AF593094

Neurodegeneration
and
Neuroprotection
in Parkinson's Disease

NEUROSCIENCE PERSPECTIVES

Titles in this series
Roger Horton and Cornelius Katona (eds), Biological Aspects of Affective Disorders
Trevor Stone (ed), Adenosine in the Nervous System
Judith Pratt (ed), The Biological Bases of Drug Tolerance and Dependence
Michel Hamon (ed), Central and Peripheral 5-HT_3 Receptors
John Waddington (ed), D_1 : D_2 Dopamine Receptor Interactions
Thomas Barnes (ed), Antipsychotic Drugs and Their Side-Effects
Eva Giesen-Crouse (ed), Peripheral Benzodiazepine Receptors
David Nicholson (ed), Anti-Dementia Agents
Alan Harvey (ed), Natural and Synthetic Neurotoxins
John Wood (ed), Capsaicin in the Study of Pain
Yossef Itzhak (ed), Sigma Receptors
John Mayer and Ian Brown (eds), Heat Shock Proteins in the Nervous System
Chris Carter (ed), Neuropharmacology of Polyamines
Roger Pertwee (ed), Cannabinoid Receptors
Steven Vincent (ed), Nitric Oxide in the Nervous System
David Latchman (ed), Genetic Manipulation of the Nervous System

Neurodegeneration and Neuroprotection in Parkinson's Disease

edited by

C. W. Olanow
Department of Neurology,
Mount Sinai Medical Center
New York, USA

Peter Jenner
Neurodegenerative Diseases Research Centre
King's College London,
London, UK

Moussa Youdim
Department of Pharmacology
Technion-Israel Institute of Technology,
Haifa, Israel

ACADEMIC PRESS
Harcourt Brace & Company, Publishers
London San Diego New York
Boston Sydney Tokyo Toronto

ACADEMIC PRESS LIMITED
24/28 Oval Road,
London NW1 7DX

United States Edition published by
ACADEMIC PRESS INC.
San Diego, CA 92101

This book is printed on acid free paper

A catalogue record for this book
is available from the British Library

ISBN 0–12–525445–8

Typeset by Servis Filmsetting Ltd, Manchester, UK
Printed and bound in Great Britain
by Hartnolls Ltd, Bodmin, Cornwall

Contents

Contributors

C. Anthony Altar Regeron Pharmaceuticals, Inc, 777 Old Saw Mill River Road, Tarrytown, NY 10591–6707, USA

Stanley H. Appel Department of Neurology, Baylor College of Medicine, Houston, TX 77030, USA

M. Flint Beal Neurology Service, Massachusetts General Hospital, Boston, MA 02114, USA

John M. Carney Free Radical Biology & Aging Research, 825 Northeast 13th Street, Oklahoma City, OK 73104, USA

Jesse Cedarbaum Regeron Pharmaceuticals, Inc, 777 Old Saw Mill River Road, Tarrytown, NY 10591–6707, USA

Robert A. Floyd Free Radical Biology & Aging Research, 825 Northeast 13th Street, Oklahoma City, OK 73104, USA

James G. Greene Departments of Neurology, Neurobiology & Anatomy, & Pharmacology, University of Rochester Medical Center, Rochester, NY 14642, USA

J. Timothy Greenamyre Departments of Neurology, Neurobiology & Anatomy, & Pharmacology, University of Rochester Medical Center, Rochester, NY 14642, USA. Present address: Department of Neurology, Emory University, Atlanta, GA 30322, USA

John M.C. Gutteridge Oxygen Chemistry Laboratory, Royal Brompton Hospital & National Heart & Lung Institute, Sydney Street, London SW3 6NP, UK

Barry Halliwell Neurodegenerative Diseases Research Centre, Pharmacology group, Biomedical Sciences Division, King's College London, Manresa Road, London SW3 6LX, UK

Peter Jenner Neurodegenerative Diseases Research Centre, Pharmacology Group, Biomedical Sciences Division, King's College London, Manresa Road, London, SW3 6LX, UK

William Y.H. Ju Departments of Physiology/Biophysics, and The Institute for Neuroscience, Dalhousie University, Halifax, Nova Scotia, Canada

Karl Kieburtz University of Rochester Medical Center, 601 Elmwood Avenue, Rochester, NY 14642, USA

Wei Dong Le Department of Neurology, Baylor College of Medicine, Houston, TX 77030, USA

Ronald M. Lindsay Regeron Pharmaceuticals, Inc, 777 Old Saw Mill River Road, Tarrytown, NY 10591–6707, USA

Pierluigi Nicotera Institute of Environmental Medicine, Division of Toxicology, Karolinska Institutet, Box 210, S–171 77 Stockholm, Sweden

C.W. Olanow Department of Neurology, Mount Sinai Medical Center, One Gustave L Levy Place, Annenberg 1494, New York, NY 10029–6574, USA

Sten Orrenius Institute of Environmental Medicine, Division of Toxicology, Karolinska Institutet, Box 210, S–171 77 Stockholm, Sweden

A.H.V. Schapira Department of Clinical Neurosciences, Royal Free Hospital School of Medicine, Rowland Hill Street, London NW3 2PF, UK

Ira Shoulson University of Rochester Medical Center, 601 Elmwood Avenue, Rochester, NY 14642, USA

R. Glenn Smith Department of Neurology, Baylor College of Medicine, Houston, TX 77030, USA

Nadine A. Tatton Departments of Anatomy/Neurobiology, and The Institute for Neuroscience, Dalhousie University, Halifax, Nova Scotia, Canada

William G. Tatton Departments of Physiology/Biophysics, Psychology and The Institute for Neuroscience, Dalhousie University, Halifax, Nova Scotia, Canada

J. Wadia Departments of Physiology/Biophysics, Dalhousie University, Halifax, Nova Scotia, Canada

Stanley J. Wiegand Regeron Pharmaceuticals, Inc, 777 Old Saw Mill River Road, Tarrytown, NY 10591–6707, USA

M.B.H. Youdim Department of Pharmacology, Technion–Israel Insitute of Technology, Haifa, Israel

Series Preface

The neurosciences are one of the most diverse and rapidly changing areas in the biological sphere. The need to understand the workings of the nervous system pervades a vast array of different research areas. By definition research in the neurosciences encompasses anatomy, pathology, biochemistry, physiology, pharmacology, molecular biology, genetics and therapeutics. Ultimately, we are striving to determine how the human brain functions under normal circumstances and perhaps more importantly how function changes in organic disease and in altered states of mind. The key to many of these illnesses will unlock one of the major therapeutic challenges remaining in this era.

The difficulty lies in the vastness of the subject matter. However I try, I find it almost impossible to keep abreast of the changes occurring in my immediate sphere of interest, let alone those exciting advances being made in other areas. The array of journals dealing with neurosciences is enormous and the flow of reprints needed to keep me updated is daunting. Inevitably piles of papers accumulate on my desk and in my briefcase. Many lie there unread until sufficient time has passed for their content to be overtaken by yet more of the ever rising tide of publications.

There are various approaches that one can take to deal with this problem. There is the blinkered approach in which you pretend that literature outside your area does not exist. There is the ignore it totally option. Indeed, one colleague of mine has ceased to read the literature in the belief that, if there is a publication of critical importance to his research, someone will tell him about it. I am not that brave and instead I arrived at what I thought was the ideal solution. I started to read critical reviews of areas of current interest. But I soon came unstuck as I realized that, for many subjects of importance to the neurosciences, such authoritative works did not exist.

Instead of simply moaning that the literature was incomplete, I conceived the idea of *Neuroscience Perspectives*. For purely selfish reasons I wanted to have available a series of individually edited monographs dealing in depth with issues of current interest to those working in the neuroscience area. Already a number of volumes have been published which have been well received and the series is thriving with books on a range of topics in preparation or in production. Each volume is designed to bring a multidisciplinary approach to the subject matter by pursuing the topic from the laboratory to the clinic. The editors of the individual volumes are producing balanced critiques of each topic to provide the reader with an up-to-date, clear and comprehensive view of the state of the art.

As with all ventures of this kind, I am simply the individual who initiates a chain of events leading to the production of the series. In reality, it is key individuals at Academic Press who are really responsible for the success of *Neuroscience Perspectives*. In particular, Dr Carey Chapman and Leona Daw had the uneviable task of recruiting editors and authors and keeping the ship on an even keel.

Finally, I hope that *Neuroscience Perspectives* will continue to be enjoyed by my colleagues in the neurosciences. Already the series is being read, understood and enjoyed by a wide audience and it is fast becoming a reference series in the field.

Peter Jenner

Preface

Parkinson's disease (PD) is an age-related neurodegenerative disorder that affects an estimated 2% of all individuals. It is characterized by a relatively selective degeneration of dopaminergic neurons in the substantia nigra pars compacta (SNc), a loss of striatal dopamine, and intracellular inclusions known as Lewy bodies. Current treatment consists of a dopamine replacement strategy, primarily using the dopamine precursor levodopa. However, after 5 to 10 years of levodopa therapy the majority of patients experience intolerable drug-related side effects, and continued disease progression is associated with the development of clinical features that do not respond to levodopa therapy. Thus, despite the best of modern medical therapy virtually all patients develop an unacceptable level of disability and mortality rates remain higher than those of age-matched control subjects. Accordingly, there has been intense interest in the development of a 'neuroprotective therapy' than can prevent neuronal degeneration and slow or stop disease progression. Towards this end, laboratory efforts have focused on attempts to understand the mechanism responsible for cell degeneration. The application of new techniques in genetics, neurochemistry, and molecular biology in animal models, PD patients, and post-mortem PD brains has provided a wealth of information that has already led to the institution of clinical trials of putative neuroprotective agents. However, there is as yet no treatment that has been definitively determined to affect the natural course of PD, and many factors have been implicated in the pathogenesis of cell degeneration. It is likely that a cascade of events involving free radicals and oxidative stress, mitochondrial dysfunction with a bioenergetic impairment, excitotoxicity, and a rise in intracytoplasmic free calcium with activation of biodestructive enzymes contributes to the neurodegenerative process. It is not clear which, if any, of these is the initiating or driving factor whose inhibition would be most likely to provide neuroprotection. There has also been recent interest in the role of apoptotic cell death in neurodegeneration and the possibility that some small molecules can provide neuroprotection or neuronal rescue by upregulating cellular defenses through the induction of translational changes and new protein synthesis. This is the first book that comprehensively reviews these and other factors implicated in the pathogenesis of cell death in PD. Each chapter examines a different factor and considers the evidence supporting its participation in the neurodegenerative process and specific strategies based on this mechanism that might lead to neuroprotection. It is likely that clinical trials will soon be instituted to test some or all of these approaches and, hopefully, at least one will lead to neuroprotection for patients with PD.

C. Warren Olanow
Mount Sinai School of Medicine, New York, USA

Peter Jenner
King's College, London, UK

Moussa Youdim
Technion, Haifa, Israel

CHAPTER 1

OXIDATIVE STRESS, BRAIN IRON AND NEURODEGENERATION. BASIC PRINCIPLES

John M.C. Gutteridge* and Barry Halliwell†

**Oxygen Chemistry Laboratory,
Unit of Critical Care, Department of Anaesthesia and Intensive Care,
Royal Brompton Hospital and National Heart and Lung Institute,
Sydney Street, London SW3 6NP, UK, and
†Neurodegenerative Diseases Research Centre,
King's College London, Manresa Road,
London SW3 6LX, UK*

Table of Contents

Neurodegeneration and Neuroprotection
in Parkinson's Disease
ISBN 0-12-525445-8

1.1 Life in oxygen: a brief comment

Although aerobes, and especially the aerobic brain, need oxygen (O_2) for survival, O_2 concentrations greater than those present in normal air have long been known to cause damage (reviewed by Balentine, 1982). The signs of O_2 toxicity depend on the organism under study, its age, physiological state, and diet. For example, pure O_2 is less toxic to adult humans than to adult rats, and less toxic to newborn rats than to adult rats. High-pressure O_2 causes acute central nervous system (CNS) toxicity in humans, leading to convulsions; this has been a problem in diving and must be considered when using hyperbaric oxygen therapy, e.g. in the treatment of gas gangrene, multiple sclerosis, and in combination with radiotherapy (O_2 aggravates the damage done to cells by ionizing radiation).

The acute effects of hyperbaric O_2 have often been attributed to direct inactivation of enzymes by O_2 (Balentine, 1982), although the evidence for this is not convincing. However, the slower-acting toxic effects of elevated O_2 have often been suggested to involve 'oxygen radicals' (Gerschmann *et al.*, 1954; Fridovich, 1978; Balentine, 1982). Such radicals may also be involved in the acute effects produced by high-pressure O_2, because administration of antioxidant enzymes has been claimed to offer some protection (Turrens *et al.*, 1984).

1.2 The basic chemistry

1.2.1 The superoxide radical

Acceptance of a single electron by an O_2 molecule forms the superoxide radical, $O_2^{\bullet -}$, which has one unpaired electron. The discovery, by McCord and Fridovich (1969), of enzymes that appear to have evolved specifically to scavenge $O_2^{\bullet -}$ (superoxide dismutases, SODs) led to the proposal that $O_2^{\bullet -}$ is a major agent responsible for O_2 toxicity and that SODs are important antioxidant defences (Fridovich, 1978, 1989). This antioxidant role of SODs is supported by a wide range of evidence, including results obtained using the techniques of modern molecular biology (e.g. see Fridovich, 1989; Touati, 1989; Chan *et al.*, 1990).

Superoxide is formed *in vivo* in a variety of ways. A major source is the activity of electron transport chains in mitochondria and endoplasmic reticulum. Some of the electrons passing through these chains 'leak' directly from intermediate electron carriers onto O_2. Because O_2 accepts electrons one at a time, $O_2^{\cdot-}$ is formed. The rate of leakage at physiological O_2 concentrations is probably very small, certainly below 5% of the total electron flow through the chains, but it rises as the O_2 concentration is increased (Fridovich, 1978, 1989). Hence, the toxicity of excess O_2 may be due to increased formation of $O_2^{\cdot-}$ (e.g. by faster electron leakage and accelerated autoxidations of such molecules as catecholamines) beyond the ability of antioxidant defenses to cope. In addition, activated phagocytic cells produce $O_2^{\cdot-}$, which plays an important part in the mechanism by which engulfed bacteria are killed (Curnutte and Babior, 1987). Phagocytes able to produce $O_2^{\cdot-}$ include monocytes, neutrophils eosinophils and macrophages of various types, including the microglial cells of the brain (Colton and Gilbert, 1987). The latter investigators have suggested that microglial $O_2^{\cdot-}$ generation helps to protect the CNS against infectious organisms. Again, excessive activation of phagocytic cells (as in chronic inflammation) can lead to free radical damage.

The evidence supporting the superoxide theory of O_2 toxicity is substantial, but the exact mechanism by which the putative excess $O_2^{\cdot-}$ generation at increased O_2 levels could exert toxic effects is not completely clear. Superoxide itself has limited reactivity. It is capable of inactivating a few enzymes directly, examples being mammalian creatine kinase (McCord and Russell, 1988) and some iron–sulphur proteins in bacteria, such as *Escherichia coli* phosphogluconate dehydratase and aconitase (Gardner and Fridovich, 1991a,b). Superoxide is also capable of inactivating the NADH dehydrogenase complex of the mitochondrial electron transport chain *in vitro* (Zhang *et al.*, 1990), although its ability to do this *in vivo* has not yet been demonstrated.

Hence, the number of targets within mammalian cells that are known to be sensitive to $O_2^{\cdot-}$ is small. Indeed, under certain circumstances, controlled $O_2^{\cdot-}$ production is a useful process, e.g. in the bacterial killing mechanisms of phagocytes referred to previously (Curnutte and Babior, 1987). Evidence is also accumulating that several cell types, such as fibroblasts, lymphocytes and vascular endothelial cells produce and release small amounts of $O_2^{\cdot-}$ in physiological reactions (e.g. Maly, 1990; Meier *et al.*, 1990; Murrell *et al.*, 1990). This $O_2^{\cdot-}$ may be involved in growth regulation and intercellular signalling, as suggested by Halliwell and Gutteridge (1986).

SOD is undoubtedly important as a major physiological antioxidant (Fridovich, 1978, 1989; Touati, 1989). For example, the familial dominant form of amyotrophic lateral sclerosis is associated with defects in the gene encoding the copper–zinc-containing (cytosolic) SOD that produce significant decreases in enzyme activity (Deng *et al.*, 1993; Bowling *et al.*, 1993). Despite this, there is evidence that *an excess* of SOD in relation to the activities of peroxide-metabolizing enzymes can sometimes be deleterious (Scott *et al.*, 1987; Avraham *et al.*, 1988; Groner *et al.*, 1990; Remacle *et al.*, 1991). For example, the gene encoding copper–zinc SOD is located on chromosome 21. Evidence consistent with the view that the excess of copper–zinc-containing SOD activity in trisomy 21 may contribute to the symptoms of Down's syndrome has come from work with transgenic mice carrying the human gene encoding copper–zinc

SOD (Avraham *et al.*, 1988; Groner *et al.*, 1990; Ceballos-Picot *et al.*, 1991). *Why* this happens is uncertain; is it due to a property of the enzyme itself, or to increased H_2O_2 production? Or perhaps some $O_2^{\cdot -}$ is needed to inactivate $NO^{\cdot}$ (Oury *et al.*, 1992), as discussed below?

1.2.2 Nitric oxide

Another free radical released by several cell types, especially vascular endothelial cells and phagocytes, is nitric oxide, $NO^{\cdot}$ (Moncada *et al.*, 1991). $NO^{\cdot}$ reacts with $O_2^{\cdot -}$ at physiological pH at a very fast rate to yield a non-radical product, peroxynitrite (equation (1)) (Saran *et al.*, 1990; Huie and Padmaja, 1993). There is considerable debate in the literature about whether this interaction of $O_2^{\cdot -}$ and $NO^{\cdot}$ might be damaging to cells. Thus, peroxynitrite may be directly cytotoxic, e.g. by oxidizing thiol groups (Beckman, 1991; Radi *et al.*, 1991). At physiological pH, peroxynitrite protonates and decomposes to a range of products including highly reactive hydroxyl radicals (see Section 2.2.4), nitrogen dioxide ($NO_2^{\cdot}$) and the nitronium ion (NO_2^{+}); the latter two can nitrate aromatic amino acid residues (Beckman, 1994; van der Vliet *et al.*, 1994).

$$O_2^{\cdot -} + NO^{\cdot} \rightarrow ONOO^{-} \qquad (1)$$

By contrast, interaction of $O_2^{\cdot -}$ and $NO^{\cdot}$ might represent a physiological regulatory process affecting vascular muscle tone. Thus, it has been suggested that vascular endothelial cells might produce both $O_2^{\cdot -}$ and $NO^{\cdot}$ as 'antagonistic agents', to give fine control of vascular tone (Halliwell, 1989). Oury *et al.* (1992) found that mice transgenic for human SOD were more sensitive to hyperbaric oxygen toxicity than control mice and this sensitivity disappeared after treatment with an inhibitor of $NO^{\cdot}$ synthetase. They postulated that, in their system, $O_2^{\cdot -}$ acts to *antagonize* neurotoxicity by removing $NO^{\cdot}$ (e.g. Laurindo *et al.*, 1991).

Nitric oxide synthetase is widespread in brain tissue (Hope *et al.*, 1991). Nitric oxide has been suggested to be involved in both the normal functioning of excitatory amino acids such as glutamate and in the damaging effects produced by their generation in excess (Dawson *et al.*, 1991; Forstermann *et al.*, 1991). It appears likely that the interaction of $O_2^{\cdot -}$ with $NO^{\cdot}$ (whether protective or dangerous) is highly relevant to normal brain metabolism and to neurodegenerative diseases (Beckman, 1991; Terada *et al.*, 1991).

1.2.3 Hydrogen peroxide

SODs remove $O_2^{\cdot -}$ by converting it into hydrogen peroxide (equation (2)). Several other enzymes that produce H_2O_2 exist in human tissues, such as L-amino acid oxidase, glycollate oxidase and monoamine oxidase. Indeed, oxidative deamination of dopamine by monoamine oxidase is the main catabolic pathway for dopamine within dopamine nerve terminals (Cohen, 1988). Cohen (1988) suggested that an accelerated turnover of dopamine in patients with Parkinson's disease, leading to increased H_2O_2 formation, may provoke an *oxidative stress* (i.e. an increase in the generation of oxygen-derived species beyond the ability of antioxidant defenses to cope with them) within surviving dopamine terminals, so accelerating their destruction. Another possibility is that the side-effects of prolonged therapy with L-DOPA in patients with Parkinson's disease might be related to increased H_2O_2 formation (Cohen, 1988; Olanow, 1990). Zhang and Piantadosi (1991) reported that the monoamine oxidase inhibitor pargyline could partially protect rats against the CNS toxicity produced by hyperbaric O_2.

$$2O_2^{\cdot -} + 2H^+ \rightarrow H_2O_2 + O_2 \quad (2)$$

H_2O_2 can act as an oxidizing agent, although it is poorly reactive. Unlike $O_2^{\cdot -}$, however, H_2O_2 crosses cell membranes easily. H_2O_2 does not qualify as a radical because it contains no unpaired electrons, and it can be removed within human cells by the action of two types of enzyme, catalases and selenium-dependent glutathione peroxidases. The latter are probably more important in the brain (Cohen, 1988; Jain *et al.*, 1991).

1.2.4 Hydroxyl radicals

One hundred years ago H.J.H. Fenton published his seminal paper 'Oxidation of tartaric acid in presence of iron' (Fenton, 1894): ferrous sulfate, hydrogen peroxide, tartaric acid and sodium hydroxide, when mixed, yield a beautiful violet color (Fenton, 1876). The reaction was shown to be driven by the mixing of a ferrous salt (Fe^{2+}) with hydrogen peroxide (H_2O_2), and the oxidation product of tartaric acid was later identified as a hydroxylacetaldehyde dimer. Although Fenton suggested that the iron was acting catalytically, the chemistry involved was not addressed until Haber and Weiss (1934) proposed that a ferrous salt reacting with H_2O_2 could yield hydroxyl radicals ($OH^{\cdot}$) (equation (3)).

$$Fe^{2+} + H_2O_2 \rightarrow Fe^{3+} + OH^- + OH^{\cdot} \quad (3)$$

By the 1950s, Gerschmann, Gilbert and colleagues had suggested that free radicals were a common link between oxygen toxicity in living systems and radiation damage. Radiation can split water molecules into a series of products, including $OH^{\cdot}$. Hydroxyl radicals react at an almost diffusion-controlled rate with all molecules found *in vivo*. Desite this, $OH^{\cdot}$ chemistry remained the domain of radiation chemists up until the discovery of the SOD enzymes. It was soon proposed that the simple inorganic chemistry used by Fenton could also take place in aerobic cells (McCord and Day, 1978; Halliwell, 1978). Although this idea has been frequently resisted, the evidence for it is now overwhelming (reviewed by Halliwell and Gutteridge, 1990, 1992; Liochev and Fridovich, 1994).

It is interesting to note that although catalase had been crystallized in 1937 by Sumner and Dounce, some 30 years earlier than the discovery of SOD, it did not trigger the same appreciation of biological Fenton chemistry. The far-sighted comments of W.A. Waters (1946) in his book *Chemistry of Free Radicals* should not go unnoticed, however. He stated: 'In fact the enzyme catalase, the destroyer of hydrogen peroxide, is to be regarded as a protective agent which prevents unduly rapid destruction of many cell components'.

Since most of our definitive knowledge concerning the formation and reactivity of $OH^{\cdot}$ radicals came from radiation chemistry, it is not surprising that many important aspects were directly transferred into biological Fenton chemistry without, in some cases, carefully considering fundamental differences between the two $OH^{\cdot}$ generating systems. The chemical reactivities of $OH^{\cdot}$ generated by radiolysis of water and those generated by Fenton chemistry are, of course, the same. However, radiolysis produces $OH^{\cdot}$ in solution, which can combine with all molecules present at rates essentially dependent on their concentrations and rate constants for reaction with $OH^{\cdot}$. Many of the simple low molecular mass 'scavengers of $OH^{\cdot}$' used in radiation experiments (such as formate, ethanol, mannitol and dimethylsulfoxide) were introduced into Fenton chemistry in attempts to identify $OH^{\cdot}$ by competition kinetics. However, 'hydroxyl radical scavengers' frequently failed to inhibit biological damage by $OH^{\cdot}$ in ways predicted from concentrations and second-order rate constants. This inevitably led to claims that the oxidizing species formed in Fenton chemistry, and reacting with added scavengers or detector molecules, was not $OH^{\cdot}$ but some oxo-iron species (reviewed by Halliwell and Gutteridge, 1990). This debate started in 1932 and periodically resurfaces (Bray and Gorin, 1932). It is certainly *possible* that oxo-iron species with considerable oxidizing power are formed during Fenton chemistry. However, they are additional to $OH^{\cdot}$, not alternative to it, and their yield will depend on the reaction conditions used (Halliwell and Gutteridge, 1992).

A ferric salt in aqueous solution at pH 7.4 rapidly forms polynuclear iron with a solubility product of 10^{-39} M and an equilibrium concentration of ferric ions of around 10^{-18} M. At similar pH values, ferrous salts in the presence of oxygen autoxidize to the ferric state, transferring an electron to oxygen which, via intermediates, forms superoxide. The rate of this oxidation depends on the ligand to the metal. In order for ferric ions to become available to aerobic life forms they have to be stabilized and solubilized by complexing to a ligand. Microbes achieve iron solubilization and capture by synthesizing siderophores (reviewed by Raymond and Carran, 1979) whereas higher life forms use more complex protein molecules for storage and transport, such as ferritin and transferrin.

The chemistry of iron in biological systems is inextricably linked to ligand binding, which makes Fenton chemistry site directed, and the iron ligand usually the first molecule to receive $OH^{\cdot}$ damage. We can therefore say that radiolysis produces $OH^{\cdot}$ with high reactivity but low specificity, whereas biological Fenton chemistry produces $OH^{\cdot}$ with high reactivity and high specificity. The $OH^{\cdot}$ is targeted to the site of iron ion binding. Hence an appreciation of the availability and location of iron (and other transition metal ions) *in vivo* is essential for understanding free radical damage.

Some ferric chelates are able to react with H_2O_2 to generate $OH^{\cdot}$, albeit at a much slower rate than ferrous chelates. Interestingly, certain ferric chelates react with H_2O_2 to form $OH^{\cdot}$ in a process involving $O_2^{\cdot -}$ that is almost completely inhibitable by superoxide dismutase (Gutteridge, 1985).

1.2.5 Iron in biological systems

Iron is essential to life, for the synthesis of important iron-containing proteins such as haemoglobin and ribonucleoside diphosphate reductase. A normal human adult contains some 4.5 g of iron, absorbs about 1 mg of iron per day, and (when in iron balance) excretes the same amount. Since the total plasma iron turnover is around 35 mg day^{-1} the body has extremely efficient mechanisms for preserving absorbed body iron. Only slight disturbances in this delicate balance of iron intake and iron loss can push the body into conditions of iron overload or iron deficiency.

Two-thirds of the iron in the body is found in the oxygen-carrying protein haemoglobin, with smaller amounts present in myoglobin, various enzymes, the iron transport proteins transferrin and lactoferrin, and the iron storage proteins ferritin and haemosiderin. Careful sequestration of iron within a protein complex ensures solubility, stability and, in the case of transferrin, lactoferrin (Gutteridge *et al.*, 1981a; Aruoma and Halliwell, 1987) and ferritin (Bolann and Ulvik, 1993) provides an antioxidant defence mechanism (see below).

1.2.6 Iron for Fenton chemistry *in vivo*

The importance of iron in mediating oxidative damage through Fenton chemistry leads to the question: what forms of iron are available *in vivo* to catalyse $OH^{\cdot}$ formation? Early claims that iron-loaded lactoferrin (2 moles of iron per mole of protein) is an effective catalyst of $OH^{\cdot}$ formation from $O_2^{\cdot-}$ and H_2O_2 have now been disproved, as have similar claims for transferrin (Aruoma and Halliwell, 1987). Iron ions, however, can become loosened from transferrin if the pH is lowered and can be released if a reductant and iron chelator are also present (Aisen, 1992). It is also likely that iron which has not correctly loaded onto the two high-affinity iron-binding sites of transferrin or lactoferrin may, under certain conditions, be labile enough to participate in Fenton chemistry (Gutteridge, 1988). Transferrin in normal healthy individuals is only one-third loaded with iron, retaining a considerable iron-binding capacity. Lactoferrin is similarly far from maximum iron-loading *in vivo*.

Ferritin is the main iron-storage protein in the body, with the ability to store some 4500 moles of iron per mole of protein. Iron enters ferritin in the ferrous state, is deposited in a core as a crystalline ferric polymer, and requires conversion back to the ferrous state before mobilization can occur. The ferric ion in ferritin is not available to directly catalyse $OH^{\cdot}$ formation. However, several studies have demonstrated that a minute amount of iron can be mobilized from ferritin by $O_2^{\cdot-}$ and other reductants to participate in $OH^{\cdot}$ formation (Biemond *et al.*, 1984). It has been suggested that this small labile fraction of 'available' iron represents iron in outer channels of ferritin, not yet loaded into the central core (Bolann and Ulvik, 1990).

Haemosiderin is a storage form of iron that increases in the liver during conditions of iron overload. Unlike ferritin, haemosiderin is highly insoluble in aqueous solution, and less reactive than ferritin in $OH^{\cdot}$ generation from H_2O_2 (O'Connell *et al.*, 1986a), presumably because iron is even less easily mobilized from it (Gutteridge and Hou,

1986). Ferritin can be converted into a haemosiderin-like molecule by incubation with proteases and/or free radicals (O'Connell *et al.*, 1986b), perhaps conferring a protective biological advantage against further oxidative damage during iron overload disease (O'Connell *et al.*, 1986a).

1.2.7 Haem proteins

Unlike the non-haem iron proteins described above, the haem proteins, haemoglobin and myoglobin, have evolved to carry and exchange oxygen. Claims that the haemoglobin molecule is a Fenton catalyst (Sadrzadeh *et al.*, 1985) are inconsistent with the chemistry of iron chelated in pyrrole rings and surrounded by a protein, making it impossible for $OH^{\cdot}$ radicals to escape into free solution.

Haemoglobin is transported in erythrocytes, which are rich in the antioxidant defence enzymes catalase, SOD, and glutathione peroxidase. However, isolated haemoglobin (and myoglobin) are degraded on exposure to excess H_2O_2, with release of iron ions, capable of catalysing free radical reactions, from the haem ring (Gutteridge, 1986; Puppo and Halliwell, 1988). In addition, haemoglobin reacts with H_2O_2 to form a protein-bound oxidizing species capable of stimulating lipid peroxidation (reviewed by Kanner *et al.*, 1987). The nature of this oxidizing species is unclear as yet. Reaction of H_2O_2 with the protein probably generates a haem ferryl species plus an amino acid radical. For both haemoglobin and myoglobin, it has been suggested that a tyrosine peroxyl radical, capable of abstracting hydrogen and initiating lipid peroxidation, is formed on exposure of the myoglobin to H_2O_2 (Davies, 1990; McArthur and Davies, 1993). However, the identity of this radical is still uncertain (Kelman and Mason, 1992).

Haemoglobin outside the erythrocyte is potentially a dangerous protein (Sadrzadeh *et al.*, 1987). Indeed, spasm of cerebral arteries can be a significant late complication of haemorrhagic stroke, and it has been proposed that release of haemoglobin from erythrocytes in the clot, and subsequent free radical reactions, are involved (Steele *et al.*, 1991). Haemoglobin also binds $NO^{\cdot}$ avidly (reviewed by Macdonald and Weir, 1991).

1.2.8 'Catalytic' iron

Clearly, protein-bound iron complexes are not efficient Fenton catalysts, although they can under certain circumstances release low molecular mass iron for Fenton chemistry. Catalytic iron is required to be redox active, and usually part of a simple complex such as iron attached to phosphate esters (ATP, ADP, etc.), organic acids (such as citrate) or even membrane lipids and DNA.

In general, however, the availability of catalytic iron or copper ions to stimulate free radical reactions in most body tissues and extracellular fluids is extremely limited, and can often control the extent of tissue damage being done (e.g. $O_2^{\cdot -}$ and H_2O_2 are unlikely to cause widespread damage unless such ions are present). However, *cell injury by almost any mechanism, including traumatic or ischaemic injury, has the potential to accelerate free radical reactions*. This is partly because injured and lysed cells release their intracellular iron into the surrounding environment, where iron is then available to accelerate formation of damaging oxygen-derived species (Halliwell and Gutteridge, 1984, 1985; Halliwell *et al.*, 1988). Other reasons include recruitment of phagocytes to the injured area with their subsequent activation, and increased 'leakage' onto O_2 of electrons from damaged electron transport chains (Ambrosio *et al.*, 1993). Bleeding as a result of injury, and subsequent haemoglobin liberation, may have a compounding effect (Gutteridge, 1986; Sadrzadeh *et al.*, 1987; Macdonald and Weir, 1991). In severe muscle injury, as in rhabdomyolysis or 'crush syndrome', the release of myoglobin into the circulation also poses a risk of oxidative damage (Odeh, 1991).

1.2.8.1 Measurement of 'catalytic iron'

In a first attempt to detect and measure iron present in biological fluids in a redox active form, the authors developed the 'bleomycin assay' (Gutteridge *et al.*, 1986). The assay is based on the ability of the antitumour antibiotic bleomycin to degrade DNA in the presence of an iron salt, oxygen and a suitable reducing agent. During the reaction, base propenals are released from DNA that rapidly break down to yield malondialdehyde, which is measured spectrophotometrically after reaction with thiobarbituric acid. Binding of the bleomycin–iron–oxygen complex to DNA makes the reaction site-specific, and biological antioxidants rarely interfere, allowing the assay to be directly applied to body fluids. By modifying the conditions and allowing both the iron and the reductant to come from the biological sample (i.e. no ascorbate is added to the iron assay) it is also possible to assess the catalytic activity of ferrous iron in a biological fluid (Gutteridge, 1992a).

1.3 Antioxidant protection

When the balance between production of oxygen-derived species, such as $O_2^{\cdot -}$, H_2O_2, hypochlorous acid (HOCl) and singlet oxygen ($^1O_2$), and antioxidant defences against them is disturbed, *oxidative stress* results (Sies, 1993). Oxidative stress can itself provide the iron necessary for Fenton chemistry to cause molecular damage (Biemond *et al.*, 1984; Gutteridge, 1986). The body has several important strategies to deal with this:

(1) removing the oxygen-derived species;
(2) sequestering iron, and other metals in non- or poorly reactive forms;
(3) scavenging formed radicals;
(4) repairing molecular damage.

Superimposed on these immediate defences is a gene-regulated protection involving the heat shock and stress proteins (reviewed by Storz *et al.*, 1990).

1.3.1 Antioxidant protection inside the cell

Oxygen metabolism occurs within cells, and it is here we expect to find antioxidants evolved to deal speedily and specifically with reduced intermediates of oxygen. The SODs rapidly promote the dismutation of superoxide into hydrogen peroxide and oxygen at a rate considerably faster than it occurs uncatalysed at physiological pH. Hydrogen peroxide, a product of the dismutation reaction, can be destroyed by two different enzymes, namely catalase and glutathione peroxidase (a selenium-containing enzyme). These enzymes function in concert during normal aerobic metabolism to eliminate toxic reduction intermediates of oxygen inside the cell, thereby allowing a small low molecular mass iron pool to safely exist to provide iron for the manufacture of iron-containing proteins (Halliwell and Gutteridge, 1986).

Prevention of radical formation inside cells is an obvious mechanism that must have evolved to restrict oxygen toxicity, and a good example of this is cytochrome oxidase, the terminal oxidase of the mitochondrial electron transport chain. This enzyme does not release reactive oxygen intermediates from its active centre whilst functioning catalytically (Chance *et al.*, 1979); even though such intermediates are produced, they remain bound to the active sites.

1.3.2 Protection of membranes

Within the hydrophobic lipid interior of membranes, different types of lipophilic radicals are formed from those seen in the intracellular aqueous milieu. Lipophilic radicals require different types of antioxidants for their removal. Vitamin E (α-tocopherol), a fat-soluble vitamin, is a poor antioxidant outside a membrane bilayer but is extremely effective when incorporated into the membrane (Gutteridge, 1978). Ubiquinols may also exert some degree of antioxidant protection within membranes, probably by cooperating with α-tocopherol (Kagan *et al.*, 1990).

An important part of membrane stability and protection is the way in which the membrane is assembled from its lipid components. Structural organization requires that the 'correct' ratios of phospholipids to cholesterol are present, and that the 'correct' types of phospholipids and their fatty acids are also attached (reviewed in Gutteridge and Halliwell, 1988).

1.3.3 Protection outside the cell

Body extracellular fluids contain little, or no, catalase activity, and extremely low levels of SOD. Glutathione peroxidases, in both selenium-containing and non-selenium-containing forms are present in plasma, but there is little glutathione in plasma ($< 1\mu M$) (Wendel and Cikryt, 1980) to satisfy enzymes with a K_m for reduced glutathione in the millimolar range. 'Extracellular' copper–zinc-containing SODs (EC-SODs) have recently been identified, and shown to be glycoproteins (Marklund, 1984). This relative lack of SOD and H_2O_2-destroying enzymes may allow the limited survival of $O_2^{\cdot -}$ and H_2O_2 in extracellular fluids to act as useful messenger, signal or trigger molecules (Halliwell and Gutteridge, 1986; Saran and Bors, 1989). A key feature of such a proposal is that $O_2^{\cdot -}$ and H_2O_2 must not meet with reactive iron or copper, suggesting that extracellular antioxidant protection has largely evolved to keep iron and copper in poorly or non-reactive forms (Halliwell and Gutteridge, 1986; Gutteridge and Halliwell, 1988).

The iron transport protein transferrin is normally one-third loaded with iron and keeps the concentration of 'free' iron in plasma effectively at nil. Iron bound to transferrin will not participate in radical reactions, and the available iron-binding capacity gives it a powerful antioxidant property towards iron-stimulated radical reactions (Gutteridge *et al.*, 1981). Similar considerations apply to lactoferrin (Gutteridge *et al.*, 1981), which, like transferrin, can bind 2 moles of iron per mole of protein, but holds onto its iron down to pH values as low as 4.0. Haemoglobin, myoglobin and other haem proteins are normally kept safely sequestered in tissues. Even if they are released from injured tissues, plasma contains proteins such as haptoglobins and haemopexin that specifically bind haemoglobin and haem iron, respectively. Binding to these proteins greatly diminishes the ability of haem and haem proteins to accelerate lipid peroxidation (Gutteridge, 1987; Gutteridge and Smith, 1988).

The major copper-containing protein of human plasma is caeruloplasmin, unique for its intense blue coloration. Apart from its known acute-phase reactant properties, its biological functions have remained an enigma. However, Stocks *et al.* (1974a) and Gutteridge and Stocks (1981) have pointed out that the ferroxidase activity of this protein makes a major contribution to extracellular antioxidant protection against lipid peroxidation and Fenton chemistry. Caeruloplasmin rapidly removes ferrous ions from solution and simultaneously reduces oxygen to water, without releasing any partially reduced oxygen intermediates into free solution (another example of protection by avoiding production of free oxygen radicals).

1.4 Brain iron and free radicals

Brain tissue contains non-haem iron, and levels of around 0.074 $\mu g\ mg^{-1}$ of protein have been estimated (Youdim and Green, 1978; Dallman *et al.*, 1975). Certain regions of the brain are particularly high in iron such as the globus pallidus, the substantia nigra and the red nucleus (Hallgren and Sowander, 1958; Harrison *et al.*, 1968; Hill and Switzer, 1984). Brain tissue slices undergo lipid peroxidation at a slow rate, but, when the tissue is homogenized, peroxidation is rapid (Stocks *et al.*, 1974b). The process of disrupting tissue structure releases and brings into contact polyunsaturated fatty acids, ascorbate, oxygen and various metal ions, particularly iron, which is normally held in safe tissue compartments. Brain homogenate peroxidation can be inhibited by adding iron chelators, enzymatically removing ascorbate or by adding chain-breaking antioxidants (Stocks *et al.*, 1974a). Recent detailed studies, using brain tissue fractions, have shown that the rate of brain homogenate peroxidation depends upon the iron content of different brain regions (Zaleska and Floyd, 1985).

There have been several reports suggesting that part of the non-haem iron present in brain tissue is in a low molecular mass form, and in the reduced (ferrous) state (reviewed by Gutteridge, 1992b). For example, the noradrenergic system may require ferrous ions to form catecholamine–Fe^{+2}–ligand complexes, the serotoninergic system may require ferrous ions for binding serotonin to serotonin-binding protein, and iron is known to be required as a cofactor for tyrosine hydroxylase and tryptophan hydroxylase enzymes. A clinical example of what can happen when iron is removed from the human brain was reported in 1985 (Blake *et al.*, 1985). Patients with rheumatoid arthritis were treated with the iron chelator desferrioxamine (which cannot cross the blood–brain barrier) in an attempt to decrease oxygen radical damage within the joint. Two patients out of seven developed severe stomach pains and were treated with prochlorperazine. Prochlorperazine, a metal complexing agent that *can* cross the blood–brain barrier, appeared to be able to chelate brain iron and transfer it outside to desferrioxamine (Blake *et al.*, 1985). Some patients receiving both desferrioxamine and prochlorperazine lost consciousness for between 48 and 72 hours. An iron-deficient animal model has confirmed the ability of these two iron chelators in concert to induce coma (Blake *et al.*, 1985). An alternative explanation for the adverse effects of desferrioxamine and prochlorperazine administered jointly is that prochlorperazine opens up the blood–brain barrier and allows desferrioxamine to enter the brain (see citation in Gutteridge, 1992b). Recently, attempts have been made to measure the low molecular mass iron fraction in rodent brains. Gerbil brain homogenates were ultrafiltered through a 10 000 Da exclusion membrane, and the iron passing through the membrane measured using the bleomycin assay. Iron levels in the range 20.5±3.5 μM were found in the filtrate (Gutteridge *et al.*, 1991), confirming the presence of low molecular mass iron in brain tissue homogenates. However, it is not clear whether this represents actual 'catalytic' iron present in the brain, or iron released from stores during tissue handling.

1.4.1 Cerebrospinal fluid iron

The iron content of cerebrospinal fluids (CSFs), using atomic absorption techniques, has been calculated to range from 0.2 to 1.1 $\mu mol\ l^{-1}$ (reviewed by Gutteridge, 1992b). Since normal CSF transferrin levels are around 0.24 $\mu mol\ l^{-1}$, and 1 mol of transferrin binds 2 moles of iron, it is likely that CSF transferrin is often at or near complete saturation with iron (Bleijenberg *et al.*, 1971). When the bleomycin iron assay is applied to normal CSF (samples taken for diagnostic purposes, but normal in respect of cell count, protein and glucose content) many show bleomycin-detectable iron at around 0.55±0.27 $\mu mol\ l^{-1}$ (Gutteridge, 1992a).

By modifying the reaction conditions so that both the iron reductant and the iron ions come solely from the CSF sample it is possible to measure ferrous ions in CSF (Gutteridge, 1992a). The reason why low molecular mass iron, present in some CSF fluids, is in the ferrous state can be ascribed to the following: ascorbate levels in CSF are higher (often 10-fold) than plasma values, whereas caeruloplasmin levels are extremely low, producing an ascorbate:caeruloplasmin ratio of around 1:25 000

(Gutteridge, 1992a). This ratio of ascorbate inhibits the ferroxidase activity of caeruloplasmin, preventing it from catalysing the oxidation of ferrous ions back to the ferric state (Gutteridge, 1991). Low levels of transferrin, at or near iron saturation, ensure that there is no iron-binding capacity; and that no ferroxidase-like activity (Harris and Aisen, 1973) can be displayed by transferrin. Any low molecular mass iron present in CSF is therefore probably maintained in the ferrous state by ascorbate.

1.4.2 Iron–aluminium synergism

Aluminium and iron are the two most abundant metals in the Earth's crust, but poor solubility at neutral pH values has maintained them as trace metals in surface water. There is, however, growing evidence that widespread industrial pollution, leading to acidification of large bodies of water, is leading to increased solubility of aluminium (and possibly iron) and to its entry into biological chains (reviewed by Martin, 1991).

Aluminium ceased to be regarded as a bland and unreactive metal when clinical links were made between aluminium toxicity and the development of an acute encephalopathy in patients with chronic renal failure on long-term haemodialysis (reviewed by Wills and Savory, 1983). Aluminium has also been implicated as a contributor to the pathogenesis of other neurological conditions (Garruto *et al.*, 1984; Perl and Brody, 1980).

Aluminium, unlike transition metal ions, has a fixed oxidation state of +3 in biological systems, and is therefore unable to promote single-electron transfer reactions. Nevertheless, there is growing evidence to suggest that aluminium ions can act synergistically with iron ions to increase free radical damage. Aluminium ingestion was shown to increase lipid peroxide formation in the brain of rats by 142% (Ohtawa *et al.*, 1983); to stimulate iron-dependent peroxidation of liposomes, micelles and red blood cells (Gutteridge *et al.*, 1985); and to increase iron-stimulated brain peroxidation (Fraga *et al.*, 1990). The iron-binding and transporting protein transferrin is also the major aluminium-binding protein in the plasma. Once bound to transferrin, aluminium can enter cells (Birchall and Chappell, 1988), and in particular the CNS (Roskama and Connor, 1990), via transferrin receptors. Aluminium hydroxide is widely used as an adjuvant for diphtheria, tetanus and pertussis (DTP) vaccines, and children can often receive up to 3.75 mg of parenteral aluminium during the first 6 months of life. When DTP vaccine is intraperitoneally injected into mice it causes a transient rise in brain tissue aluminium levels, peaking around the second and third day after injection (Redhead *et al.*, 1992).

1.5 Oxidative stress: the molecular targets

Ischaemia/reoxygenation injury and traumatic damage to the brain can produce liberation of catalytic metal ions and an increase in the formation of reactive radicals. How then could further cellular damage occur? Lipid peroxidation is one possibility, and several groups have found evidence consistent with increased CNS lipid peroxidation after ischaemia (e.g. Michel *et al.*, 1987; Watson and Ginsberg, 1988), trauma (Hall and Braughler, 1988), iron salt injection (Willmore *et al.*, 1980) or damage by methylmercury (Sarafian and Verity, 1991). It must be emphasized, however, that *damage caused by oxidative stress need not necessarily involve lipid peroxidation* (Halliwell, 1987; Orrenius *et al.*, 1989; Cochrane, 1991). Hence, the failure of some groups to find lipid peroxides in injured nervous tissue does not rule out the occurrence of oxidative damage; damage to DNA and proteins may be of equal, or even greater, importance *in vivo* (Halliwell, 1987). Early events in human cells subjected to oxidative stress include DNA damage and consequent activation of poly (ADP-ribose) synthetase (this enzyme polymerizes ADP-ribose residues derived from NAD^+ and can lead to depletion of cellular NAD^+), decreases in ATP content, and increases in intracellular free Ca^{2+}, with consequent activation of Ca^{2+}-stimulated proteases that can cause such phenomena as 'bleb' formation on the plasma membrane of the cells (Orrenius *et al.*, 1989). The changes that can occur in mammalian cells subjected to oxidative stress are extremely complex, and it is very difficult to disentangle the sequence of events. Increases in intracellular free Ca^{2+} might be particularly damaging to neurones (reviewed by Siesjö, 1990), because several neurotoxins are thought to act in this way (Komulainen and Bondy, 1988; Siesjö, 1990). An additional mechanism that has been suggested to explain the damage produced by ischaemia/reoxygenation in brain is the generation within the tissue of excitatory amino acids such as glutamate (reviewed by Meldrum, 1985; Siesjö, 1988). It is therefore interesting to note claims that excitatory amino acids can be released from rat hippocampal slices as a result of exposure to $O_2^{\cdot -}$ and H_2O_2 (Pelligrini-Girampietro *et al.*, 1988), and that the toxicity of the excitatory neurotoxin kainic acid to cerebellar neurones in culture may involve increased oxygen radical formation (Dykens *et al.*, 1987). Further work is required in this area.

Protein damage is frequently an important consequence of oxidative stress (Halliwell, 1987; Orrenius *et al.*, 1989; Cochrane, 1991). Indeed, radicals can damage brain proteins (Oliver *et al.*, 1990), including the enzyme glutamine synthetase (Schor, 1988). Thus, in gerbil brain, glutamine synthetase, the major enzyme responsible for glutamate removal, may be inactivated by oxidative stress (Oliver *et al.*, 1990; Carney *et al.*, 1991). Increased levels of end-products of oxidative damage to proteins, and decreased glutamine synthetase activities, have been reported in old (>70 years) human brains when compared with young (~30 years) controls (Smith *et al.*, 1991).

The acidosis resulting from ischaemia has been suggested to aggravate free radical injury by making released iron more soluble (Siesjö, 1988; Bralet *et al.*, 1992); it might also interfere with the activity of antioxidant defence enzymes (Link, 1988). However, there are also reports that acidosis may protect certain cell types, such as liver endothelial cells, against oxidative damage (e.g. Bronk and Gores, 1991). Further work is needed for clarification.

1.6 Oxidative stress in neurodegenerative diseases

In a recent survey of over 100 diseases in which free radicals have been implicated (Gutteridge, 1993) the author concluded that in most cases free radicals were a consequence of tissue damage, at worst exacerbating and amplifying disease pathology. It seems likely that similar considerations apply to most of the neurodegenerative diseases. Compared with other tissues, the CNS may be particularly susceptible to oxidative damage (reviewed by Halliwell and Gutteridge, 1985; Halliwell, 1992; Evans, 1993) due to its high rate of oxidative metabolic activity, high concentration of polyunsaturated lipids, low levels of catalase, glutathione peroxidase and possibly SOD, neurochemical oxidations and auto-oxidations, regions of high iron and ascorbate content, and the ease with which iron appears to be mobilized in a catalytic form. The specific diseases are considered here only very briefly, as they are presented in detail elsewhere in this volume.

1.6.1 CNS trauma and ischaemia

Damage to the CNS is a major clinical problem, often resulting in more extensive injury than equivalent insults to other tissues. Free radical reactions have been implicated in CNS damage (Demopoulos *et al.*, 1982; Halliwell and Gutteridge, 1985), especially when ischaemic tissue is reoxygenated (Patt *et al.*, 1988), with suggestions that xanthine oxidase-generated radicals contribute to increased oxidative damage. The role of this enzyme in human brain reoxygenation injury, however, remains unclear, and the importance of free radical generation may be critically dependent on the extent and period of oxygen deprivation (Agardh *et al.*, 1991).

Trauma to the CNS leads to a rapid amplification of tissue damage and this probably involves free radical reactions stimulated by released iron and haem proteins. There is therefore considerable current interest in the possibility that antioxidants and chelating agents might be clinically useful in arresting the spread of tissue injury from the site of damage (Hall *et al.*, 1988a,b, 1990).

1.6.2 Parkinson's disease

In Parkinson's disease there is a selective and progressive destruction of the nigrostriatal dopaminergic neurones. Cohen (1988) suggested that the compensatory increase in dopamine turnover in the remaining dopamine neurones could lead to increased generation of H_2O_2 by monoamine oxidase. It has been proposed that iron-dependent free radical reactions contribute to damage of the substantia nigra seen in Parkinson's disease patients (reviewed by Halliwell, 1992). Consistent with this, increased nigral iron and decreased brain ferritin levels have been reported (Dexter *et al.*, 1989, 1991), as well as increased nigral iron, increased ferritin and decreased reduced glutathione (GSH) in parkinsonian brains (Riederer *et al.*, 1989).

CSFs taken from Parkinsonian patients show normal levels of total iron. However, Pall *et al.* (1987) have found increased levels of copper in CSF of some patients, although copper levels are not elevated in parkinsonian substantia nigra (Dexter *et al.*, 1989). When an iron salt is injected intranigrally into rats, it produces a 'parkinsonian-type' behaviour (Ben-Shachar and Youdim, 1991), consistent with the proposal that iron-driven oxidative reactions play some role in Parkinson's disease. However, insufficient data are available concerning the chemical nature and reactivity of iron present in human brain. Observations of oxidative damage in Parkinson's disease may be a consequence of a primary injury, perhaps due to the action of a toxin, which lead to important secondary events that contribute significantly to the disease pathology. The effects of DOPA treatment on brain oxidative damage are also unclear.

1.6.3 Alzheimer's disease

Controversy concerning the role of environmental exposure to aluminium and the development of Alzheimer's disease continues, based on contested findings of aluminosilicates at the cores of senile plaques and within neurones bearing neurofibrillary tangles (Garruto *et al.*, 1984; Perl and Brody, 1980; Birchall and Chappell, 1988). Aluminium ions are *known* to be neurotoxic, and can enter the brain attached to transferrin via transferrin receptors (Roskama and Connor, 1990). It has been reported that iron deposition is increased within or close to plaques and in neurofibrillary tangles (Youdim, 1988; Good *et al.*, 1992). This is particularly interesting in view of the oxidant synergism shown by iron and aluminium (Gutteridge *et al.*, 1985) and the reported interaction between superoxide and aluminium to form a 'stronger oxidizing agent' (Kong *et al.*, 1992). Aluminium salts can also stimulate phagocytes to generate reactive oxygen species (Evans *et al.*, 1989) increasing the potential for oxidative damage, and possibly promoting the release of iron from storage sites. Some plasma antioxidants have been reported to be low in Alzheimer's patients (Zaman *et al.*, 1992) and skin fibroblasts from such patients are reported to be more susceptible to oxidative damage (Piersanti *et al.*, 1992) consistent with a compromised defense system. It is far from clear, however, whether these observations are only a consequence of the disease pathology or whether they contribute substantially to the development of the disease.

1.6.4 Batten's disease

The neuronal ceroid–lipofuscinoses (NCLs) are a group of recessively inherited neurodegenerative lysosomal storage diseases which have a particularly high incidence in the Nordic countries. In all types of NCLs there is a widespread deposition of ceroid and lipofuscin pigments in the body, which has, in the past, led to oversimplistic claims that lipid peroxidation was a key event in disease progression. Recent studies in sheep, however, have shown that the deposited pigment is mostly made of a single protein with an M_r of 3500 which is the c subunit of ATP synthase (Fearnley *et al.*, 1990).

CSFs from NCL patients show increased levels of bleomycin-detectable iron under certain assay conditions (Heiskala *et al.*, 1988). Concentrations of 'catalytic' iron and copper have been claimed to increase in CSF of NCL patients as a function of age, which may represent increasing tissue destruction accompanying progressive degeneration (Heiskala *et al.*, 1988).

1.6.5 Abetalipoproteinaemia

In abetalipoproteinaemia, patients have a rare inborn error of lipid metabolism in which dietary fat is ingested and absorbed, but not transported out of the intestinal mucosal cells, because of an inherited inability to synthesize apoprotein B. Patients with abetalipoproteinaemia have negligible plasma vitamin E concentrations and eventually show neuropathy, retinal degeneration and abnormally shaped erythrocytes (acanthocytes). Recent studies have shown that both the neuropathy and the retinopathy can be prevented by giving patients very large oral doses of vitamin E (Runge *et al.*, 1986), supporting the proposal that vitamin E is an important brain antioxidant (Gross-Sampson and Muller, 1987).

1.7 Conclusions

Free radical reactions are a part of normal human metabolism. When produced in excess, radicals can cause tissue injury. However, tissue injury can cause more radical reactions, which may (or may not) contribute to a worsening of the injury. The careful use of a range of antioxidants, combined with new methods for measuring free radical generation in humans, is, at long last, enabling the exact contribution of free radical reactions to human disease to be evaluated and may allow the development and effective testing of new therapeutic agents.

Acknowledgements

JMCG is a BLF/BOC Senior Research Fellow in Respiratory Critical Care, and thanks the British Lung Foundation and the British Oxygen Group for their generous support. The authors also thank the British Heart Foundation, the Medical Research Council, the MAFF, the Cancer Research Campaign, the NIH and the Wellcome Trust for research support.

References

Agarh, C.D., Zhang, H., Smith, M.L. & Siesjö, B.L. (1991) *Int. J. Dev. Neurosci.* **9**, 127–138.
Aisen, P. (1992) *Ann. Neurol.* **32**, 562–568.

Ambrosio, G., Zweier, J.L. & Duilio, C. *et al.* (1993) *J. Biol. Chem.* **268**, 18532–18541.
Aruoma, O.I. & Halliwell, B. (1987) *Biochem. J.* **241**, 273–278.
Avraham, K.B., Shickler M., Sapoznikov D., Yarom, R. & Groner Y. (1988) *Cell* **54**, 823–839.
Balentine, J.D. (1982) *Pathology of Oxygen Toxicity*. Academic Press, New York.
Beckman, J.S. (1991) *J. Dev. Physiol.* **15**, 53–59.
Beckman, J.S., Chem, J., Ischiropoulos, H. & Crow, J.C. (1994) *Meth. Enzymol.* **233**, 229–240.
Ben-Shachar, D. & Youdim, M.B.H. (1991) *J. Neurochem.* **57**, 2133–2135.
Biemond, P., Van Eijk, H.G., Swaak, A.J.G. & Koster, J.F. (1984) *J. Clin. Invest.* **73**, 1576–1579.
Birchall, J.D. & Chappel, J.S. (1988) *Lancet* **ii**, 1008–1010.
Blake, D.R., Winyard, P., Lunec, J. *et al.* (1985) *Quart. J. Med.* **56**, 345–355.
Bleijenberg, B.G., Van Eljk, H.G. & Eijnse, B. (1971) *Clin. Chim. Acta* **31**, 277–281.
Bolann, B.J. & Ulvik, R.J. (1990) *Eur. J. Biochem.* **193**, 899–904.
Bolann, B.J. & Ulvik, R.J. (1993) *FEBS Lett.* **328**, 263–267.
Bowling, A.C., Schulz, J.B., Brown, R.H. Jr & Beal, M.F. (1993) *J. Neurochem.* **61**, 2322–2325.
Bralet, J., Schreiber, L. & Bouvier, C. (1992) *Biochem. Pharmacol.* **43**, 979–983.
Bray, W.G. & Gorin, M. (1932) *J. Am. Chem. Soc.* **54**, 2124–2125.
Bronk, S.F. & Gores, G. (1991) *Hepatology* **14**, 150–157.
Carney, J.H., Starke-Reed, P.E., Oliver, C.N., Landum, R.W., Cheng, M.S., Wu, J.F. & Floyd R.A. (1991) *Proc. Natl Acad. Sci. USA* **88**, 3633–3636.
Ceballos-Picot, I., Nicole, A., Briand, P., Grimber, G., Delacourte, A., Defossez, A., Javoy-Agid, F., Lafon, M., Blovin, J.L. & Sinet P.M. (1991) *Brain Res.* **552**, 198–214.
Chan, P.H., Chu, L., Chen, S.F., Carlson, E.J. & Epstein, C.J. (1990) *Stroke* **21** (Suppl. III), 80–82.
Chance, B., Sies, H. & Boveris, A. (1979) *Physiol. Rev.* **59**, 527–605.
Cochrane, C. (1991) *Mol. Aspects Med.* **12**, 137–147.
Cohen, G. (1988) In *Oxygen Radicals and Tissue Injury* (ed. Halliwell, B.), pp. 130–135, *FASEB*, Bethesda, Maryland.
Colton, C.A. & Gilbert, D.L. (1987) *FEBS Lett.* **223**, 284–288.
Curnutte, J.T. & Babior, B.M. (1987) *Adv. Hum. Genet.* **16**, 229–297.
Dallman, P.R., Simes, M.A. & Monies, E.L. (1975) *Br. J. Haematol.* **31**, 209–215.
Davies, M.J. (1990) *Free Radical Res. Commun.* **10**, 361–370.
Dawson, V.L., Dawson, T.M., London, E.D., Bredt, D.S. & Snyder, S.M. *Proc. Natl Acad. Sci. USA* **88**, 6368–6371.
Demopoulos, H.B., Flamm, E., Selligman, M. & Pietronigro, D.D. (1982) In *Pathology of Oxygen* (ed. Autor, A.P.), pp. 127–155. Academic Press, New York.
Deng, H.X. *et al.* (1993) *Science* **261**, 1047–1051.
Dexter, D.T., Wells, F.R., Lees, A.J., Agid, F., Agid, Y., Jenner, P. & Marsen, C.D. (1989) *J. Neurochem.* **52**, 1830–1836.
Dexter, D.T., Carayon, A., Javoy-Agid, F., Agid, Y., Wells, F.R., Daniel, S.E., Lees, A.J., Jenner, P. & Marsden, C.D. (1991) *Brain* **114**, 1953–1975.
Dykens, J.A., Stern, A. & Trenkner, E (1987) *J. Neurochem.* **49**, 1222–1228.
Evans, P.H. (1993) *Br. Med. Bull.* **49**, 577–587.
Evans, P.H., Klinowski, J., Yano, E. & Urano, N. (1989) *Free Radical Res. Commun.* **6**, 317–321.
Fearnley, I.M., Walker, J.E., Martins, R.D. *et al.* (1990) *Biochem. J.* **268**, 751–758.
Fenton, H.J.H. (1876) *Chem. News*, 25 April.
Fenton, H.J.H. (1894) *J. Chem. Soc.* **65**, 899–909.
Fostermann, U., Schmidt, H.H.H.W., Pollock, J.S., Sheng, H., Mitchell, A.J., Warner, T.D., Nakane, M. & Murad, F. (1991) *Biochem. Pharmacol.* **42**, 1849–1857.
Fraga, C.G., Oteiza, P.I., Golub, M.S. *et al.* (1990) *Toxicol. Lett.* **51**, 213–219.
Fridovich, I. (1978) *Science* **209**, 875–877.
Fridovich, I. (1989) *J. Biol. Chem.* **264**, 7761–7764.
Gardner, P.R. & Fridovich, I. (1991a) *J. Biol. Chem.* **266**, 1478–1483.
Gardner, P.R. & Fridovich, I. (1991b) *J. Biol. Chem.* **266**, 19328–19333.

Garruto, R.M., Fakatsu, R., Yanagihare, R. *et al.* (1984) *Proc. Natl Acad. Sci. USA* **81**, 1875–1879.
Gerschmann, R., Gilbert, D.L., Nye, S.W., Dwyer, P. & Fenn, W.O. (1954) *Science* **119**, 623–626.
Good, P.F., Perl, D. P., Bierer, L.M. & Schmeidler, J. (1992) *Ann. Neurol.* **31**, 286–292.
Groner, Y., Elroy-Stein, O., Avraham, K.B., Yarom, Y., Shickler, M., Knobler, H. & Rotman, G. (1990) *J. Physiol. (Paris)* **84**, 53–77.
Gross-Sampson, M.A. & Muller, D.P.R. (1987) *Neuropathol. Appl. Neurobiol.* **13**, 289–296.
Gutteridge, J.M.C. (1978) *Res. Commun. Chem. Pathol. Pharmacol.* **22**, 563–571.
Gutteridge, J.M.C. (1985) *FEBS Lett.* **185**, 19–23.
Gutteridge, J.M.C. (1986) *FEBS Lett.* **201**, 291–295.
Gutteridge, J.M.C. (1987) *Biochim. Biophys. Acta* **917**, 219–223.
Gutteridge, J.M.C. (1988) In *Free Radicals: Methodology and Concepts* (eds Rice-Evans, C. & Halliwell, B.), pp. 429–446. Richelieu Press, London.
Gutteridge, J.M.C. (1991) *Clin. Sci.* **81**, 413–417.
Gutteridge, J.M.C. (1992a) *Clin. Sci.* **82**, 315–320.
Gutteridge, J.M.C. (1992b) *Ann. Neurol.* **32**, S16–S21.
Gutteridge, J.M.C. (1993) *Free Radical Res. Commun.* **19**, 141–158.
Gutteridge, J.M.C. & Halliwell, B. (1988) In *Cellular Antioxidant Defense Mechanisms* (ed. Chow, C.K.), pp. 1–23. CRC Press, Boca Raton.
Gutteridge, J.M.C. & Hou, Y. (1986) *Free Radical Res. Commun.* **2**, 143–151.
Gutteridge, J.M.C. & Smith, A. (1988) *Biochem. J.* **256**, 861–865.
Gutteridge, J.M.C. & Stocks, J. (1981) *CRC Crit. Rev. Clin. Lab. Sci.* **14**, 257–329.
Gutteridge, J.M.C., Paterson, S.K., Segal, A.W. & Halliwell, B. (1981a) *Biochem. J.* **199**, 259–261.
Gutteridge, J.M.C., Rowley, D.A. & Halliwell, B. (1981b) *Biochem. J.* **199**, 263–265.
Gutteridge, J.M.C., Quinlan, G.J., Clark I.A. & Halliwell, B. (1985) *Biochim. Biophys. Acta* **835**, 441–447.
Gutteridge, J.M.C., Cao, W. & Chevion, M. (1991) *Free Radical Res. Commun.* **11**, 317–320.
Haber, F. & Weiss, J. (1934) *Proc. R. Soc. London* **147**, 332–351.
Hall, E.D. & Braughler, J.M. (1988) In *Oxygen Radicals and Tissue Injury* (ed. Halliwell, B.), pp. 92–98. *FASEB*, Bethesda, Maryland.
Hall, E.D., Yonkers, P.A., McCall, J.M. & Braughler, J.M. (1988a) *J. Neurosurg.* **68**, 456–461.
Hall, E.D., Yonkers, P.A. & McCall, J.M. (1988b) *Eur. J. Pharmacol.* **147**, 299–303.
Hall, E.D., Pazara, K.E., Braughler, J.M., Linemon, K.L. & Jacobsen, E.J. (1990) *Stroke* **21** (Suppl. III), 83–87.
Hallgren, B. & Sowander, P. (1958) *J. Neurochem.* **3**, 41–51.
Halliwell, B. (1978) *FEBS Lett.* **92**, 321–326.
Halliwell, B. (1987) *FASEB J.* **1**, 358–364.
Halliwell, B. (1989) *Free Radical Res. Commun.* **5**, 315–318.
Halliwell, B. (1992) *J. Neurochem.* **59**, 1609–1623.
Halliwell, B. & Gutteridge, J.M.C. (1984) *Lancet* **i**, 1396–1398.
Halliwell, B. & Gutteridge, J.M.C. (1985) *Trends Neurosci.* **8**, 22–26.
Halliwell, B. & Gutteridge, J.M.C. (1986) *Arch. Biochem. Biophys.* **246**, 501–514.
Halliwell, B. & Gutteridge, J.M.C. (1989) *Free Radicals in Biology and Medicine*, 2nd edn. Clarendon Press, Oxford.
Halliwell, B. & Gutteridge, J.M.C. (1990) *Methods Enzymol.* **186**, 1–85.
Halliwell, B. & Gutteridge, J.M.C. (1992) *FEBS Lett.* **397**, 108–112.
Halliwell, B., Aruoma, O.I., Mufti, G. & Bomford, A. (1988) *FEBS Lett.* **241**, 202–204.
Harris, D.C. & Aisen, P. (1973) *Biochim. Biophys. Acta* **329**, 156–158.
Harrison, W.W., Netsky, M.G. & Brown, M.D. (1968) *Clin. Chim. Acta* **21**, 55–60.
Heiskala, H., Gutteridge, J.M.C., Westermarck, T. *et al.* (1988) *Am. J. Med. Genet. S.* **5**, 193–202.
Hill, J.M. & Switzer, R.C. (1984) *Neuroscience*, **11**, 595–603.
Hope, B.T., Michael, G.J., Knigge, K.M. & Vincent, S.R. (1991) *Proc. Natl Acad. Sci. USA* **88**, 2811–2814.

Huie, R.E. & Padmaja, S. (1993) *Free Radical Res. Commun.* **18**, 195–199.
Jain, A., Martensson, J., Stole, E., Auld, P.A.M. & Meister, A. (1991) *Proc. Natl Acad. Sci. USA* **88**, 1913–1917.
Kagan, V.E., Serbinova, E.A. & Packer, L. (1990) *Biochem. Biophys. Res. Commun.* **169**, 851–857.
Kanner, J., German, J.B. & Kinsella, J.E. (1987) *CRC Crit. Rev. Food Sci. Nutr.* **25**, 317–374.
Kelman, D.J. & Mason, R.P. (1992) *Free Radical Res. Commun.* **16**, 27–33.
Komulainen, H. & Bondy, S.C. (1988) *Trends Pharmacol. Sci.* **9**, 154–156.
Kong, S., Liochev, S. & Fridovich, I. (1992) *Free Radical Biol. Med.* **13**, 79–81.
Laurindo, F.R.M., da Luz, P.L., Uint, L., Rocha, T.F., Jaeger, R.G. & Lopes, E.A. (1991) *Circulation* **83**, 1705–1715.
Link, E.M. (1988) *Arch Biochem. Biophys.* **265**, 362–372.
Liochev, S.I. & Fridovich, I. (1994) *Free Radical Biol. Med.* **16**, 29–33.
Macdonald, R.L. & Weir, B.K.A. (1991) *Stroke* **22**, 971–982.
McArthur, K.M. & Davies, M.J. (1993) *Biochem. Biophys. Acta* **1202**, 173–181.
McCord, J.M. & Day, E.D. Jr (1978) *FEBS Lett.* **86**, 139–142.
McCord, J.M. & Fridovich, I. (1969) *J. Biol. Chem.* **244**, 6049–6055.
McCord, J.M. & Russell, W.J. (1988) In *Oxy Radicals in Molecular Biology and Pathology* (eds Cerutti, P.A., Fridovich, I. & McCord, J.M.), pp. 27–35. Alan R. Liss, New York.
Maly, F.E. (1990) *Free Radical Res. Commun.* **8**, 143–148.
Marklund, S.L. (1984) *Biochem. J.* **222**, 649–655.
Martin, R.B. (1991) In *Aluminium in Chemistry, Biology and Medicine* (eds Nicolini, M., Zalta, P.F. & Corain, B.), pp. 3–20. Raven Press, New York.
Meier, B., Radeke, H., Selle, S., Raspe, H.H., Sies, H., Resch, K. & Habermehl, G.G. (1990) *Free Radical Res. Commun.* **8**, 149–160.
Meldrum, B. (1985) *Trends Neurosci.* **8**, 47–48.
Michel, H.S., Vaishnav, Y.B.N., Kempski, O., von Lubitz, D., Weiss, J.F. & Feuerstein, G. (1987) *Stroke* **18**, 426–430.
Moncada, S., Palmer, R.M.J. & Higgs, E.A. (1991) *Pharmacol. Rev.* **43**, 109–142.
Murrell, G.A.C., Francis, M.J.O. & Bromley, L. (1990) *Biochem. J.* **265**, 659–665.
O'Connell, M.J., Halliwell, B., Moorhouse, C.P., Aruoma, O.I., Baum, H. & Peters, T.J. (1986a) *Biochem. J.* **234**, 724–731.
O'Connell, M.J., Baum, H. & Peters, T.J. (1986b) *Biochem. J.* **240**, 297–300.
Odeh, M. (1991) *N. Engl. J. Med.* **324**, 1417–1422.
Ohtawa, M., Seko, M. & Takayama, F. (1983) *Chem. Pharm. Bull.* **31**, 1415–1418.
Olanow, C.W. (1990) *Neurology* **40** (Suppl. 3), 32–37.
Oliver, C.N., Starke-Reed, P.E., Stadtman, E.R., Liu, G.J., Carney, J.M. & Floyd, R.A. (1990) *Proc. Natl Acad. Sci. USA* **87**, 5144–5147.
Orrenius, S., McConkey, D.J., Bellomo, G. & Nicotera, P. (1989) *Trends Pharmacol. Sci.* **10**, 281–285.
Oury, T.D., Ho, Y.S., Piantadosi, C.A. & Crapo, J.D. (1992) *Proc. Natl Acad. Sci. USA* **89**, 9715–9719.
Pall, H.S., Williams, A.C., Blake, D.R., Lunec, J., Gutteridge, J.M.C., Hall, M. & Taylor, A. (1987) *Lancet* **ii**, 238–241.
Patt, A., Harkam, A.H., Burton, L.K., Rodell, T.C., Piermattei, D., Schorr, W.J., Parker, N.B., Berger, E.M., Horesh, I.R., Terada, L.S., Linas, S.L., Cheronic, J.C. & Repine, J.E. (1988) *J. Clin. Invest.* **81**, 1556–1562.
Pellegrini-Giampietro, D.E., Cherichi, G., Alesiani, M., Carla, V. & Moroni, F. (1988) *J. Neurochem.* **51**, 1960–1963.
Perl, D.P. & Brody, A.R. (1980) *Science* **208**, 297–299.
Piersanti, P., Tesco, G., Latorraca, S. *et al.* (1992) *Aging* **13**, S1–S111.
Puppo, A. & Halliwell, B. (1988) *Biochem. J.* **249**, 185–190.
Radi, R., Beckman, J.S., Bush, K.M. & Freeman, B.A. (1991) *J. Biol. Chem.* **266**, 4244–4250.
Raymond, V.N. & Carran, C.J. (1979) *J. Am. Chem. Soc.* **12**, 183–190.

Redhead, K., Quinlan, G.J., Das, R.G. & Gutteridge, J.M.C. (1992) *Pharm. Toxicol.* **70**, 278–280.
Remacle, C., Raes, M., Houbion, A. & Remacle, J. (1991) *Free Radical Res. Commun.* **14**, 323–334.
Riederer, P., Sofic, E., Rausche, W.D., Schmidt, B., Reynolds, G.P., Jallinger, K. & Youdim, M.B.H. (1989) *J. Neurochem.* **52**, 515–520.
Roskama, A.J. & Connor, J.R. (1990) *Proc. Natl Acad. Sci.* **87**, 9024–9027.
Runge, P., Muller, D.P.R., McAllister, J. *et al.* (1986) *Br. J. Ophthalmol.* **70**, 166–173.
Sadrzadeh, S.M.H., Graf, E., Panter, S.S., Hallaway, P.F. & Eaton, J.W. (1985) *J. Biol. Chem.* **259**, 14354–14356.
Sadrzadeh, S.M., Anderson, D.K., Panter, S.S., Hallaway, P.E. & Eaton, J.W. (1987) *J. Clin. Invest.* **79**, 662–664.
Sarafian, T. & Verity, M.A. (1991) *Int. J. Dev. Neurosci.* **9**, 147–153.
Saran, M. & Bors, W. (1989) *Free Radical Res. Commun.* **7**, 213–220.
Saran, M., Michel, C. & Bors, W. (1990) *Free Radical Res. Commun.* **10**, 221–226.
Schor, N.F. (1988) *Brain Res.* **456**, 17–21.
Scott, M.D., Meshnick, S.R. & Eaton, J.W. (1987) *J. Biol. Chem.* **262**, 3640–3645.
Sies, H. (1993) *Eur. J. Biochem.* **215**, 213–219.
Siesjö, B.K. (1988) *Crit. Care Med.* **16**, 954–963.
Siesjö, B.K. (1990) *News Physiol. Sci.* **5**, 120–125.
Smith, C.D., Carney, J.M., Starke-Reed, P.E., Oliver, C.N., Stadtman, E.R., Floyd, R.A. & Markesbery, W.R. (1991) *Proc. Natl Acad. Sci. USA* **88**, 10540–10543.
Steele, J.A., Stockbridge, N., Malijkovic, G. & Weir, B. (1991) *Circ. Res.* **68**, 416–423.
Stocks, J., Gutteridge, J.M.C., Sharp, R.J. & Dormandy, T.L. (1974a) *Clin. Sci.* **47**, 223–233.
Stocks, J., Gutteridge, J.M.C., Sharp, R.J. & Dormandy, T.L. (1974b) *Clin. Sci. Mol. Med.* **47**, 215–222.
Storz, G., Tartaglia, L.A. & Ames, B.N. (1990) *Science* **248**, 189–194.
Terada, L.S., Willingham, I.R., Rosandich, M.E., Leff, J.A., Kindt, G.W. & Repine, J.E. (1991) *J. Cell Physiol.* **148**, 191–196.
Touati, D. (1989) *Free Radical Res. Commun.* **8**, 1–9.
Turrens, J.F., Crapo, J.D. & Freeman, B.A. (1984) *J. Clin. Invest.* **73**, 87–95.
van der Vliet, M., O'Neill, C.A., Halliwell, B., Cross, C.E. & Kaur, H. (1994) *FEBS Lett.* **339**, 89–92.
Waters, W.A. (1946) *The Chemistry of Free Radicals.* Oxford University Press, Oxford.
Watson, B.D. & Ginsberg, M.D. (1988) In *Oxygen Radicals and Tissue Injury* (ed. Halliwell, B.), pp. 81–91. *FASEB*, Bethesda, Maryland.
Wendel, A. & Cikryt, P. (1980) *FEBS Lett.* **120**, 209–211.
Wills, M.R. & Savory, J. (1983) *Lancet* **ii**, 29–33.
Willmore, L.J., Balinger, W.E, Boggs, W., Sypert, G.W. & Rubin, J.J. (1980) *Neurosurgery* **7**, 142–146.
Willmore, L.J., Triggs, W.J. & Gray, J.D. (1986) *Brain Res.* **382**, 422–426.
Youdim, M.B.H. (1988) *Mt Sinai J. Med.* **55**, 97–101.
Youdim, M.B.H. & Green, A.R. (1978) *Proc. Nutr. Soc.* **37**, 173–179.
Zaleska, M.M. & Floyd, R.A. (1985) *Neurochem. Res.* **10**, 397–410.
Zaman, Z., Roche, S., Fielden, P. *et al.* (1992) *Age Ageing* **10**, 125–132.
Zhang, J. & Piantadosi, C.A. (1991) *J. Appl. Physiol.* **71**, 1057–1061.
Zhang, Y., Marcillat, O., Giulivi, C., Ernster, L. & Davies, K.J.A. (1990) *J. Biol. Chem.* **265**, 16330–16336.

CHAPTER 2

PATHOLOGICAL EVIDENCE FOR OXIDATIVE STRESS IN PARKINSON'S DISEASE AND RELATED DEGENERATIVE DISORDERS

P. Jenner* and C.W. Olanow†

*Neurodegenerative Diseases Research Centre, Pharmacology Group, Biomedical Sciences Division, King's College London, Manresa Road, London SW3 6LX, UK, and
†Department of Neurology, Mount Sinai Medical Center, One Gustave L Levy Place, Annenberg 1494, New York, NY 10029–6574, USA

Table of Contents

Neurodegeneration and Neuroprotection in Parkinson's Disease
ISBN 0-12-525445-8

2.1 Introduction

Parkinson's disease (PD) is one of the best understood of the neurodegenerative disorders which afflict man. Its core pathology and biochemistry have been established and there is effective treatment for the symptoms of the disorder, at least in its early stages. However, the cause of cellular destruction in PD remains a mystery. Primary loss of neuromelanin-containing dopamine cells in the zona compacta of the substantia nigra (SNc) is known to account for the motor symptoms of PD. But why selective destruction of these neurones occurs has proven difficult to unravel. Even less is known about pathological changes occurring in other areas of the brain in PD. Little attention has been focused on the degeneration of other pigmented brain stem nuclei, such as the locus coeruleus, as well as non-pigmented non-catecholamine-containing areas, such as the substantia innominata. What is common to the pathological process occurring in all these brain regions is the presence of Lewy bodies. Yet the significance of the Lewy body is unknown and it lacks disease specificity in much the same way as the plaques and tangles that have been identified with Alzheimer's disease (AD). So, a core hypothesis of the process leading to nigral cell death has been missing.

This situation has been altered dramatically in the last few years with the advent of a body of information suggesting that excessive free radical formation and oxidative stress form a common mechanism of cellular destruction underlying PD and perhaps other neurodegenerative illnesses (Olanow, 1993; Jenner, 1994). It is not known whether oxidative stress is a primary part of the pathology of such diseases or whether it is a secondary phenomenon that occurs subsequent to an alternate aetiological process. Nonetheless, it may still contribute to the progression of neuronal degeneration, and it provides a testable concept of cellular degeneration against which novel therapies can be evaluated.

The notion of free radical involvement in PD arose from the recognition that both the auto-oxidation and enzymatic metabolism of dopamine produces potentially toxic free radicals and reactive oxidant species (Halliwell and Gutteridge, 1985; Olanow, 1990). Dopamine auto-oxidation generates semiquinones ($SQ^{\cdot}$), the superoxide radical ($O_2^{\cdot -}$) and hydrogen peroxide (H_2O_2). Enzymatic metabolism of dopamine by monoamine oxidase (MAO) results in the formation of H_2O_2. Dopamine metabolism can thus generate H_2O_2 capable of reacting with ferrous iron and forming the highly reactive, cytotoxic hydroxyl radical ($OH^{\cdot}$). These reactions are catalysed by transition metals such as iron which have been shown to selectively accumulate in neuromelanin granules within dopaminergic neurones. Dysregulation of dopamine metabolism could thus lead to an increased steady state concentration of H_2O_2 capable of reacting with iron in dopaminergic neurones and generating excess free radicals with consequent cell loss due to oxidative damage. In addition, 1-methyl-4-phenyl-1,2,3,6-tetrahydropyridine (MPTP), which can induce selective degeneration of nigral neurones via its metabolite 1-methyl-4-phenylpyridinium (MPP^+), has been shown to inhibit complex I of the mitochondrial respiratory chain and to induce secondary oxidative stress (Singer *et al.*, 1987). Such evidence has provoked an examination of post-mortem brain material from patients dying with PD

Table 1 Iron and Parkinson's disease

Iron levels increased in the midbrain	Earle (1968)
Increased iron on magnetic resonance imaging	Olanow (1992)
Increase in iron and total iron in the substantia nigra	Sofic *et al.* (1988) Riederer *et al.* (1989)
Increased total iron in the substantia nigra but not other brain regions	Dexter *et al.* (1989b) Dexter *et al.* (1991)
Increase in iron selective to the zona compacta	Sofic *et al.* (1991)
Altered distribution of iron in the substantia nigra	Jellinger *et al.* (1990) Hirsch *et al.* (1991) Good *et al.* (1992) Morris and Edwardson (1994)
No increase in iron	Uitti *et al.* (1989)
Iron infusion induced model of parkinsonism	Sengstock (1994) Ben-Schacher and Youdim (1991)

and related basal ganglia degenerative illnesses to establish whether evidence for free radical damage or mitochondrial change can be detected. The basic hypothesis has been supported by the discovery of increased iron levels, alterations in mitochondrial function and changes in the levels of reduced glutathione in the SNc of PD patients (for reviews, see Jenner *et al.*, 1992; Olanow *et al.*, 1992; Jenner, 1993). Many of these observations are examined in more detail in subsequent chapters in this volume. The object of this chapter is to review the pathological data which has led to the conclusion that free radical damage may, indeed, be an important part of the pathogenesis of PD.

2.2 Alterations in iron in PD

Several lines of evidence support a role for iron in the pathogenesis of PD (Table 1). Iron is a transition metal that has the capacity to stimulate free radical formation. In particular, iron is able to convert H_2O_2 into the highly toxic $OH^{\cdot}$ by way of the Fenton reaction (Olanow *et al.*, 1992; Youdim *et al.*, 1993; Gerlach *et al.*, 1994). The concept of altered iron metabolism in PD is not new. In the 1960s, Earle used X-ray spectrometry to show that iron levels were increased in the midbrain in fixed material from PD although the precise area affected was not delineated (Earle, 1966). More recently, a host of investigations have been undertaken using a variety of imaging and analytical techniques to study iron changes within various brain regions in PD and related degenerative disorders. There is general agreement that there is a 30–40% increase in the levels of iron in the substantia nigra (particularly in the SNc) of patients dying with PD (Sofic *et al.*, 1988; Dexter *et al.*, 1989; Jellinger *et al.*, 1990; Hirsch *et al.*, 1991; Good *et al.*, 1992; Jellinger *et al.*, 1992; Olanow *et al.*, 1992). Iron accumulation does not appear to be due to drug treatment as early studies were undertaken prior to the introduction of levodopa therapy.

Recent studies have concerned themselves with the localization of iron changes at a regional and subcellular level. Alterations in iron levels in PD appear to be specific to the SNc as they are not found in other areas of basal ganglia that have been studied including the substantia nigra pars reticularis (SNr) (Sofic *et al.*, 1988; Dexter *et al.*, 1989b). Using a semiquantitative histological technique to evalute the distribution of brain iron in PD, Jellinger and colleagues showed increased iron in astrocytes, macrophages, reactive microglia, non-pigmented neurones, and in damaged areas devoid of pigmented neurones (Jellinger *et al.*, 1990). In agreement, Hirsch *et al.*, using X-ray microanalysis, showed that in comparison to control individuals iron levels in the SNc of PD patients were increased in regions devoid of neuromelanin and reduced in neuromelanin aggregates (Hirsch *et al.*, 1991). However, these techniques utilize thick sections that do not permit precise histological identification. Using laser microprobe mass analysis (LAMMA), which utilizes semithin sections and permits precise histological identification, we showed that within the SNc in PD, iron does significantly accumulate within neuromelanin granules (Good *et al.*, 1992). No evidence of increased iron was detected in regions of non-melanized cytoplasm of nigral neurones or in the adjacent neuropil. Similarly, Jellinger and colleagues using energy-dispersive X-ray microanalysis demonstrated significant iron peaks only in the highly electron-dense regions of neuromelanin granules of SNc neurones in PD patients (Jellinger *et al.*, 1992). A massive increase in aluminium levels within neuromelanin granules of SNc neurones has also been detected (Good *et al.*, 1992). As aluminium can promote the oxidative damage associated with iron-induced oxidative stress (Gutteridge *et al.*, 1985), these studies provide further support for the notion that oxidative damage contributes to the pathogenesis of cell death in PD.

It should be pointed out that not all studies have shown iron levels to be increased (Uitti *et al.*, 1989). The reason for this is not clear but may relate to the technique employed, the stage of the illness, the manner in which the post-mortem tissue was handled, or the number of patients studied. Indeed, it should be pointed out that as with many of the biochemical changes found in post-mortem brain from PD, there is no absolute separation between normal individuals and those with PD. Thus, while the average group values are increased, there is considerable overlap between the normal population and the parkinsonian subjects.

Iron is more reactive when it is unbound and present in the ferrous (Fe^{2+}) form and is less reactive when it is in the ferric (Fe^{3+}) form complexed to an iron-binding protein such as transferrin or ferritin (Halliwell and Gutteridge, 1988). It has been claimed that the excess iron in the SNc is present in the Fe^{3+} form (Sofic *et al.*, 1988). However, it has not been established that the analytical techniques employed can accurately determine the form in which iron existed during life. There is also controversy over whether iron in the SNc exists in a free or sequestered form. This is an important issue since it affects the degree of reactivity of the excess iron deposited in PD. Normally, it would be anticipated that an increase in iron would stimulate the formation of ferritin into which iron can be sequestered and rendered relatively inactive. However, in our own study of brain ferritin levels in PD, the ferritin content in the SNc and selected other brain areas was shown to be decreased (Dexter *et al.*, 1990). In contrast, Riederer and colleagues showed an increase in the ferritin content of the SNc in PD (Riederer *et al.*, 1989). This finding was confirmed by a subsequent histological analysis (Jellinger *et al.*, 1990). There are several reasons why the results of these two studies may have differed. Antibodies recognizing the L isoform of ferritin used in these studies were raised in different species and from different organs (liver and spleen) and thus may recognize different subunits of brain ferritin. There have been questions as to whether tissues employed in one of the investigations (Riederer *et al.*, 1989) came from patients with Parkinson's plus syndromes (multiple system atrophy, progressive supranuclear palsy) rather than idiopathic PD. There were also differences in the methodologies employed for measuring the ferritin content and in the way in which ferritin was extracted. We have subsequently found that repeated freeze–thawing of tissues can alter dramatically the apparent tissue content of ferritin (unpublished observations). Thus, there can be a switch from a decrease in ferritin levels to an apparent increase using the same tissue preparations. This may reflect the ability of repeated freeze–thawing to break ferritin into its component subunits which are then individually recognized by the antibodies employed. Indeed, in a further study carried out by our own group using a different analytic technique, we were unable to confirm the decrease in ferritin levels shown in our initial studies but still could show no rise (Mann *et al.*, 1994). More recently, Connor and colleagues have demonstrated that H-chain or neuronal ferritin does not rise in response to increased iron levels in the SNc of PD patients as might normally be anticipated (Connor *et al.*, 1995).

Where the excess iron comes from in PD is unknown. Iron gains access to the brain and to neurones via transferrin receptor-mediated endocytosis (Aisen, 1992; Roberts *et al.*, 1992). There is evidence of a decrease in the number of transferrin receptors in membrane fractions from the putamen of patients with PD and in the striatum of MPTP-treated animals (Mash *et al.*, 1991). However, these findings contrast with a more recent study suggesting that the density of transferrin receptors is unchanged throughout the brain in PD (Faucheux *et al.*, 1993). An alteration in the blood brain barrier could account for increased iron in PD. Some support for this concept comes from preliminary ^{52}Fe studies utilizing positron emission tomography (PET) (Leenders *et al.*, 1993). Riederer and colleagues also reported that transferrin levels in the cerebrospinal fluid (CSF) in PD were decreased (Riederer *et al.*, 1988). A variety of additional studies measuring serum levels of iron, transferrin and ferritin and CSF levels of transferrin, ferritin and caerulplasmin have been unable to find any alterations to support this concept (Cabrera-Valdivia *et al.*, 1994; Kuiper *et al.*, 1994; Loeffler *et al.*, 1994a).

Iron is known to be toxic to dopaminergic cells. Iron alone or in combination with levodopa, other catecholamines or neuromelanin can kill *in vitro* preparations of dopaminergic neurones (Tanaka *et al.*, 1991; Michel *et al.*, 1992; Mochizuki *et al.*, 1993). In addition, we and others have shown that intranigral infusion of iron induces a dose-related loss of nigral dopaminergic cells and striatal dopamine (Ben-Schachar and Youdim, 1991; Sengstock *et al.*, 1992, 1993). However, the relationship between iron and the pathology of PD remains unproven. There is no doubt that iron accumulates in the substantia nigra but it is not clear in which form it exists and the extent to which it is reactive. There is also doubt as to the stage in the illness when iron accumulation occurs (see later). Further, iron accumulation is not specific to PD and occurs in a number of other neurodegenerative disorders. For example, we have demonstrated increased iron in areas of neurodegeneration in progressive supranuclear palsy (PSP), multiple system atrophy (MSA) and Huntington's chorea (HC) (Dexter *et al.*, 1991; Olanow and Hauser, 1994). In these conditions, increased iron is noted in areas of pathological damage. For example, iron levels are increased in the SNc in PD, MSA and PSP, and in the striatum in MSA and HC, consistent with the sites of pathological damage in these conditions. In contrast, iron levels are not increased in the SNc in HC or in the striatum in PD where there is little or no pathology. Increased iron levels have also been noted in the Hallevorden–Spatz syndrome, AD, multiple sclerosis, spastic paraplegia, pallidonigro-Luysian degeneration and guamanian neurodegeneration (Valberg *et al.*, 1989; Arena *et al.*, 1992; Connor *et al.*, 1992a,b; Kawai *et al.*, 1993; Yasui *et al.*, 1993; Olanow, 1994). Further, we and others have demonstrated that increased iron levels in the SNc can occur as a consequence of 6-hydroxydopamine (6-OHDA) lesions of the nigrostriatal tract (Oestreicher *et al.*, 1994) or MPTP administration (Temlett *et al.*, 1994).

These studies indicate that iron accumulation is not selectively associated with nigral cell damage or with PD. It would seem, therefore, that alterations in iron metabolism in PD must presently be viewed as a probable secondary phenomenon common to many degenerative illnesses and that a primary change in iron-mediated systems cannot be directly implicated in the aetiopathogenesis of PD. This does not mean that alterations in iron levels are not an important component of the neurodegenerative process. Iron deposition, regardless of the aetiology, may accelerate the pathological process and the rate of neuronal degeneration. These concepts are discussed in a later chapter.

2.3 Alterations in mitochondrial function

The possibility of a mitochondrial defect in PD was suggested by evidence that mitochondrial dysfunction is the mechanism responsible for MPTP parkinsonism (reviewed by Singer *et al.*, 1987). Within the brain, MPTP is oxidized to MPP^+, which is actively accumulated by mitochondria where it acts to inhibit mitochondrial respiratory activity at the level of complex I. MPP^+ also inhibits the non-respiratory enzyme α-ketoglutarate dehydrogenase (α-KGDH), which may also inhibit mitochondrial function and cellular energy production (Mizuno *et al.*, 1987). The actions of MPP^+ are shared by a range of isoquinoline derivatives which have been proposed as endogenous neurotoxins (McNaught *et al.*, 1995). These also inhibit complex I and α-KGDH with potencies even greater than that of MPP^+.

Schapira and colleagues (1990a) and Mizuno *et al.* (1989) were the first to show that there was a specific decrease in the activity of complex I in homogenates of substantia nigra from patients with PD. No alterations in the levels of enzyme activity associated with complexes II, III or IV were detected. This work has subsequently been extended to show that complex I activity is preserved in other areas of the basal ganglia in PD patients (Schapira *et al.*, 1990b). Inhibition of complex I also appears to be restricted to PD since similar changes are not found in the SNc of MSA patients or in the caudate–putamen in patients with HC (Schapira, 1994). Complex I inhibition does not appear to be related to drug treatment since MSA patients had been treated with levodopa in similar doses and for similar periods of time as PD patients. In addition, no changes in complex I were detected in the striatum of PD patients despite levodopa accumulation. There have been various investigations of complex I activity in platelets and muscle in PD. This issue has been comprehensively reviewed by Schapira in Chapter 7.

The localization of the change in complex I activity in the SNc in PD is disputed. Using an immunohistochemical technique, reduced complex I immunoreactivity in PD was localized to melanized nigral neurones (Hattori *et al.*, 1991). However, it is difficult to account for a 30% decrease in complex I activity in SNc by a selective change in the neuronal population which comprises only approximately 2% of the total number of cells within that tissue sample. The magnitude of decrease in complex I activity observed suggests that at least part of the change must occur in non-neuronal elements.

The cause of the complex I deficiency is also unclear. Changes in subunits of complex I have been linked to a common deletion of mitochondrial DNA (Ikebe *et al.*, 1990). However, these results were obtained using areas such as the striatum in which no overall complex I defect is apparent (Lestienne *et al.*, 1990; Schapira *et al.*, 1990c). Others have found no differences in the extent to which the common mitochondrial deletion occurs in PD other than that expected with the normal aging process. At present, there appears to be no obvious alteration in encoding for complex I which would explain the enzyme deficiency observed. An alternative possibility would be the presence of a neurotoxin with actions equivalent to that of MPP^+. However, using a range of antibodies raised against MPTP- and MPP^+-like substances, Markey and colleagues were unable to show immunoreactivity within the substantia nigra of patients dying with PD.

Changes in non-respiratory mitochondrial enzymes in PD have largely been ignored. These may be important since a 30% decrease in complex I activity is likely to be insufficient to alter cellular energy metabolism. A recent study notes that SNc immunostaining for α-KGDH is decreased in PD (Mizuno *et al.*, 1994). Inhibition of α-KGDH would decrease the reducing equivalents available to complex II. A

Table 2 Antioxidant systems in PD

Glutathione peroxidase and catalase decreased in the substantia nigra	Ambani *et al.* (1975)
Small decrease in glutathione peroxidase in the substantia nigra	Kish *et al.* (1985)
Catalase, and glutathione reductase normal in PD	Marttila *et al.* (1988)
Glutathione peroxidase, glutathione reductase and glutathione transferase are normal in substantia nigra	Sian *et al.* (1994a)
Superoxide dismutase increased but dispute over isoenzyme involved	Marttila *et al.* (1988), Saggu *et al.* (1989)
Ascorbic acid levels normal in the substantia nigra	Riederer *et al.* (1989)
α-Tocopherol levels normal in the substantia nigra	Dexter *et al.* (1992)

combined complex I and complex II impairment would be more likely to reduce ATP production than a defect in complex I alone. The alterations in α-KGDH immunostaining in PD appear to be localized to melanized dopaminergic neurones in the SNc. Attempts to measure enzyme activity using nigral homogenates have so far been unsuccessful, presumably reflecting post-mortem degeneration. Such changes, however, would not influence the immunoreactivity of antibodies to α-KGDH.

2.4 Alterations occurring in protective enzyme systems

Free radical activity can induce damage to brain if normal defence mechanisms are overwhelmed or defective. Free radical activity is normally limited by the actions of superoxide dismutase (SOD), glutathione peroxidase (GPX), catalase, α-tocopherol (vitamin E) and ascorbic acid (vitamin C). These have all been studied in post-mortem tissues in patients with PD (Table 2).

In a single study, catalase activity in the brain of PD patients appeared to be lower than that in age-matched controls (Ambani *et al.*, 1975). Indeed, in the substantia nigra and putamen, catalase activity was significantly decreased by 30–50%. However, no recent study of catalase has been undertaken to confirm these findings. Early studies also suggested that GPX activity was lower in the brain of patients with PD compared to age matched controls, and a significant decrease was reported in the substantia nigra, caudate nucleus and putamen (Kish *et al.*, 1985, 1986). However, these results are disputed since another early study of GPX showed only a 19% reduction in substantia nigra (Ambani *et al.*, 1975). Recent studies failed to show any alteration in GPX activity (Marttila *et al.*, 1988; Sian *et al.*, 1994a). No alterations in GPX activity have been detected in any brain area in HC, MSA or PSP. Further investigations are warranted.

There have been limited studies of the antioxidant vitamin levels in the SNc in PD. Riederer and colleagues reported that the levels of vitamin C were unchanged in PD (Riederer *et al.*, 1989). We have similarly reported that levels of α-tocopherol were also unaltered (Dexter *et al.*, 1992). However, post-mortem studies may not provide a true indication of antioxidant activity. First, the post-mortem stability of vitamin C and vitamin E, and the effect of post-mortem delay and freezing have not been studied. Second, the rate of conversion between the oxidized and reduced forms of these vitamins has not been measured and may be considerably increased in the presence of excess free radical formation.

Changes in SOD activity in PD have been measured in two recent studies. In our own study, we found an increase in the inducible Mn-dependent mitochondrial form of SOD in the SNc but not in the cerebellum of PD patients (Saggu *et al.*, 1989). No changes were detected in the constituent Cu/Zn-dependent cytosolic form of the enzyme. In contrast, Marttila and colleagues noted an increase in the cytosolic form of SOD but not in mitochondrial levels of the enzyme (Marttila *et al.*, 1988). Solubilization of mitochondria with leakage of enzyme into the cytosol might account for the differences observed. Whatever the explanation, there is agreement that SOD activity is increased in the SNc in PD. Interestingly, Mn-dependent SOD activity but not the Cu/Zn-dependent isoenzyme was increased in the CSF of PD patients (Yoshida *et al.*, 1994).

Why alterations in SOD activity occur in PD is not clear but a number of explanations are possible. First, an increase in SOD may contribute to pathogenesis by increasing levels of H_2O_2, which may itself be toxic or may be transformed to the cytotoxic $OH^{\cdot}$ radical. This is an unlikely sequence as $O_2^{\cdot -}$ spontaneously decomposes to H_2O_2 at rates which are almost equal to those catalysed by SOD. More likely, an increase in SOD represents a compensatory change. A reduction in the glutathione system is often associated with reciprocal changes in SOD activity. An increase in SOD activity may also reflect mitochondrial dysfunction or auto-oxidation of dopamine.

2.5 Alterations in glutathione content

Reduced glutathione (GSH) detoxifies H_2O_2 to water in a reaction catalysed by GPX. Oxidized glutathione (GSSG) generated by this reaction is reduced to GSH by the enzyme glutathione reductase. These reactions constitute one of the principal defenses of the brain against oxidant stress.

Thomas Perry first reported alterations in GSH in the substantia nigra in PD (Perry, 1982, 1986, 1988). These studies indicated that GSH was markedly reduced in the substantia nigra but the investigations were considered to be flawed. There was criticism of the reported ratio of GSH to GSSG which was attributed to either post-mortem artefacts or to the analytical technique employed (Slivka *et al.*, 1987a).

Table 3 Glutathione and PD

Levels of GSH reduced in the substantia nigra in PD	Perry *et al.* (1982) Perry and Yong (1986) Perry *et al.* (1988)
Doubts due to methodology and post-mortem artefacts	Slivka *et al.* (1987a)
Total glutathione levels decreased in the substantia nigra parallel with disease severity	Riederer *et al.* (1989)
Decrease in GSH levels in the substantia nigra confirmed	Jenner (1992) Sofic *et al.* (1992) Sian *et al.* (1994)
Increase in γ-glutamyltranspeptidase in the substantia nigra	Sian *et al.* (1994)

Consequently, the possibility of an alteration in GSH in PD was largely ignored. Subsequently, Riederer and colleagues showed a reduction in total glutathione levels in the SNc in PD that paralleled the severity of disease (Riederer *et al.*, 1989). Still, this investigation did not resolve the issue of the relative alteration in the levels of GSH and GSSG.

Recently, we and others have returned to the issue of GSH in PD and related disorders, using modern analytical techniques and paying particular attention to the potential of post-mortem artefacts. These studies showed a decrease in levels of GSH without an accompanying alteration in levels of GSSG in the SNc which was not found in other brain regions examined (Sofic *et al.*, 1992; Sian *et al.*, 1994). These changes appear to be relatively specific for PD since they are not seen in other degenerative disorders affecting the SNc such as MSA and PSP (Sian *et al.*, 1994b).

The precise cause of the GSH defficiency in PD is unknown. Studies using sections of nigral tissue stained with mercury orange to detect thiol groups indicate that the majority of GSH is in neuronal processes and glia (Slivka *et al.*, 1987b). In PD, mercury orange histofluorescence is dramatically reduced and appears to reflect the loss of neuronal processes. Consequently, the decrease in GSH might merely reflect the extent of neuronal degeneration. However, the lack of decrease in GSH levels in the SNc in MSA or PSP despite extensive neuronal loss makes this interpretation unlikely. Further, the extent of the decrease in GSH that occurs in PD is not readily accounted for by neuronal degeneration and implies involvement of glia which are another major source of GSH in SNc. Further, alterations in GSH closely parallel changes in mitochondrial function, suggesting that these phenomena may be interrelated. This raises the possibility that oxidant stress consequent to GSH deficiency damages mitochondria and/or that a bioenergetic defect due to mitochondrial dysfunction leads to diminished GSH synthesis.

There is no apparent alteration in the rate-limiting enzyme for GSH formation (γ-glutamylcysteine synthetase), the enzymes responsible for its oxidation and reduction (glutathione peroxidase and glutathione reductase), or the enzyme utilized in the formation of mercapturic acids (glutathione transferase). However, changes have been detected in the level of γ-glutamyltranspeptidase (γ-GTP) activity that normally clears excess GSSG. In the SNc in PD there is an approximate doubling of γ-GTP activity which is not seen in the SNc of patients with MSA or PSP (Sian *et al.*, 1994). γ-GTP is a cell membrane enzyme concerned with conservation of the peptide precursors of glutathione. Why γ-GTP activity is altered in PD is not clear but a number of hypotheses can be proposed. γ-GTP activity may increase as an adaptive response to the loss of GSH in an attempt to conserve the intracellular glutathione content. Alternatively, alterations in γ-GTP activity may occur as a result of GSH leakage from cells due to disruption of cell membranes resulting from the disease process. Increased γ-GTP activity may even contribute to the pathological process occurring in PD, since cysteinyl residues resulting from its actions on glutathione may themselves exert free radical-mediated toxicity (Zhang and Dryhurst, 1994). An attractive hypothesis is that γ-GTP activity increases in response to a rise in GSSG as a result of its increased formation from GSH due to oxidative stress. As GSSG can itself be toxic, its removal by γ-GTP may be protective. One could speculate that under conditions of continuing oxidative stress, GSH is converted to GSSG which is then removed from the cell by γ-GTP. This would result in a loss of GSH without a corresponding increase in GSSG as has been observed in PD.

Alternatively, a bioenergetic defect secondary to impaired mitochondrial function could result in decreased GSH synthesis. There are some clues as to the relationship between GSH and GSSG from impaired mitochondrial function in hepatocytes. The mitochondrial inhibitors MPP^+, potassium cyanide and antimycin A all inhibit the mitochondrial respiratory chain and cause a loss of GSH from hepatocytes without a corresponding increase in the level of GSSG (Mithofer *et al.*, 1992). This also mimics what occurs in PD and raises the possibility that alterations in GSH levels could be a reflection of altered mitochondrial activity.

A summary of glutathione and related changes is presented in Table 3.

2.6 Oxidative damage in PD

The changes in brain iron concentrations, mitochondrial function, protective enzymes and GSH are indicative of increased free radical activity. However, they do not prove that free radical damage occurs. The finding of oxidative damage would support the notion that oxidant stress was a contributing factor to cell death in PD. There is some evidence to suggest that oxidative damage does occur in the SNc in PD. Lipid peroxidation is thought to occur based on the finding of increased levels of malondialdehyde (MDA) (Dexter *et al.*, 1989a). This would suggest free radical damage to lipid membranes. However, MDA levels may have been falsely elevated as a consequence of post-mortem lipid peroxidation due to release of bound iron during homogenization (Olanow and Arendash, 1994). Consequently, we recently examined another index of lipid peroxidation, namely the formation of lipid hydroperoxides (Dexter *et al.*, 1994a). A 10-fold increase in lipid hydroperoxide formation was found in the SNc of PD patients compared to control individuals. In this studies, post-mortem lipid peroxidation was inhibited by iron chelators. So, increased lipid peroxidation indicative of a free radical-mediated mechanism may indeed occur in PD.

Very recently, evidence of oxidative damage to DNA has also been observed in the SNc in PD in the form of increased levels of 8-hydroxy-2-deoxyguanosine (Sanchez-Ramos *et al.*, 1994). While these studies need to be corroborated, 8-hydroxy-2-deoxyguanosine is a specific marker of oxidative damage to DNA and provides further evidence of free radical-mediated damage in PD. However, it remains necessary to undertake a comprehensive analysis of oxidative damage to lipids, proteins and DNA in order to ascertain the extent to which free radicals contribute to neurodegeneration in PD.

2.7 Role of levodopa in alterations in indices of oxidative stress

Post-mortem studies in PD are complicated by concern over the possibility that pathological changes and evidence of oxidative stress might be due to drug treatment, particularly levodopa. The metabolism of levodopa to dopamine and its subsequent enzymatic and auto-oxidation have the potential to promote oxidative stress and damage nigral neurones. So, it is impossible to ignore the possibility that levodopa contributes to the markers of oxidative damage observed in PD. Levodopa is toxic to neuronal cells in culture and synergizes with iron in producing cell toxicity, presumably through the production of $OH^{\cdot}$ radicals (Tanaka *et al.*, 1991; Mytilineou *et al.*, 1993; Ogawa *et al.*, 1993). Levodopa has been shown to increase GSSG levels in striatal synaptosomes (Spina and Cohen, 1988), although a recent *in vivo* study failed to find any levodopa effect on brain GSSG content (Loeffler *et al.*, 1994b). In other studies, chronic administration of levodopa to rats and to neuroblasts decreased complex I activity (Przedborski *et al.*, 1993). However, the extent of complex I inhibition was less than occurs in the SNc in PD despite the high doses of levodopa employed.

There is also evidence for the *in vivo* formation of free radicals from levodopa. In a recent *in vivo* microdialysis study, levodopa administration caused a pronounced increase in $OH^{\cdot}$ production (Spencer Smith *et al.*, 1994a), and levodopa administration to 6-OHDA-lesioned rats increased the degree of lipid peroxidation (Tanaka *et al.*, 1992; Ogawa *et al.*, 1994). Oxidative damage produced by levodopa may be particularly prominent in the presence of specific metal ions. Levodopa, dopamine, and 3-O-methyldopa cause extensive oxidative damage to DNA which is increased in the presence of H_2O_2 and traces of copper ions (Spencer *et al.*, 1994). This was associated with the formation of 8-hydroxy-2-deoxyguanosine, which is also elevated in the SNc in PD.

The results of these studies emphasize the importance of determining precisely if levodopa contributes to the oxidative stress observed in PD. At present, we resort to comparisons of findings in PD, MSA and PSP, since these illnesses are associated with nigral degeneration and treated with equivalent amounts of levodopa. In addition, we compare findings in the SNc with other brain areas such as the caudate and putamen, in which levodopa also accumulates. Such comparisons are clearly unsatisfactory, and more direct evidence of the precise effect of levodopa on indices of oxidative stress is required.

Very recently, we obtained samples of brain material from normal monkeys chronically treated with high doses of levodopa for a 3 month period. We are presently examining these to determine whether indices of oxidative stress are affected by chronic levodopa treatment. Preliminary studies using LAMMA show no change in nigral iron content (unpublished observations). Further studies are aimed at detecting any alterations in glutathione content or mitochondrial activity. Still, the effect of levodopa may be different in the normal SNc in comparison to the PD nigra which is in a state of oxidant stress and may have reduced defense capabilities. A determination of the effect of levodopa on nigral neurones in PD is of crucial importance as levodopa remains the primary symptomatic treatment of PD.

2.8 Oxidative stress and incidental Lewy body disease

Studies of oxidative stress in PD have largely been undertaken in brain material from patients at the end of their illness. This raises the question as to whether the alterations observed are primary or secondary and whether similar changes would be present early in the illness. Ideally, this problem could be solved by studying brain tissue from patients at different stages of the disease, but this material is not readily available. To overcome this problem, we have turned to another condition, namely, incidental Lewy body (ILB) disorder. ILBs and nigral degeneration are found at post mortem in 10–15% of normal individuals over the age of 60 years, and has been considered to represent presymptomatic PD (Gibb and Lees, 1988; Fearnley and Lees, 1991). This presumes that had these individuals lived longer they would have developed clinical features of PD. Of course, this cannot be established with certainty. Others have viewed ILB disease as a mild form of stable pathology which does not progress into clinical PD. Whatever the interpretation, ILB disease provides an opportunity to study the association of mild nigral pathology with indices of oxidative stress.

Iron levels in the SNc in ILB disease were normal (Dexter *et al.*, 1994b). In addition, no alterations in ferritin levels were found. To the extent that ILB disease represents a preclinical form of PD, this suggests that the iron changes observed in PD occur later in the disease and are non-specific or secondary. However, studies in ILB utilized bulk analytical techniques which may have failed to recognize a modest increase confined to subcellular areas such as neuromelanin granules, the site of maximal iron accumulation in PD. More sensitive studies using LAMMA are pending. Measurements of complex I activity were not significantly decreased in patients with

ILBs. However, values were halfway between those found in normal controls and patients with advanced PD. In contrast, SNc levels of GSH in ILB disease were reduced to the same extent as in PD. In addition, there was no increase in GSSG. The finding of equivalent decreases in GSH levels in PD and ILB disease argues against this reduction being merely a marker of neuronal loss. Interestingly, the induction of oxidant stress in cultured dopamine neurones is associated with a compensatory rise in GSH levels and protection against the effects of toxins such as MPP^+ (Mytilineou *et al.*, 1993). Thus, the finding that patients who may be in the early stages of PD have decreased GSH levels without an accompanying rise in iron raises the possibility that a defect in GSH is a primary pathogenetic feature of PD.

At this time, indices of free radical formation and oxidative damage such as lipid peroxidation, DNA damage and protein oxidation have not been assessed in ILB disease. Further study of ILB cases may provide much information on the role of oxidative stress in the pathogenesis of PD. In particular, since these patients have no neurological deficits in life and have not been treated with levodopa, the effect of the disease process can be assessed in the absence of drug-mediated changes.

2.9 Other related studies

Studies of oxidative stress in PD have largely concentrated on changes in the SNc. It remains to be determined whether similar changes occur in other affected brain regions. There have been no systematic studies of pigmented brainstem nuclei, such as the locus coeruleus, or non-pigmented brainstem nuclei, such as the substantia innominata, which are also affected by Lewy body pathology. Preliminary studies suggest that there is no alteration in iron, GSH, or mitochondrial complex I activity (unpublished observations). However, the degree of pathology in these regions is small and may be insufficient for changes to be detected.

Cases with cortical Lewy bodies (so-called Lewy body disease) have been examined separately (unpublished information). In the cingulate cortex there is an elevation of iron content and a reduction in GSH but no change in complex I activity. Patients with AD also show evidence of oxidant stress with increased iron and decreased GSH levels (Adams *et al.*, 1991; Conner, 1992a,b). One of the problems in these studies is the criteria for defining PD, Lewy body disease and AD which appear to form a spectrum of neurodegenerative disease. Whatever the case, there is evidence that oxidative stress contributes to the neurodegenerative process in a variety of clinically or pathologically defined entities.

PD is generally thought of as a primary degeneration of the nigral dopamine neurones. However, it may be that in our search for a cause of nigral cell death the possibility of involvement of other cell types within substantia nigra, namely the glial cells, has been overlooked. Evidence for such an involvement comes from a number of sources. In MPTP toxicity the conversion of MPTP to MPP^+ is a glial-mediated process (Ransom *et al.*, 1987). Reactive microglia have been demonstrated in the

SNc of PD patients (McGeer *et al.*, 1988), and recent studies have shown that reactive microglia can induce oxidative damage in cultured neurones. In PD, many of the indices of oxidative stress which are altered in post-mortem tissues are either directly known to involve glial cells or are implicated as a result of the extent of the changes which have occurred. Changes in iron, iron proteins, GSH, GPX, mitochondria complex I and monoamine oxidase B, have all been identified in the SNc in PD and are all present in glial cells. The magnitude of change observed in PD is not readily accounted for by dopamine neurones, which constitute only 5–10% of nigral neurones and suggest that SNc glia may be affected by the degenerative process.

Glial cell involvement in the pathological process can be viewed in one of two ways. First, glial cells may themselves be the primary target of oxidative stress and thereby be unable to fulfil their normal role of supporting neuronal activity such that nigral cells degenerate. Increasing evidence demonstrates the dependence of neurones on glia and glial-derived trophic factors. Several studies demonstrate enhanced survival of cultured dopamine neurones in the presence of glia or in the presence of a variety of glial trophic factors including brain-derived neurotrophic factor (BDNF), basic fibroblast growth factor (bFGF), epidermal growth factor (EGF) and glial cell line-derived neurotrophic factor (GDNF) (see Chapter 10). Similar results have been found with respect to transplanted embryonic dopamine neurones. Alternatively, damage to SNc neurones may result in secondary glial cell damage due to the state of oxidant stress or the release of toxins such as dopamine. This could, in turn, contribute to ongoing neuronal damage. This is an important issue which warrants further investigation.

2.10 Which comes first? The horse or the cart?

Post-mortem studies provide a snapshot of the processes underlying nigral cell death in PD. However, it is impossible to tell whether the various indices of oxidative stress and damage are the cause or result of the degenerative process or for that matter, if they are causally connected to one another. Unfortunately, there are few experimental animal models of altered indices of oxidative stress in which this can be assessed. Models of nigral iron overload lead to progressive cell degeneration (Sengstock *et al.*, 1995; Riederer and Wasserman, 1995), but there has been insufficient biochemical analysis to determine whether the pattern of change resembles that found in PD. Nonetheless, this model demonstrates the capacity of iron, which is increased in PD, to induce oxidant stress and a model of neurodegeneration which has histological, biochemical and behavioral features of parkinsonism. We are presently attempting to develop models of parkinsonism secondary to brain glutathione deficiency. The nigral toxin MPP^+ probably represents the most appropriate means of producing *in vivo* inhibition of complex I. Measures of complex I activity are not affected by acute *in vivo* MPTP or MPP^+ toxicity as changes are reversible and inhibition of mitochondrial function is lost during tissue preparation. MPP^+ does produce destruction of nigral neurones associated with increased iron and increased lipid peroxidation (Temlett *et al.*, 1994). These changes can be blocked by antioxidants or over-expression of SOD (Przedborski *et al.*, 1992). As far as we are aware there have been no *in vivo* studies of the effect of MPP^+ or other mitochondrial inhibitors on the levels of GSH or SOD. *In vivo* elevation of SOD as a result of drug treatment has not been accompanied by assessment of mitochondrial function, iron metabolism or GSH levels.

In vitro models of oxidant stress using cells in culture or tissue preparations have also been studied but do not clarify the situation. Oxidative stress with lipid peroxidation and neuronal degeneration can be generated *in vitro* using iron and a variety of free radical generator systems. SOD activity and GSH levels have not been assessed. Iron can lead to inhibition of mitochondrial function, but this differs from PD in that it affects both complexes I and IV (A. Schapira, personal communication). To date, it has not been possible to produce a selective decrease in complex I activity. Inhibition of mitochondrial function at the level of complex I leads to a depletion of GSH and the development of lipid peroxidation, but its effect on iron levels and SOD activity are unknown. Depletion of GSH inhibits complexes I and IV of the mitochondrial respiratory chain and may lead to reciprocal changes in SOD activity, but the effect on lipid peroxidation and iron levels has not been studied.

The recent finding of point mutations in the gene that encodes for Cu/Zn SOD in patients with familial amyotrophic lateral sclerosis (Rosen *et al.*, 1993) suggests that alterations in SOD can represent the initial event in the cascade of events leading to neurodegeneration (Olanow, 1993). This is supported by the finding that transgenic mice with the mutant SOD gene develop a neurodegenerative syndrome with an amyotrophic lateral sclerosis (ALS) phenotype (Gurney *et al.*, 1994).

So, at the present time there is insufficient evidence on which to determine the relationship between the key post-mortem parameters seen in PD, namely increased iron, reduced GSH, mitochondrial inhibition, and SOD activation, and which of these, if any, drives the neurodegenerative process. There is still no satisfactory hypothesis to explain the precise mechanism of how oxidative stress leads to the pattern of neurodegeneration and biochemical alterations found in PD and whether it represents a primary or secondary phenomenon.

2.11 Which free radical species are involved in PD?

Post-mortem studies of oxidative stress in PD have not yet determined the specific free radical species involved. Evidence of lipid peroxidation suggests that it could be $OH^{\cdot}$, which is the most highly reactive radical and is thought to be the prime mediator of oxidative damage. Studies using electron paramagnetic spin resonance (EPR) to determine the precise free radical species and the pattern of oxidative damage to DNA may provide further clues. Recently, there has also been interest in the possibility that nitric oxide (NO˙) might play a role in producing oxidative damage in PD.

NO˙ is a free radical that is formed by the conversion of arginine to citrulline in a reaction catalysed by calcium-activated nitric oxide synthase (NOS). Physiologically, NO˙ acts as a vasodilator and neuromodulator through its ability to stimulate guanylate cyclase (Moncada *et al.*, 1991; Dawson *et al.*, 1992). However, NO˙ can also exert neurotoxicity, possibly through its ability to combine with superoxide (Beckman *et al.*, 1990). This reaction forms peroxynitrite, which is itself toxic and can also decompose to produce the OH˙. NO˙-mediated neuronal death has been implicated in the pathogenesis of acute stroke and excitotoxicity (Dawson *et al.*, 1991). In this regard, it is noteworthy that cell damage in this model can be prevented by agents that scavenge or inhibit the formation of NO˙ (Buisson *et al.*, 1992; Izumi *et al.*, 1992; Nagafuji *et al.*, 1992, 1994). In addition, NO˙ can displace iron from its binding sites on ferritin, inhibit mitochondrial respiration and stimulate lipid peroxidation (Lancaster and Hibbs, 1990; Reif and Simmons, 1990; Radi *et al.*, 1991; Bolanos *et al.*, 1994). These properties would seem highly relevant to PD although NO˙ appears to affect mitochondrial complex IV rather than complex I. This may still be significant in view of the recent finding that inhibition of complex IV is necessary to induce irreversible inhibition of complex I (Cleeter *et al.*, 1994).

The potential relevance of NO˙ to chronic progressive neurodegenerative disorders, such as PD, is derived from several sources. Firstly, mutations in the SOD gene as seen in familial amyotrophic lateral sclerosis (FALS) may permit superoxide radical to interact with NO˙ and generate peroxynitrite and OH˙ with consequent cell damage. It is noteworthy that neurones which stain positively for NADPH diaphorase are relatively spared in neurodegenerative disorders such as HC, AD and PD (Kowall and Beal, 1988; Mufson and Brandabur, 1994). These cells contain NOS and are thought to be relatively spared because they are enriched in manganese SOD, which can clear the superoxide radical and prevent its interaction with NO˙. The sparing of these cells suggests that NO˙ toxicity may be a contributing factor in neurodegenerative disorders. There is additional evidence to suggest that NO˙ may contribute to the pathology of PD. Kaur and Halliwell (1994) have shown that 3-nitrotyrosine (3-NT) is a product of oxidative damage induced by NO˙. Recently, we have examined post-mortem PD brain and have shown the presence of 3-NT within the SNc (unpublished observation). The levels of 3-NT appear to be increased over those found in age-matched control tissues, but this remains to be confirmed. This reaction may be relevant to neurodegeneration as nitration of tyrosine receptors may prevent their phosphorylation and limit trophic factor-induced signalling necessary for neuronal support and maintenance (Ischiropoulos *et al.*, 1992). In addition, NO˙ activity may be diminished in PD as a thiol carrier may be required to allow diffusion to the target site. The major thiol in the brain is GSH which is markedly decreased in PD and, accordingly, NO˙ activity may be truncated.

There is also experimental evidence to suggest that NO˙ contributes to MPTP toxicity. Inhibition of NOS has been shown to reduce MPP^+-induced $OH^{\cdot}$ formation in the striatum of rats and MPTP-induced nigrostriatal dopamine terminal damage (Spencer-Smith *et al.*, 1994b). MPTP-induced striatal dopamine depletion in mice is also reduced with the immunosuppressant SK506, which interferes with dephosphorylation–activation of NOS (Kitamura *et al.*, 1994). However, others have been unable to demonstrate that inhibition of NOS prevents MPP^+ toxicity in the rat striatum (Santiago *et al.*, 1993). Indeed, we were unable to prevent MPTP toxicity in primates by administration of N^6-nitro-L-arginine methyl ester (L-NAME). Thus, there is reason to consider the possibility that $NO^{\cdot}$ is involved in PD neurodegeneration but, as with other possible contributing factors, its precise role remains undefined.

2.12 Interrelationship of oxidant stress with other mechanisms of cell death

Oxidative stress and oxidative damage to critical biomolecules appears to be an important process mediating cell death in PD. However, it has not yet been proven that this is the primary event which initiates nigral cell degeneration. There are a number of other mechanisms of cellular degeneration including mitochondrial damage, excitatory amino acids (EAAs) and calcium cytotoxicity that could also be involved in PD and which will be reviewed individually in subsequent chapters in this book. Indeed, oxidant stress can damage mitochondria, induce the release of EAAs, and promote a rise in intracellular calcium. In turn, EAAs, mitochondrial damage and increased cytosolic calcium can result in free radical generation. An interaction between free radicals, mitochondrial dysfunction, EAAs, and a rise in intracellular free calcium could comprise a cascade of events leading to cell degeneration consequent to different aetiological events (Figure 1).

Apoptotic cell death may be the manner by which cells die in neurodegenerative disease. Cultured dopaminergic cells exposed to low concentrations of neurotoxins, oxidant stress or trophic factor withdrawal die by an apoptotic mechanism (Mesner *et al.*, 1992; Troy and Shelanski, 1994; Walkinshaw and Waters, 1994; Ratan *et al.*, 1994a; Ziv *et al.*, 1994). Interestingly, anti-apoptotic proteins such as *Bcl-2* inhibit neuronal death and decrease free radical formation (Kane *et al.*, 1993). This may provide a common link between oxidative stress and apoptosis in PD. Further, neuronal apoptotic death resulting from cysteine deprivation, GSH loss, and oxidative stress can be prevented by inhibitors of macromolecular synthesis. The protection afforded in this model is associated with enhanced availability of cysteine and restoration of cellular GSH levels. Deprenyl-induced protection of cultured dopaminergic neurones similarly involves up-regulation of antioxidant proteins and the *Bcl-2* protein (Tatton *et al.*, 1994). So, there are a number of ways in which oxidative mechanisms may link to apoptotic cell death in PD. There is no direct evidence at the present time that apoptosis is the basis of cellular degeneration in PD or the other

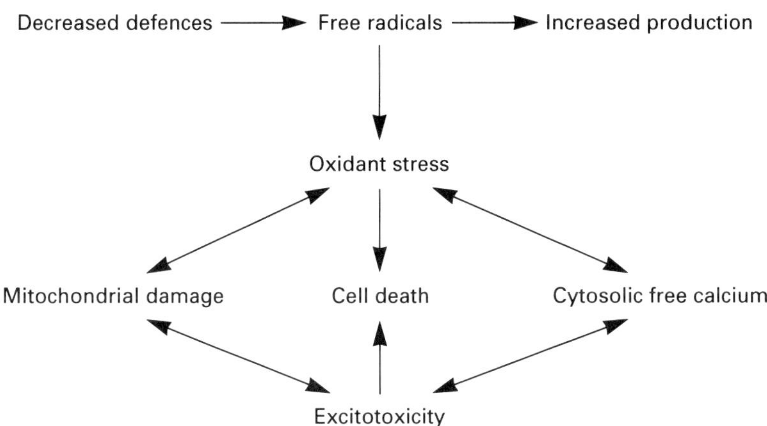

Figure 1 Aetiological events leading to cell degeneration.

neurodegenerative disorders. However, in end-stage degeneration, traces of apoptotic cell death may be gone.

2.13 Conclusions

There is increasing evidence that oxidative stress and oxidative damage to critical biomolecules are an important component of the pathology of PD. As yet, there is no clear explanation for the pattern of biochemical alterations which take place in PD such that unequivocal mechanisms can be proposed. Some of the changes which occur, such as the alterations in GSH and mitochondrial function appear to be earlier and more specific components of the disease process than others such as iron which is common to many different neurodegenerative diseases. Studies pursuing alterations in mitochondrial function and GSH might thus lead to further understanding of the primary mechanism responsible for nigral cell death in PD. An interesting hypothesis is that free radical-mediated neurodegeneration is common to a range of neurodegenerative diseases caused by a variety of aetiological mechanisms (Olanow, 1993; Olanow and Arendash, 1994). For example, there is evidence of oxidant stress in PD, AD, ALS and stroke. Similarly, most, if not all, models of cell death involve free radical species and oxidative stress. It may thus be possible to interfere with cell death in the neurodegenerative diseases by devising therapeutic strategies aimed at stopping or slowing free radical-mediated oxidative damage. In this regard it is anticipated that a variety of antioxidants will be tested in the neurodegenerative diseases in the coming years. There is still much to be learned about the nature of the neurodegenerative process, and it is important to recall that other causes of nigral cell death may ultimately turn out to play a more important role. For the present, substantial direct and indirect information suggests that oxidant stress plays a major role in the pathogenesis of cell degeneration in PD and other neurodegenerative disorders, raising the possibility that agents that interfere with these mechanisms will one day provide a neuroprotective therapy.

References

Adams, J., Klaidman, L., Odunze, I. (1991) *Molec. Chem. Neuropathol.* **14**, 213–226.

Aisen, P. (1992) *Ann. Neurol.* **32**, 62–68.

Ambani, L.M., Van Woert, M.H. & Murphy, S. (1975) *Arch. Neurol.* **32**, 114–118.

Arena J.F., Schwartz, C., Stevenson, R., Lawrence, L., Carpenter, A., Duara, R., Ledbetter, D., Huang, T., Lehner, T., Ott, J. & Lubs, H.A. (1992) *Am. J. Med. Genet.* **43**, 479–490.

Beckman, J.S., Beckman, T.W., Chen, J. (1990) *Proc. Natl Acad. Sci. USA* **87**, 1620–1624.

Ben-Schachar, D. & Youdim, M.B.H. (1991) *J. Neurochem.* **57**, 2133–2135.

Bolanos, J.P., Peuchen, S., Heales, S.J.R. (1994) *J. Neurochem.* **63**, 910–916.

Buisson, A., Plotkine, M. & Boulu, R.G. (1992) *Br. J. Pharmacol.* **106**, 766–767.

Férnandez-Calle, P., Vázquez, A., Cañizares-Liébana, F., Larumbe-Lobalde, S., Ayuso-Peralta, L., Rabasa, M. & Codoceu, R. (1994) *J. Neurol. Sci.* **125**, 82–86.

Cleeter, M.W.J., Cooper, J.M., Darley-Usmar, V.M., Moncada, S. & Schapira, A.H.V. (1994) *FEBS Lett.* **345**, 50–54.

Connor, J.R., Menzies, S.L., St Martin, S.M. & Mufson, E.J. (1992a) *J. Neurosci. Res.* **31**, 75–83.

Connor, J.R., Snyder, B.S., Beard, J.L., Fine, R.E. & Mufson, E.J. (1992b) *J. Neurosci. Res.* **31**, 327–335.

Connors, J.R., Snyder, B.S., Arosio, P. (1995) *J. Neurochem.* (in press).

Dawson, T.M., Dawson, V.L. & Snyder, S.H. (1992) *Ann. Neurol.* **32**, 297–311.

Dawson, V.L., Dawson, T.M., London, E.D. (1991) *Proc. Natl Acad. Sci. USA* **88**, 6368–6371.

Dexter, D.T., Carter, C.J., Wells, F.R. (1989a) *J. Neurochem.* **52**, 381–389.

Dexter D.T., Wells, F.R. & Lees, A.J. (1989b) *J. Neurochem.* **52**, 1830–1836.

Dexter, D.T., Carayon, A., Vidaihet, M., Ruberg, M., Agid, F., Agid, Y., Lees, A.J., Wells, F.R., Jenner, P. & Marsden, C.D. (1990) *J. Neurochem.* **55**, 16–20.

Dexter, D.T., Carayon, A., Javoy-Agid, F., Agid, Y., Wells, F.R., Daniel, S.E., Lees, A.J., Jenner, P. & Marsden, C.D. (1991) *Brain* **114**, 1953–1975.

Dexter, D.T., Ward, R.J., Wells, F.R., Daniel, S.E., Lees, A.J., Peters, T.J., Jenner, P. & Marsden, C.D. (1992) *Ann. Neurol.* **32**, 591–593.

Dexter, D.T., Holley, A.E., Flitter, W.D., Slater, J.F., Wells, F.R., Daniel, S.E., Lees, A.J., Jenner, P. & Marsden, C.D. (1994a) *Movement Disorders* **9**, 92–97.

Dexter, D.T., Sian, J., Rose, S., Hindmarsh, J.G., Mann, V.M., Cooper, J.M., Wells, F.R., Daniel, S.E., Lees, A.J., Schapirá, A.H.V., Jenner, P. & Marsden, C.D. (1994b) *Ann. Neurol.* **35**, 38–44.

Earle, K.M. (1966) *J. Neuropathol. Exp. Neurol.* **27**, 1–14.

Earle, K.M. (1968) *J. Neuropathol. Exp. Neurol.* **27**, 1–14.

Faucheux, B.A., Hirsch, E.C. & Villares, J. (1993) *J. Neurochem.* **60**, 2338–2341.

Fearnley, J.M. & Lees, A.J. (1991) *Brain* **114**, 2283–2301.

Gerlach, M., Ben-Schachar, D., Riederer, P. & Youdim, M.B.H. (1994) *J. Neurochem.* **63**, 793–807.

Gibb, W.R.G. & Lees, A.J. (1988) *J. Neurol. Neurosurg. Psychiatry* **51**, 745–752.

Good, P., Olanow, C.W. & Perl, D.P. (1992) *Brain Res.* **593**, 3343–3346.

Gurney, M.E., Pu H., Chiu, A.Y., Dal Canto, M.C., Polchow, C.Y., Alexander, D.D. Caliendo, J., Hentati, A., Kwon, Y.W., Deng, H.-X., Chen, W., Zhai, P., Sufit, R.L. & Siddique, J. (1994) *Science* **264**, 1772–1775.
Gutteridge, J.M., Quinlan, G.J., Clark, I. & Halliwell, B. (1985) *Biochem. Biophys. Acta* **835**, 441–447.
Halliwell, B. & Gutteridge, J. (1985) *Trends Neurol. Sci.* **8**, 22–29.
Halliwell, B. & Gutteridge, J. (1988) *ISI Atlas Sci. Biochem.* **I**, 48–52.
Hattori, N., Tanaka, M., Ozawa, T. & Mizuno, Y. (1991) *Ann. Neurol.* **30**, 563–571.
Hirsch, E.C., Brandel, J.P., Galle, P., Javoy-Agid, F. & Agid Y. (1991) *J. Neurochem.* **56**, 446–451.
Ikebe, S-I., Tanaka, M., Ohno, K., Sato, W., Hattori, K., Kondo, T., Mizuno, Y. & Ozawa, J. (1990) *Biochem. Biophys. Res. Commun.* **170**, 1044–1048.
Ischiropoulos, H., Zhu, L. & Chen, J. (1992) *Arch. Biochem. Biophys.* **298**, 431–437.
Izumi, Y., Benz, A.M., Clifford, D.B. & Zorumski, C.F. (1992) *Neurosci. Lett.* **135**, 227–230.
Jellinger, K., Paulus, W., Grundke-Iqbal, I., Riederer, P. & Youdim, M.B.H. (1990) *J. Neural Transm.* **2**, 327–340.
Jellinger, K., Kienzl, E. & Rumpelmair, B. (1992) *J. Neurochem.* **59**, 1168–1171.
Jenner, P., Schapira, A.H.V. & Marsden, C.D. (1992) *Neurology* **42**, 2241–2250.
Jenner, P. (1993) *Acta Neurol. Scand.* **87** (Suppl. 146), 6–13.
Jenner, P. (1994) *Lancet* **344**, 796–798.
Jenner, P., Schapira, A.H.V. & Marsden, C.D. (1992) *Neurology* **42**, 2241–2250.
Kane, D.J., Sarafian, T.A. & Anton, R. *et al.* (1993) *Science* **262**, 1274–1277.
Kaur, H. & Halliwell, B. (1994) *FEBS Lett.* **350**, 9–12.
Kawai, J., Sasahara, M., Hazama, F., Kumo, S., Komure, O., Nomura, S. & Yamaguchi, M. (1993) *Acta Neuropathol.* **86**, 609–616.
Kish, S.J., Morito, C. & Hornykiewicz, O. (1985) *Neurosci. Lett.* **58**, 343–346.
Kish, S.J., Morito, C.L.H. & Hornykiewicz, O. (1986) *Neurochem. Path.* **4**, 23–28.
Kitamura, Y., Itano, Y. & Kubo, T. (1994) *J. Neuroimmunol.* **50**, 221–224.
Kowall, N.W. & Beal, M.F. (1988) *Ann. Neurol.* **23**, 105–114.
Kuiper, M.A., Mulder, C., van Kamp, G.J., Scheltens, P.H. & Wolters, E. Ch. (1994) *J. Neural Transm.* **7**, 109–114.
Lancaster, J.R. Jr. & Hibbs, J.B. Jr (1990) *Proc. Natl Acad. Sci USA* **87**, 1223–1227.
Leenders, K.K., Antonini, A., Schwarzbach, P., Smith-Jones, P., Reist, H., Ben-Shachar, D., Youdim, M. & Henn, V. (1993) In *Quantifications of Brain Function. Tracer Kinetics and Image Analysis in Brain PET.* (ed. Uemurd, K.), pp. 145–146. Elsevier, Amsterdam.
Lestienne, P., Nelson, J., Riederer, P., Jellinger, K. & Reichmann, H. (1990) *J. Neurochem.* **55**, 1810–1812.
Loeffler, D.A., DeMaggio, A.J., Juneau, P.L., Brickman, C.M., Mashour, G.A. Finkelman, J.H., Pomara, N. & le Witt, P.A. (1994a) *Alzheimer Dis. Assoc. Disorders* **8**, 190–197.
Loeffler, D.A., DeMaggio, A.J., Juneau, P.L., Haraich, M.K. & le Witt, S.A. (1994b) *Clin. Neuropharmacol.* **17**, 370–379.
McGeer, P.L., Itagaki, S., Boyes, B.E. & McGeer, E.G. (1988) *Neurology* **38**, 1285–1291.
McNaught, K.St.P., Thull, U., Carrupt, P.-A., Altomare, C., Cellamare, S., Carotti, S., Testa, B., Jenner, P. & Marsden, C.D. (1995) *Biochem. Pharmacol.* (in press).
Mann, V.M., Cooper, J.M., Daniel, S.E., Srai, K., Jenner, P., Marsden, C.D. & Schapira, A.H.V. (1994) *Ann. Neurol.* **3**, 876–881.
Marttila, R.J., Lorentz, H. & Rinne, U.K. (1988) *J. Neurol. Sci.* **86**, 321–331.
Mash, D.C., Pablo, J. & Buck, B.E. (1991) *Exp. Neurol.* **114**, 73–81.
Mesner, P.W., Winters, T.R. & Green, S.H. (1992) *J. Cell Biol.* **119**, 1669–1680.
Michel, P.P., Vyas, S. & Agid, Y. (1992) *J. Neurochem.* **59**, 118–127.
Mithofer, K., Sandy, M.S., Smith, M.T. & DiMonte, D. (1992) *Arch. Biochem. Biophys.* **295**, 132–136.
Mizuno, Y., Saitoh, T. & Sone, N. (1987) *Biochem. Biophys. Res. Commun.* **143**, 971–976.

Mizuno, Y., Ohta, S., Tanaka, M., Takamiya, S., Suzuki, K., Sato, T., Oya, H., Ozawa, T. & Kagawa, Y. (1989) *Biochem. Biophys. Res. Commun.* **163**, 1450–1455.
Mizuno, Y., Matuda, S., Yoshino, H., Mori, H., Hattori, N. & Ikebe, S.-J. (1994) *Ann. Neurol.* **35**, 204–210.
Mochizuki, H., Nishi, K. & Mizuno, Y. (1993) *Neurodegeneration* **2**, 1–7.
Moncada, S., Palmer, R.M.J. & Higgs, E.A. (1991) *Pharmacol. Rev.* **43**, 109–142.
Mufson, E.J. & Brandabur, M.M. (1994) *Neurol. Rep.* **5**, 705–708.
Mytilineou, C., Han, S.-K. & Cohen, G. (1993) *J. Neurochem.* **61**, 1470–1478.
Nagafuji, T., Matsui, T., Koide, T. & Asano, T. (1992) *Neurosci. Lett.* **147**, 159–162.
Nagafuji, T., Sugiyama, M., Matsui, T. & Koide, T. (1993) *Eur. J. Pharmacol – Environ. Toxicol. Pharmacol. Sect.* **248**, 325–328.
Oestreicher, E., Sengstock, G.J., Riederer, P., Olanow, C.W., Dunn, A.J. & Arendash, G. (1994) *Brain Res.* **660**, 8–18.
Ogawa, N., Edamatsu, R., Mizukawa, K., Asanuma, M., Kohno, M. & Mori, A. (1993) *Adv. Neurol.* **60**, 242–250.
Ogawa, N., Asanuma, M., Kondo, Y., Kawada, Y., Yamamoto, M. & Mori, A. (1994) *Neurosci. Lett.* **171**, 55–58.
Olanow, C.W. (1990) *Neurology* **40**, 32–37.
Olanow, C.W. (1992) *Neurol. Clin. N. Am.* 405–420.
Olanow, C.W. (1993) *TINS* **16**, 439–444.
Olanow, C.W. (1994) In *Neurodegenerative Disorders* (ed. Calne, D.B.), pp. 807–824. W.B. Saunders, Philadelphia.
Olanow, C.W. & Arendash, G. (1994) *Curr. Opin. Neurol.* **7**, 548–558.
Olanow, C.W. & Hauser, R. (1994) In *Neurodegenerative Disorders* (ed. Calne, D.B.). W.B. Saunders, Philadelphia.
Olanow, C.W., Cohen, G., Perl, D.P. & Marsden, C.G. (1992) *Ann. Neurol.* **32**, 1–145.
Perry, T. & Yong, V.W. (1986) *Neurosci. Lett.* **67**, 269–274.
Perry, T.L., Godin, D.V. & Hansen, S. (1982) *Neurosci. Lett.* **33**, 305–310.
Perry, T.L., Hansen, S. & Jones, K. (1988) *Neurology* **38**, 943–946.
Przedborski, S., Kostivc, V., Jackson-Lewis, V., Muthane, H., Jiang, H., Ferreira, M., Naini, A.B. & Fahn, S. (1992) *J. Neurosci.* **12**, 1658–1667.
Przedborski, S., Jackson-Lewis, V. & Muthane, U. (1993) *Ann. Neurol.* **34**, 715–723.
Radi, R., Beckman, J.S., Bush, K.M. & Freeman, B.A. (1991) *Arch. Biochem. Biophys.* **288**, 481–487.
Ransom, B.R., Kunis, D.M., Irwin, I. & Langston, J.W. (1987) *Neurosci. Lett.* **75**, 323–328.
Ratan, R.R., Murphy, T.H. & Baraban, J.M. (1994a) *J. Neurochem.* **62**, 376–379.
Ratan, R.R., Murphy, T.H. & Baraban, J.M. (1994b) *J. Neurosci.* **14**, 4385–4392.
Reif, D.W. & Simmons, R.D. (1990) *Arch. Biochem. Biophys.* **283**, 537–541.
Riederer, P. & Wasserman, A. (1995) *J. Neural Transm.* (in press).
Reider, P., Rausch, W.D., Schmidt, B. & Youdim, M.B.H. (1988) *Mt Sinai J. Med.* **55**, 21–29.
Riederer, P., Sofic, E., Rausch, W.-D., Schmidt, B., Reynolds, G.P., Jellinger, K. & Youdim, M.B.H. (1989) *J. Neurochem.* **52**, 515–520.
Roberts, R., Sandra, A. & Siek, G.C. (1992) *Ann. Neurol.* **32**, 43–50.
Rosen, D.R., Siddique, T. & Patterson, D. (1993) *Nature* **362**, 59–62.
Saggu, H., Cooksey, J., Dexter, D., Wells, F.R., Lees, A., Jenner, P. & Marsden, C.D. (1989) *J. Neurochem.* **53**, 692–697.
Sanchez-Ramos, J.R., Overvik, E. & Ames, B.N. (1994) *Neurodegeneration* **3**, 197–204.
Santiago, M., Machado, A. & Cano, J. (1993) *Br. J. Pharmacol.* **111**, 837–842.
Schapira, A.H.V. (1994) *Movement Disorders* **9**, 125–138.
Schapira, A.H.V., Cooper, J.M., Dexter, D., Clark, J.B., Jenner, P. & Marsden, C.D. (1990a) *J. Neurochem.* **54**, 823–827.
Schapira, A.H.V., Mann, V.M., Cooper, J.M., Dexter, D., Daniel, S.E., Jenner, P., Clark, J.B. & Marsden, C.D. (1990b) *J. Neurochem.* **55**, 2142–2145.

Schapira, A.H.V., Holt, I.J., Sweeney, M., Harding, A.E., Jenner, P. & Marsden, C.P. (1990c) *Movement Disorders* **5**, 294–297.
Sengstock, G., Olanow, C.W., Dunn, A.J. & Arendash, G.W. (1992) *Brain Res. Bull.* **28**, 645–649.
Sengstock, G.J., Olanow, C.W., Menzies, R.A., Dunn, A.J. & Arendash, G. (1993) *J. Neurosci. Res.* **35**, 67–82.
Sengstock, G., Olanow, C.W., Dunn, A.J. *et al.* (1995) *Exp. Neurol.* (in press).
Sian, J., Dexter, D.T., Less, A.J., Daniel, S., Jenner, P. & Marsden, C.D. (1994a) *Ann. Neurol.* **36**, 356–361.
Sian, J., Dexter, D.T., Lees, A.J., Daniel, S., Agid. Y., Javoy-Agid, F., Jenner, P. & Marsden, C.D. (1994b) *Ann. Neurol.* **36**, 348–355.
Singer, T.P., Castagnoli, N. Jr., Ramsay, R.R. & Trevor, A.J. (1987) *J. Neurochem.* **49**, 1–8.
Slivka, A., Spina, M.B. & Cohen, G. (1987a) *Neurosci. Lett.* **74**, 112–118.
Slivka, A., Mytilineou, C. & Cohen, G. (1987b) *Brain Res.* **409**, 275–284.
Sofic, E., Riederer, P. & Heinsen, H. *et al.* (1988) *J. Neural Transm.* **74**, 199–205.
Sofic, E., Lange, K.W., Jellinger, K. & Riederer, P. (1992) *Neurosci. Lett.* **142**, 128–130.
Spencer, J.P.E., Jenner, A., Aruoma, O.I., Evans, P.J., Kaur, H., Dexter, D.T., Jenner, P., Lees, A.J., Marsden, C.D. & Halliwell, B. (1994) *FEBS Lett.* **353**, 246–250.
Spencer Smith, T., Parker, W.D. Jr. & Bennett J.P. (1994a) *NeuroReport* **5**, 1009–1011.
Spencer Smith, T., Swerdlow, R.H., Davis Parker W. Jr & Bennett, J.P. Jr. (1994b) *NeuroReport* **5**, 2598–2600.
Spina, M.B. & Cohen, G. (1988) *J. Pharmacol. Exp. Ther.* **247**, 502–507.
Tanaka, M., Sotomatsu, A., Kinai, H. & Hirai, S. (1991) *J. Neurosci.* **101**, 198–203.
Tanaka, M., Sotomatsu, A., Kanai, H. & Hirai, S. (1992) *Neurosci. Lett.* **140**, 42–46.
Tatton, W.G., Ju, W.Y.L. & Holland, D.P. (1994) *J. Neurochem.* **63**, 1572–1575.
Temlett, J.A., Landsberg, J.P., Watt, F. & Grime, G.W. (1994) *J. Neurochem.* **62**, 134–146.
Troy, C.M. & Shelanski, M.L. (1994) *Proc. Natl Acad. Sci. USA* **91**, 6384–6387.
Uitti, R.J., Rajput, A.H., Rozdilsky, B., Bickis, M., Wollin, T. & Yuen, W.K. (1989) *Can. J. Neurol. Sci.* **16**, 310–314.
Valberg, L.S., Flanagan, P.R., Kertesz, A. & Ebers, G.C. (1989) *Can. J. Neurol. Sci.* **16**, 184–186.
Walkinshaw, G. & Waters, C.M. (1994) *Neuroscience* **63**, 975–987.
Yasui, M., Ota, K. & Garruto, R.M. (1993) *Neuro Toxicology* **14**, 445–450.
Yoshida, E., Mokuno, K., Aoki, S.-I., Takahashi, A., Riku, S., Murayama, T., Yanagi, T. & Kato, K. (1994) *J. Neurol. Sci.* **124**, 25–31.
Youdim, M.B.H., Ben-Shachar, D. & Riederer, P. (1993) *Movement Disorders* **8**, 1–12.
Zhang, Fa. & Dryhurst, G. (1994) *J. Med. Chem.* **37**, 1084–1098.
Ziv, I., Melamed, E. & Nardi, N. (1994) *Neurosci. Lett.* **170**, 136–140.

CHAPTER 3

TREATMENT OF PARKINSON'S DISEASE WITH DEPRENYL (SELEGILINE) AND OTHER MONOAMINE OXIDASE INHIBITORS

Karl Kieburtz and Ira Shoulson

University of Rochester Medical Center,
601 Elmwood Avenue,
Rochester, NY 14642, USA

Table of Contents

3.1 Introduction

The development of a 'neuroprotective' intervention that retards or prevents the death of vulnerable neurones in the substantia nigra will be a significant advance in the treatment of Parkinson's disease (PD). Such an intervention would be expected to forestall or slow the clinical progression of the illness. Currently available 'symptomatic' treatments that relieve symptoms and temporarily improve motor function in PD have not been demonstrated to slow the underlying neuronal dysfunction and degeneration. Circumstantial preclinical and clinical data have suggested that deprenyl (selegiline) and other monoamine oxidase (MAO) inhibitors may exert neuroprotective effects in PD.

Neurodegeneration and Neuroprotection
in Parkinson's Disease
ISBN 0-12-525445-8

3.2 Characteristics of MAOs

'MAO' refers to a class of flavin-containing enzymes that promote the oxidative deamination of monoamines (Kalaria *et al.*, 1988). MAOs located in the outer mitochondrial membrane of neuronal and glial cells are categorized by A (MAO-A) and B (MAO-B) forms based on distribution, substrate selectivity and inhibitor specificity. MAO-A is found in exocrine pancreas, liver, kidney, intestine and heart, while MAO-B is found in the endocrine pancreas, liver, lung, platelets and epididymis (Johnson, 1968; Denney *et al.*, 1983; Saura *et al.*, 1992). In primate brain, MAO-A is found in the locus coeruleus, substantia nigra and periventricular thalamus, whereas MAO-B localizes predominantly to the raphé nucleus, ependyma and lateral hypothalamus (Westlund *et al.*, 1985). In human brain, MAO-A is synthesized in catecholamine-containing neurones of the locus coeruleus, and MAO-B is synthesized in serotonin-containing neurones of the raphé nucleus (Richards *et al.*, 1992). There is no direct evidence of MAO-A or -B synthesis within neurones of the substantia nigra (Richards *et al.*, 1992), although enzyme autoradiography detects MAO-A and -B within glial cells of the substantia nigra (Saura *et al.*, 1992).

MAO-A preferentially oxidizes hydroxylated amines such as adrenaline, tyramine and serotonin while MAO-B preferentially oxidizes non-hydroxylated amines such as phenylethylamine and benzylamine (Johnson, 1968). In humans, dopamine appears to be oxidized by both enzymes, although there may be a preference for MAO-B (Glover *et al.*, 1977).

Some compounds can inhibit both MAO-A and -B (non-selective inhibitors), whereas others exert preferential inhibition of the A or B forms. MAO inhibitors may also act irreversibly and reversibly. Irreversible MAO inhibitors bind covalently to inactivate the enzyme. After discontinuing irreversible inhibitors, *de novo* synthesis of MAO is required for restitution of enzymatic activity. Clorgyline can induce selective and irreversible inhibition of MAO-A in human locus coeruleus neurones, while deprenyl can produce selective and irreversible inhibition of MAO-B in human raphé neurones (Konradi *et al.*, 1989). In contrast, reversible inhibitors of MAO do not bind covalently to the enzyme and enzymatic activity is restored within hours after discontinuation of the inhibitor. Brofaromine and meclobamide are selective, reversible inhibitors of MAO-A; lazabemide is a selective, reversible inhibitor of MAO-B (Gerlach *et al.*, 1993).

3.3 Development of MAO inhibitors

Several non-selective MAO inhibitors that block the deamination of monoamines, particularly tyramine, carry the risk of inducing a hypersympathetic syndrome (hypertension, tachycardia, headaches) (Blackwell, 1963). The syndrome, which can be induced by even small amounts of food rich in tyramine (e.g. red wine and chocolate) in individuals taking MAO inhibitors, has been referred to as 'the cheese effect'.

Because primarily MAO-A is found in the intestine, inhibitors of MAO-A permit excess tyramine to be absorbed which in turn displaces noradrenaline from nerve terminals and induces the hypersympathetic state. In contrast, selective inhibitors of MAO-B do not inhibit the oxidation of tyramine or produce sympathomimetic effects.

In 1965, Knoll *et al.* first reported the synthesis of deprenyl (phenylisopropylmethylpropynylamine, selegiline), a selective MAO-B inhibitor at dosages of less than 20 mg day^{-1} which was not attended by the cheese effect (Knoll *et al.*, 1965). At dosages higher than 20 mg day^{-1}, deprenyl begins to inhibit MAO-A and lose its MAO-B selectivity.

L-Deprenyl acts as an irreversible inhibitor of MAO-B and is metabolized primarily in the liver to the laevo forms of amphetamine and methamphetamine, and the inactive metabolite 1-*N*-desmethyldeprenyl (Kalir *et al.*, 1981). The laevo forms of these amphetamine metabolites do not exert the strong sympathomimetic actions of their corresponding dextrorotatory forms.

Deprenyl readily crosses the blood–brain barrier and has a plasma half-life of approximately 40 hours (Elsworth *et al.*, 1978). The pharmacodynamic action of deprenyl has not been studied extensively, but inhibition of platelet MAO activity, which is exclusively MAO-B, begins within 2 hours of deprenyl administration. MAO inhibition persists for at least 5 days after cessation of therapy (Lee *et al.*, 1989). Since the half-life of platelets is only about 2 days, the effect of deprenyl on platelet MAO-B is not an accurate reflection of the effect on brain MAO activity. Treatment with deprenyl 10 mg daily for at least 1 week in patients with very advanced PD was found to produce a 90% decrease in post-mortem striatal MAO-B activity, a threefold increase in striatal and nigral dopamine levels, and a 60% reduction in MAO-A activity that was not associated with increased serotonin levels (Riederer and Youdim, 1986).

Lazabemide (Ro 19–6327) is a short-acting, reversible, competitive inhibitor of MAO-B which is not metabolized to active compounds. Lazabemide is approximately 100-fold more selective than deprenyl in inhibiting MAO-B compared with MAO-A (DaPrada *et al.*, 1988). Over 90% of platelet MAO-B is inhibited for 24 hours by a single oral dose of lazabemide 200 mg.

3.4 Deprenyl in PD

Deprenyl has been studied as an adjunctive therapy in levodopa-treated PD patients, and has been shown to be beneficial in patients experiencing motor fluctuations (Csanda and Tarczy, 1983). In double-blind controlled trials (for a review, see Golbe, 1988), deprenyl was found to prolong the beneficial motor effects of levodopa, but had little or no effect on unpredictable motor fluctuations or loss of levodopa efficacy. Taken together, these observations suggest that the benefits of deprenyl in levodopa-treated patients probably relate to inhibition of brain dopamine metabolism.

In 1983, Birkmayer *et al.* reported their retrospective experience involving more than 900 patients who had been treated with either levodopa alone or levodopa in combination with deprenyl (Birkmayer *et al.*, 1983). They found that the 'evolution' of Parkinson's disease was slower in patients receiving both levodopa and deprenyl and also suggested an increased life expectancy in the combined treatment group (Birkmayer *et al.*, 1983, 1986). This uncontrolled study suggested that deprenyl and other MAO-B inhibitors might function to slow the progression of PD by possibly slowing neuronal death. This suggestion was supported by the later report that deprenyl treatment was associated with more nigral neurones and fewer Lewy bodies in the post-mortem brains of PD patients (Rinne *et al.*, 1991).

3.5 Basis for neuroprotective action

The interaction of MAO-B and exogenous neurotoxins may play a role in the dysfunction and death of nigral neurones. The recognition of young intravenous drug users with parkinsonism led to the identification of 1-methyl-4-phenyl-1,2,3,6-tetrahydropyridine (MPTP) as a toxin which can induce the neuropathological and clinical characteristics of PD in humans and experimental animals (Langston *et al.*, 1983; Ballard *et al.*, 1985). MPTP is metabolized by MAO-B to the active neurotoxin 1-methyl-4-phenylpyridinum (MPP^+), which is actively taken up by the dopamine transporter system and concentrated within mitochondria (Langston *et al.*, 1984; Markey *et al.*, 1984; Chiba *et al.*, 1984). MPP^+ is a potent inhibitor of complex I, which may interfere with oxidative phosphorylation and lead to neuronal death (Youngster *et al.*, 1989).

MAO-B inhibitors such as deprenyl, nialamide and tranylcypromine can prevent the oxidative biotransformation of MPTP to MPP^+ and prevent the development of parkinsonism in pretreated experimental animals exposed to MPTP (Heikkila *et al.*, 1984; Cohen *et al.*, 1985). While MAO-B inhibitors can prevent the transformation of some MPTP-like protoxins to neurotoxins that cause nigral injury, other nigrotoxic MPTP analogues (e.g. 2′-ethyl-MPTP) are good substrates for MAO-A as well as MAO-B (Heikkila *et al.*, 1990).

Oxidative stress describes the conditions that promote the damage of cellular constituents, such as DNA, and the initiation of harmful reactions, such as lipid peroxidation (Fahn and Cohen, 1992). There are several ways by which oxidative stress is increased in the substantia nigra of patients with PD. The synthetic turnover of dopamine is increased in surviving nigrostriatal neurones in patients with PD, which may lead to dopamine auto-oxidation (i.e. a non-enzymatic reaction with oxygen) and promote oxidative stress (Graham, 1978; Fornstedt *et al.*, 1989). The increase in brain iron and reduction in glutathione activity found in the post-mortem PD brain may further impair the ability of cells to handle oxidative stress (Dexter *et al.*, 1989; Riederer *et al.*, 1989). Dopamine metabolism by MAO-B may also lead to the formation of hydrogen peroxide, a cellular oxidant, and other oxidants and toxins such as superoxide, hydroxyl and peroxyl radicals, quinones and 6-hydroxydopamine (Cohen, 1990). Thus, deprenyl and other MAO inhibitors may retard oxidative stress in nigral neurones and protect cells from endogenous toxins (Cohen and Spina, 1989).

3.6 Clinical trials with MAO inhibitors in early PD

The multicentre, controlled clinical trial DATATOP (Deprenyl and Tocopherol Antioxidative Therapy of Parkinsonism), conducted by the Parkinson Study Group, was designed to determine if deprenyl 10 mg daily and/or α-tocopherol (the active component of vitamin E that quenches hydroxyl radical reactions) 2000 IU daily would prolong the time until levodopa therapy was judged necessary to treat the disability of patients with early PD (Parkinson Study Group, 1989a,b, 1993a). This randomized double-blind trial was planned for 2 years of observation including a baseline evaluation, follow-up evaluations at 3 month intervals, a 1 or 2 month withdrawal of experimental medications, and a final evaluation.

The primary response variable for DATATOP was the time from randomization until sufficient disability developed to require levodopa therapy. This end-point was readily quantifiable in days and is an important milestone of PD progression relevant to the care of patients. In determining the need for levodopa therapy, investigators considered (1) the threat to employability, (2) the threat to management of domestic or financial affairs, and (3) the worsening of PD, especially impairments in activities of daily living, gait and balance. Other standardized clinical assessments including the Unified Parkinson's Disease Rating Scale (UPDRS) were performed (Parkinson Study Group, 1989b).

Eight hundred eligible subjects, including 530 men and 270 women, were enrolled in DATATOP and followed for 14±6 months (mean±SD) of observation. The four treatment groups did not differ significantly at baseline with respect to age, gender or clinical ratings of PD. The experimental medications were generally well tolerated.

α-Tocopherol treatment did not influence the need for levodopa treatment or ameliorate the features of PD, and there was no interaction between deprenyl and α-tocopherol. Deprenyl treatment significantly reduced the risk of disability requiring levodopa treatment. The projected median time to the primary end-point of disability was 719 days for subjects who were assigned to deprenyl and 454 days for subjects not assigned to deprenyl, a difference of approximately 9 months.

Subjects treated with deprenyl also showed a modest improvement in clinical features of PD as assessed by the UPDRS, during their first 3 months of therapy. After withdrawal of deprenyl for 1 month, there were no significant treatment differences (deprenyl versus no deprenyl) in clinical PD features among subjects who had reached the end-point and required levodopa. However, withdrawal of deprenyl for 2 months in 'surviving' subjects who had not reached the end-point resulted in a small but significant deterioration of PD features. For the entire duration of the study, including the 2 month period of deprenyl withdrawal, subjects treated with deprenyl showed a significantly slower overall rate of clinical decline than subjects not treated with deprenyl.

Other controlled trials have produced similar results. Tetrud and Langston conducted a trial involving 54 subjects with early PD who did not yet require treatment with levodopa (Tetrud and Langston, 1989). Subjects were randomly assigned to receive deprenyl 5 mg twice daily or a matching placebo, and were followed by blinded investigators for up to 3 years to determine if and when they had reached the primary end-point of PD disability sufficient to warrant levodopa therapy. Clinical evaluations included the UPDRS.

During the 3 years of follow-up, all but one of the 54 subjects reached the primary end-point or withdrew from the study. The time to the end-point of disability was 312±44.5 days (mean±SD) for the placebo group and 550±61 days in the deprenyl group ($p<0.002$), a difference of about 8 months. The clinical features of PD did not change significantly by 1 month after initiation or withdrawal of deprenyl. Deprenyl-treated subjects showed an overall 50% slowing in their UPDRS motor scores compared with subjects not treated with deprenyl. There were no significant adverse experiences. The investigators concluded that deprenyl as monotherapy in PD was safe, delayed the need for treatment with levodopa, and appeared to slow the clinical progression of illness.

Myllyla and his Finnish colleagues conducted a randomized, double-blind, controlled trial of deprenyl in 52 patients with early untreated PD (Myllyla *et al.*, 1992). They found that deprenyl prolonged the time to initiation of levodopa therapy by about 6 months (545±90 days versus 372±28 days with placebo) and was associated with mild improvement in the motor features of PD. Allain and colleagues in the French Selegiline [Deprenyl] Multicentre Trial enrolled 93 patients in a 3 month double-blind, placebo-controlled multicentre study of deprenyl 10 mg day^{-1} (Allain *et al.*, 1991). They found 18.4% of placebo-treated patients required levodopa therapy in comparison with 4.5% of deprenyl-treated patients. They also found a 15% improvement in the UPDRS scores compared with the baseline, related to deprenyl therapy.

Clinical trials have been conducted in patients with early PD to examine other MAO-B inhibitors such as lazabemide. A total of 201 patients were randomized to lazabemide 100, 200 or 400 mg day^{-1} or matching placebo in an 8 week, multicentre, clinical trial. Lazabemide at dosages of 100 and 200 mg day^{-1} was generally well tolerated (Parkinson Study Group, 1993b). Subjects treated with lazabemide for 4–6 weeks showed a significant improvement in their activities of daily living and a trend towards improvement in other UPDRS ratings. Further trials of lazabemide are under way to determine the long-term safety of this MAO-B inhibitor and its influence on the progression of disability in patients with early untreated PD.

3.7 Neuroprotection versus symptomatic benefit

There has been considerable debate as to whether deprenyl acts through neuroprotective mechanisms to retard the degeneration of nigral neurones or as symptomatic treatment to temporarily ameliorate the features of PD (Olanow and Calne, 1991). The results of at least four controlled trials have consistently demonstrated that deprenyl delays the progression of disability as measured by the need for levodopa in patients with early untreated PD. The DATATOP study also shows a modest but detectable improvement in motor performance among deprenyl-treated patients during the initial 3 months of therapy and a comparable worsening after withdrawal of deprenyl for 2 months. However, the rate of PD progression as measured by the UPDRS was significantly slower overall in the deprenyl-treated group.

The most parsimonious conclusion from these trials is that deprenyl as monotherapy in early PD produces modest improvement in motor function. This benefit hampers the detection of a clear-cut influence of deprenyl on the underlying progression of PD. A critical methodological obstacle in trials aimed at identifying neuroprotective benefits is the lack of well-validated biological markers of PD progression (Shoulson, 1992). Clinical assessment scales such as the UPDRS and the end-point of disability requring levodopa therapy are useful measures of the progression of clinical features, but do not necessarily indicate a concurrent slowing of nigral degeneration. The development of well-validated biological markers of the presence (trait) and extent (state) of PD would aid appreciably in achieving the goal of neuroprotective therapy.

References

Allain, H., Cougnard, J., Neukirch, H.C. & the FSMT members (1991) *Acta Neurol. Scand.* **84** (Suppl. 136), 73–78.

Ballard, P.A., Tetrud, J.W. & Langston, J.W. (1985) *Neurology* **35**, 949–956.

Birkmayer, W., Knoll, J., Riederer, P. & Youdim, M.B. (1983) *Mod. Probl. Pharmacopsychiatr.* **19**, 170–176.

Birkmayer, W., Knoll, J., Riederer, P., Youdim, J.B., Ilars, V. & Martan, J. (1986) *J. Neural Trans.* **64**, 113–127.

Blackwell, B. (1963) *Lancet* **ii**, 849–851.

Chiba, K., Trevor, A.J. & Castagnoli, N. Jr (1984) *Biochem. Biophys. Res. Commun.* **120**, 574–578.

Cohen, G. (1990) *J. Neural Transm.* **32**, 229–238.

Cohen, G. & Spina, M.B. (1989) *Ann. Neurol.* **26**, 689–690.

Cohen, G., Pasik, P. Cohen, B., Leist, A., Mytilineou, L. & Yahr, M.D. (1985) *Eur. J. Pharmacol.* **106**, 209–210.

Csanda, E. & Tarczy, M. (1983) *Acta Neurol. Scand.* **95** (Suppl.), 117–122.

DaPrada, M., Kettler, R., Keller, H.H. & Burkard, W.P. (1988) In *Progress in Catecholamine Research. Part B: Central Aspects* (ed. Sandler, M.), pp. 359–363. Alan R. Liss, New York.

Denney, R.M., Fritz, R.R., Patel, N.T., Widen, S.G. & Abell, C.W. (1983) *Molec. Pharmacol.* **24**(1), 60–68.

Dexter, D.R., Wells, F.R., Lees, A.J., Agid, Y., Jenner, P. & Marden, C.D. (1989) *J. Neurochem.* **52**, 1830–1836.
Elsworth, J.D., Glover, V., Reynolds, G.P., Sandler, M., Lees, A.J. *et al.* (1978) *Psychopharmacology* **57**, 33–38.
Fahn, S. & Cohen, G. (1992) *Ann. Neurol.* **32**, 804–812.
Fornstedt, B., Brun, A., Rosengren, E. & Carlsson, A. (1989) *J. Neural Transm.* **1**, 29–295.
Gerlach, M., Riederer, P. & Youdim, M.B.H. (1993) In *Inhibitors of MAOB: Pharmacology and Clinical Use in Neurodegenerative Disorders* (ed. Szelenyi, I.). Birkhäuser, Basel. pp. 183–200.
Glover, V., Sandler, M., Owen, F. & Riley, G. (1977) *Nature* **265**, 80–81.
Golbe, L.I. (1988) *Clin. Neuropharmacol.* **5**, 387–400.
Graham, D.G. (1978) *Mol. Pharmacol.* **14**, 633–643.
Heikkila, R.E., Manzino, L., Cabbat, F.S. & Duvoisin, R.C. (1984) *Nature* **311**, 467–469.
Heikkila, R.E., Terleckyj, I. & Sieber, B.A. (1990) *J. Neural Transm.* **32** (Suppl.), 217–227.
Johnson, J.P. (1968) *Biochem. Pharmacol.* **17**, 1285–1297.
Kalaria, R.N., Mitchell, M.J. & Harik, S.I. (1988) *Brain* **111**, 1441–1445.
Kalir, A., Sabbagh, A. & Youdim, M.B.H. (1981) *Br. J. Pharmacol.* **73**, 55–64.
Knoll, J., Esceri, Z., Kelemen, K., Nievel, J. & Knoll, B. (1965) *Arch. Int. Pharmacodyn. Ther.* **155**, 154–164.
Konradi, C., Kornhuber, J., Froelich, L., Fritze, J., Heinsen, H., Beckmann, J.H., Schulz, E. & Riederer, P. (1989) *Neuroscience* **33**, 383–400.
Langston, J.W., Ballard, P.A., Tetrud, J.W. & Irwin, I. (1983) *Science* **219**, 979–980.
Langston, J.W., Irwin, I., Langston, E.B. & Forno, L.S. (1984) *Neurosci. Lett.* **48**, 87–92.
Lee, D.H., Mendoza, M., Dvorozniak, M.T., Chung, E., van Woert, M.H. & Yahr, M.D. (1989) *J. Neural Trans.* **1**, 189–194.
Markey, S.P., Johannessen, J.N., Chiueh, C.C., Burns, R.S. & Herkenhaw, M.A. (1984) *Nature* **311**, 464–467.
Myllylä, V.V., Sotaniemi, K.A., Vuorinen, J.A. & Heinonen, E.H. (1992) *Neurology* **42**, 339–343.
Olanow, C.W. & Calne, D. (1991) *Neurology* **42** (Suppl. 4), 13–26.
Parkinson Study Group (1989a) *Arch. Neurol.* **46**, 1052–1060.
Parkinson Study Group (1989b) *N. Engl. J. Med.* **321**, 1364–1371.
Parkinson Study Group (1993a) *N. Engl. J. Med.* **328**, 176–183.
Parkinson Study Group (1993b) *Ann. Neurol.* **33**, 350–356.
Richards, J.G., Saura, J., Ulrich, J. & Da Prada, M. (1992) *Psychopharmacol.* **106**, S21–S232.
Riederer, P. & Youdim, M.B.H. (1986) *J. Neurochem.* **46**, 1359–1365.
Riederer, P., Sofic, E., Rausch, W.D. *et al.* (1989) *J. Neurochem.* **52**, 515–520.
Rinne, J.O., Röyttä, M., Paljärvi, L., Rummukainen, J. & Rinne, U.K. (1991) *Neurology* **41**, 859–861.
Saura, J., Kettler, R., DePrada, M. & Richard, J.G. (1992) *J. Neurosci.* **12**(5), 1977–1999.
Shoulson, I. (1992) *Ann. Neurol.* **32**, S143–S145.
Tetrud, J.W. & Langston, J.W. (1989) *Science* **245**, 519–522.
Westlund, K.N., Denney, R.M., Kochersperger, L.M., Rose, R.M. & Abell, C.W. (1985) *Science* **230**, 181–183.
Youngster, S.K., Nicklas, W.J. & Heikkila, R.E. (1989) *J. Pharmacol. Exp. Ther.* **249**, 829–835.

CHAPTER 4

IRON AND NEURODEGENERATION: PROSPECTS FOR NEUROPROTECTION

C.W. Olanow* and M.B.H. Youdim†

**Department of Neurology, Mount Sinai Medical Center, New York, NY 10029-6574, USA and †Department of Pharmacology Technion—Israel Institute of Technology, Haifa, Israel*

Table of Contents

4.1 Introduction

Free radicals and oxidative stress have been implicated in the pathogenesis of neurodegeneration (Olanow and Arendash, 1994; Gotz *et al.*, 1994). Free radicals are atoms or molecules that contain one or more orbitals housing a single unpaired electron (Halliwell and Gutteridge, 1985). They tend to be unstable and highly reactive molecules capable of extracting an electron from, and thereby damaging, neighbouring molecules through this oxidation process. Oxidative damage is thought to be primarily mediated by the hydroxyl radical ($OH^{\cdot}$), which is extremely reactive and capable of damaging a variety of critical biological molecules including cellular proteins, membrane lipids and DNA (Wolff *et al.*, 1986; Richter *et al.*, 1988). More recently, evidence has accumulated indicating that free radicals are also associated with excitotoxicity, a rise in intracellular calcium, and apoptotic cell death (Dawson *et al.*, 1991; Orrenius *et al.*, 1992; Ratan *et al.*, 1994).

Oxidation and reduction (redox) reactions that generate free radicals and other reactive oxidant species (ROSs) are catalysed by transition metals such as iron, copper and manganese. Transition metals exist in more than one valence state and contain a loosely bound electron in their outer shell. An electron can thus be donated or accepted by a transition metal, which can thereby promote redox reactions and the

Neurodegeneration and Neuroprotection in Parkinson's Disease
ISBN 0-12-525445-8

formation of free radicals and other ROS (Halliwell and Gutteridge, 1988). Transition metal-catalysed reactions are particularly important for the reduction of molecular oxygen (O_2). Molecular oxygen is a diradical containing two orbitals with unpaired electrons. In the ground state, these electrons have *parallel spin.* In contrast, molecules with covalent bonds have orbitals containing two electrons with *opposite rotational spin.* For this reason, O_2 is relatively unreactive, and spontaneous auto-oxidation of most molecules rarely occurs. However, O_2 will readily accept a single electron from a transition metal. The oxidation of iron from its ferrous (Fe^{2+}) to its ferric (Fe^{3+}) form promotes the reduction of O_2 with the generation of the superoxide radical ($O_2^{\cdot -}$) as shown in equation (1). Similarly, the conversion of iron from Fe^{2+} to Fe^{3+} catalyses the reduction of hydrogen peroxide (H_2O_2) to the hydroxyl radical ($OH^{\cdot}$) in a reaction known as the Fenton reaction (equation (2)). While H_2O_2 can spontaneously generate $OH^{\cdot}$, the importance of the iron catalysis is illustrated by studies showing that co-infusion of H_2O_2 and Fe^{2+} results in significantly greater $OH^{\cdot}$-induced neurodegeneration than occurs with infusion of H_2O_2 alone (Liu *et al.*, 1994).

$$O_2 + Fe^{2+} \rightarrow O_2^{\cdot -} + Fe^{3+} \tag{1}$$

$$H_2O_2 + Fe^{2+} \rightarrow OH^{\cdot} + OH^{-} + Fe^{3+} \tag{2}$$

Iron is most likely to be reactive and to promote oxidation reactions when it is unbound or when it exists in a low molecular weight formulation complexed to citrate or ATP. In contrast, iron is less reactive when it is bound to an iron chelator or an iron-binding protein. Reactive iron promotes oxidative damage in a dose-related fashion while iron-induced oxidative damage can be reduced or prevented by iron chelators (Gutteridge *et al.*, 1983). In general, the greater the concentration of the transition metal, the more likely oxidation reactions are to occur. Redox cycling of iron between its Fe^{2+} and Fe^{3+} valence states can lead to the formation of a cascade of free radicals and consequent tissue damage. The Fe^{5+} and Fe^{6+} valence states may also be generated through the Fenton reaction and are thought to be particularly toxic, but they have not been well studied in man (Bielski, 1992).

Transition metals are essential for normal cellular function. Iron is particularly important because it is involved in a number of critical biological functions including protein synthesis, DNA replication, membrane receptor formation, and numerous metabolic enzymes (Youdim, 1988). It is now apparent that maintenance of brain homeostasis is essential for normal brain function and that abnormalities of iron metabolism can be associated with devastating neurological disorders (Lauffer, 1992). Brain iron deficiency is associated with cognitive impairment and subsensitivity of dopamine transmission (Yeuda and Youdim, 1989; Youdim *et al.*, 1989). Excess iron has been implicated in the cellular damage that occurs in a variety of neurodegenerative disorders including Parkinson's disease (PD), Alzheimer's disease (AD) and Hallervorden–Spatz's disease (Ben-Shacher *et al.*, 1991a; Connor *et al.*, 1992a; Olanow *et al.*, 1992; Olanow, 1994; Youdim, 1994). Neurodegeneration has also been associated with accumulation of other transition metals including copper in Wilson's disease (Lewitt and Brewer, 1994) and manganese in parkinsonism (Mena *et al.*, 1967;

Olanow *et al.*, 1995). These findings suggests that dysregulation of transition metals in the brain can contribute to selective neuronal damage.

It is now apparent that iron is increased at the site of neurodegeneration in a variety of disorders. Under normal circumstances iron does not readily cross the blood–brain barrier, and excess peripheral iron in disorders such as haemochromatosis or following multiple transfusions does not result in increased brain iron. Thus, the finding of increased iron accumulation or deposition in areas undergoing neurodegeneration suggests the possibility that iron-induced oxidative stress may contribute to the pathogenesis of neurodegeneration process and the development of disease pathology. This chapter will review brain iron with an aim to understanding its possible role in the pathogenesis of the neurodegenerative process and the potential of developing a neuroprotective therapy based on the concept of iron chelation.

4.2 Iron and the brain

Zaleski first measured iron in the brain using Turnbull blue staining (Zaleski, 1887). Subsequently, the pioneering work of Spatz in 1920 and Hallgren and Sourander (1958) demonstrated that, normally, iron specifically accumulates in the globus pallidus (GP), the substantia nigra pars reticularis (SNr), the red nucles (RN), and the dentate nucleus of the cerebellum (DN). A similar pattern of iron distribution has been documented on high field strength magnetic resonance imaging (MRI) based on the capacity of iron to induce signal hypointensity on *T2*-weighted scans (Olanow, 1992). Iron is not detected in the brain at birth, gradually accumulates in the GP, RN and SNr during the first few years of life, and reaches adult levels by the third decade. Iron accumulation in the DN lags behind these structures and is not detected until the teenage years. During the sixth and seventh decades, iron accumulation increases in these structures and becomes particularly pronounced in the posterolateral striatum. It is noteworthy that iron accumulation in these regions exceeds that in the liver on a per milligram weight basis. It is not known why iron accumulates in these specific areas in concentrations which substantially exceed known physiological requirements.

Iron in a free (ionic) form is highly reactive and capable of inducing cytotoxicity by way of oxidative damage as described above. To avoid risks associated with iron toxicity, iron is normally maintained in a relatively unreactive state by iron-binding proteins such as transferrin or ferritin. In plasma, iron is primarily bound to transferrin. Transferrin is a Y-shaped molecule capable of binding two atoms of iron. The abundance of transferrin in the plasma serves to ensure that free iron is not available to induce local tissue damage. In adults, there is an effective blood–brain barrier for iron, and excess peripheral iron is excluded from the brain. For example, brain iron levels are normal in haemochromatosis despite the massive increase in peripheral iron concentration. Recent studies indicate that iron enters the brain coupled to transferrin by way of transferrin receptor-mediated endocytosis (Aisen, 1992; Roberts *et al.*, 1992). The iron–transferrin complex binds to transferrin receptors located on blood–brain barrier capillary endothelial cells. Within the acid pH of the endothelial cell, iron dissociates from the transferrin–transferrin receptor complex and is absorbed into the brain through the abluminal membrane. The apotransferrin–transferrin receptor complex does not enter the brain and is recycled to the luminal surface. At the more alkaline pH of the blood, they dissociate, making apotransferrin available to bind additional molecules of iron, and transferrin receptor free to bind another iron–transferrin complex. Iron entry into the brain is regulated by the expression of transferrin receptor mRNA. Increased brain iron levels induce a reduction in the expression of transferrin receptor mRNA thus limiting formation of transferrin receptor and the ability of iron to be transported into the brain.

Once in the brain, iron binds to a transferrin molecule for transport to sites of utilization and storage (possibly by axonal transport). Iron enters brain cells through a mechanism similar to that which it uses to cross the blood–brain barrier, and is normally stored as ferritin. This contrasts with liver, where iron is also stored as haemosiderin, another iron storage protein. Ferritin is a 20 kDa protein comprised of four α helices with a hollow core capable of storing up to 4500 atoms of iron as polynuclear aggregates of hydrated ferric oxide (Drysdale *et al.*, 1977). Recent studies suggest that iron is primarily bound to H chain ferritin within neurones and to L chain ferritin within oligodendroglia and, to a lesser degree, astrocytes (Connor *et al.*, 1995).

Transferrin receptors in the brain are primarily localized in the frontal cortex, hippocampus and striatum. In contrast, the specific areas where iron is stored appear to be relatively deficient in transferrin receptors. This raises the possibility that within the brain, iron is transported from sites that are rich in transferrin receptors to the GP, RN, SNr and DN, where it is deposited or stored in high concentrations. In support of this concept, Dwork *et al.* (1990) have shown that in newborn rats, in whom the blood–brain barrier has not yet fully developed, labelled iron initially accumulates in cortical regions and is subsequently transported to the GP, SNr and RN. Similar patterns of brain iron uptake and transport are seen in adult rodents and non-human primates if the blood–brain barrier for iron is artificially disrupted with neuroleptic agents such as chlorpromazine and haloperidol or by iron deficiency (Ben-Shachar *et al.*, 1988, 1994).

It is noteworthy, that the absorption, transport and storage of iron is tightly regulated. Throughout these processes, iron is maintained in a bound and unreactive state in order to protect critical cell molecules from iron toxicity. Accordingly, alterations in brain iron regulation resulting in an increased availability of reactive or low molecular weight iron could lead to free radical formation, tissue damage, and a neurodegenerative disorder.

4.3 Iron and neurodegeneration

Increased levels of iron have been noted in association with neurodegeneration in both PD and AD. Increased iron in the PD brain was first noted by Lhermitte in

1924. In 1966, Earle reported a generalized increase in brain iron in PD using X-ray fluorescence spectroscopy. It is noteworthy that these findings were described in the pre-levodopa era and, accordingly, iron accumulation was not influenced by administration of this drug. The topic remained relatively obscure until MRI studies indicated that iron could be detected in living patients and was increased in parkinsonism (Drayer *et al.*, 1986; Olanow and Drayer, 1987). Subsequently, numerous studies using a variety of analytical techniques have demonstrated that iron levels are increased in PD, primarily within the SNc (Sofic *et al.*, 1988; Dexter *et al.*, 1989; Hirsch *et al.*, 1991). Ben-Shachar *et al.* (1991a) studied the binding of iron to dopamine– melanin and speculated that neuromelanin might account for the accumulation of iron within the SNc. Further, they speculated that semiquinones derived from the auto-oxidation of dopamine could reduce iron from Fe^{3+} to Fe^{2+}, the form in which it can react with H_2O_2 derived from dopamine metabolism and generate the cytotoxic $OH^{\cdot}$ radical. This concept has received substantial support by recent laser microprobe mass analysis (LAMMA) and X-ray microanalytical studies which demonstrated that iron accumulation in the SNc of PD patients is primarily within neuromelanin granules (Good *et al.*, 1992a; Jellinger *et al.*, 1992). Neuromelanin might thus account for the site-specific accumulation of iron and aluminium, within SNc neurones (Ben-Shachar *et al.*, 1991a; Youdim *et al.*, 1994). Interestingly, LAMMA studies also demonstrate a massive accumulation of aluminium within SNc neurones in PD (Good *et al.*, 1992a). While aluminium exists in only a single valence state and is not a transition metal, oxidant damage consequent to iron-induced oxidative stress is substantially increased by the addition of low concentrations of aluminium salts (Gutteridge *et al.*, 1985). We have speculated that aluminium may displace iron from its binding site on neuromelanin and thereby permit it to interreact with H_2O_2 derived from neighbouring dopamine metabolism, leading to the formation of cytotoxic radicals and neurodegeneration (Olanow, 1993a). As iron and aluminium can promote free radical formation, these findings have supported the hypothesis that iron-induced oxidative stress contributes to the development of cell degeneration in PD (Ben-Shacher *et al.*, 1991a; Olanow *et al.*, 1992; Olanow and Arendash, 1994; Gotz *et al.*, 1994).

Experimental models of parkinsonism can also involve iron-induced neurodegeneration. 6-Hydroxydopamine (6-OHDA) has been shown to release iron from ferritin and to promote ferritin-dependent lipid peroxidation (Monteiro and Winterbourn, 1989). Further, iron chelation prevents the development of 6-OHDA-induced parkinsonism (Ben-Schachar *et al.*, 1991b). An animal model of PD can be generated secondary to iron-induced oxidant stress by injecting iron directly into the SNc. In separate studies, the authors have stereotactically infused iron into the SNc of rodents and induced a model of neurodegeneration with behavioural, histological and neurochemical features of parkinsonism (Ben-Shachar and Youdim, 1991; Sengstock *et al.*, 1992). Neurodegeneration in these animals is dose related, associated with lipid peroxidation, and attenuated by co-administration of the iron-binding protein transferrin (Sengstock *et al.*, 1993; Arendash *et al.*, 1994). Further, it has now been demonstrated that changes in neurochemical markers observed in this model are progressive

Table 1 Neurodegenerative conditions associated with increased iron

Condition	*References*
Parkinson's disease	Olanow and Drayer (1987), Sofic *et al.* (1988), Dexter *et al.* (1989), Good *et al.* (1992a), Jellinger *et al.* (1992), Hirsch (1991)
Alzheimer's disease	Connor *et al.* (1992a,b), Good *et al.* (1992b), Richardson (1993)
Amyotrophic lateral sclerosis	Unpublished observation (CWO)
Multiple system atrophy	Drayer *et al.* (1986), Dexter *et al.* (1991)
Progressive supranuclear palsy	Olanow (1992), Dexter *et al.* (1991)
Hallervorden–Spatz's disease	Olanow (1994)
Wilson's disease	Lewitt and Brewer (1994)
Huntington's chorea	Olanow and Hause (1994)
Multiple sclerosis	Valberg *et al.* (1989)
Tardive dyskinesia	
Guam neurodegeneration	Yasui *et al.* (1993)
Manganese intoxication	Olanow *et al.* (1993)
6-OHDA-induced parkinsonism	Oestreicher *et al.* (1995)
MPTP-induced parkinsonism	Temlett *et al.* (1994)

(Riederer and Wasserman, 1995; Sengstock *et al.*, 1994). In this way, the iron-induced oxidant stress model of parkinsonism more closely resembles PD than do models based on 6-OHDA or 1-methyl-4-phenyl-1,2,3,6-tetrahydropyridine (MPTP).

Iron levels are also increased at sites of degeneration in AD. Increased iron has been detected in the hippocampus, nucleus basalis, frontal cortex and motor cortex of AD patients (Conner *et al.*, 1992a; Richardson, 1993). Histologically, iron has also been shown to concentrate in neuritic plaques within the neocortex and hippocampus (Conner *et al.*, 1992b). Using the LAMMA technique, iron has also been shown to selectively accumulate in the cytoplasm, nucleus and neurofibrillary tangles of hippocampal neurones in AD patients (Good *et al.*, 1992b).

Increased iron has also been observed in the lesion site of a number of other neurodegenerative disorders and models of neurodegeneration (Table 1). Iron accumulation is most pronounced in Halervorden Spatz disease where there is massive iron deposition in the GP and SNr. This is accompanied by the development of axonal spheroids and pigmentary formation – pathological findings that can be induced experimentally by iron-induced oxidant stress (Dooling *et al.*, 1974). The finding of increased iron at sites of cell degeneration in these conditions raises the possibility that iron and oxidant stress contribute to the pathogenesis of cell death in human neurodegenerative disorders. There is substantial evidence of oxidant stress in disorders such as PD and AD (Gotz *et al.* 1994; Jenner *et al.*, 1992; Olanow *et al.*,

Table 2 Evidence of oxidative stress in PD and AD

	PD	AD
Iron	Increased	Increased
Monoamine oxidase B	Increased	Increased
Reduced glutathione	Decreased	Increased
Oxidized glutathione	Unchanged	
Superoxide dismutase	Increased	Increased/decreased
Zinc	Increased	
Uric acid	Decreased	
Mitochondrial complex I	Decreased	
Lipid peroxidation	Increased	Increased
Lipid hydroperoxides	Increased	
Lipofuscin	Increased	
Carbonyl proteins	Increased	
8-Oxyguanidine	Increased	
Ubiquitin	Increased	Increased
Calcium-binding proteins	Decreased	Decreased

1992; Richardson, 1993) This evidence is reviewed in Chapter 2 and summarized in Table 2. Prominent among these findings are increased iron, alterations in the body's naturally occurring antioxidant defence mechanisms, and an increase in markers of oxidative damage. These findings are not in themselves conclusive, but they suggest the possibility that the increased iron found in conditions such as PD and AD contributes to the pathogenesis of the neurodegenerative process. The recent finding that familial amyotrophic lateral sclerosis is associated with mutations in the gene that encodes for copper/zinc superoxide dismutase (Rosen *et al.*, 1993) raises the possibility that oxidant stress may also contribute to the pathogenesis of amyotrophic lateral sclerosis (Olanow, 1993b).

On the other hand, the finding of increased iron in the area of neurodegeneration in such a wide array of conditions also raises the possibility that it is a secondary phenomenon that occurs consequent to cell death of any aetiology. In support of this latter interpretation are two recent studies demonstrating an increase in SNc iron following neuronal degeneration induced by either MPTP or a 6-OHDA injection into the median forebrain bundle (Temlett *et al.*, 1994; Oestreicher *et al.*, 1994). This does not, however, exclude the possibility that secondary iron contributes to the neurodegenerative process.

4.3.1 Unresolved issues

While there are many reasons to consider the possibility that pathological accumulation of iron is a factor that contributes to the cell death that characterizes neurodegeneration, a number of important issues remain to be resolved.

4.3.1.1 Is the iron that accumulates in the neurodegenerative disorders in a reactive form?

There are currently no studies that have determined the precise form in which iron deposits exist in the neurodegenerative disorders. The only direct way to measure iron in its different valence forms is with the technique of *Mössbauer spectroscopy*, which has not yet been applied to the study of neurodegeneration nor fully evaluated in the study of biological samples. However, there is indirect information about the state of iron that can be gleaned through the study of iron-binding proteins. Normally, an increase in cellular iron would be expected to be associated with a compensatory rise in the storage protein ferritin and a down-regulation of the transport protein transferrin. Increased iron coupled with unanticipated changes in iron-binding proteins might imply an altered capacity of the brain to regulate iron and suggest that iron exists in a reactive form. In PD, ferritin levels have been reported to be both decreased (Dexter *et al.*, 1990) and increased (Riederer *et al.*, 1989). The conflicting results in these studies may relate to differences in the source of ferritin (liver vs spleen) used to generate monoclonal antibodies. More recently, immunocytological studies using antibodies directed at the H and L chains of brain ferritin showed no change in ferritin levels in the SNc despite elevated levels of iron (Mann *et al.*, 1994; Connor *et al.*, 1995), suggesting an impaired compensatory mechanism. Other iron regulatory proteins have been even less well studied. A decrease in transferrin receptor density has been observed in the putamen (Mash *et al.*, 1991), and reduced levels of transferrin were detected in the cerebrospinal fluid of PD patients (Riederer *et al.*, 1988). In AD, Connor *et al.* (1992b) have reported increased ferritin levels in the motor cortex, decreased transferrin levels in cerebral cortex white matter, and decreased transferrin receptors in the hippocampus and temporal cortex. They interpret these alterations to be consistent with a reaction to increased iron, but the findings do not imply that the increased iron is in a reactive form. In summary, there is evidence of alterations in iron-binding proteins in PD and AD but the information is insufficient to permit a determination of whether iron deposits are in a reactive form. New studies aimed at correlating sites of iron accumulation and oxidative damage at a subcellular level may provide further insights into the role of iron in the production of neurodegenerative pathology.

4.3.1.2 Does increased iron drive the neurodegenerative process or does iron accumulate secondary to an alternative aetiological process?

Iron is capable of inducing oxidative damage and causing an experimental model of PD. However, we and others have shown that 6-OHDA and MPTP lesions of the nigrostriatal tract induce a secondary increase in iron at the level of the SNc (Temlett *et al.*, 1994; Oestreicher *et al.*, 1995). This increase was not accounted for by degradation of iron storage proteins. Indeed, our studies indicated that new iron had accumulated at the site of neurodegeneration. These studies demonstrate that iron can accumulate within a degenerating region as a consequence of a remote and unrelated lesion. Jenner and colleagues (1992) argue that iron accumulation in PD is likely to be secondary based on their studies of patients with incidental Lewy bodies whom they consider to have preclinical PD. These are individuals who at pathology exhibit neuronal degeneration and Lewy bodies in the SNc but who were free of parkinsonian features during life. In these individuals, reduced glutathione (GSH) levels are decreased to a level comparable to that seen in PD, suggesting that the nigra is in a state of oxidant stress even at this early stage of the putative disease (Dexter *et al.*, 1994). Interestingly, iron levels in the SNc were within normal limits in these patients. The authors interpret these findings to suggest that alterations in GSH may be primary in PD, and iron accumulation a secondary event that occurs at a later time point in the neurodegenerative process. However, iron measurements in these studies were performed using bulk analysis which may fail to recognize a significant iron increase in a specific subcellular region such as neuromelanin granules. Further studies to assess iron levels using microprobe techniques such as LAMMA are awaited. All the same, an increase in iron if it is in a reactive form may still contribute to the neurodegenerative process and extend the degree of cell damage even if it accumulates secondary to an alternative aetiological process. Indeed, it has now been shown that iron can be released from its storage proteins by free radicals such as $O_2^{\cdot -}$, $OH^{\cdot}$ and $NO^{\cdot}$ and promote a continuing cycle of free radical formation and oxidative damage (Monteiro and Winterbourn, 1988, 1989).

4.3.1.3 What is the source of increased iron in areas that undergo neurodegeneration?

Some considerations include:

(1) Redistribution of iron from other brain regions. Iron is transported to brain specific regions as part of its normal metabolism. An alteration in this distribution pattern could result in pathological iron accumulation in affected areas.
(2) Impaired clearance. Decreased iron efflux could result in iron accumulation.
(3) Protein degradation. Cell degeneration could result in denaturation of proteins such as ferritin with release of stored iron. This interpretation is unlikely as total iron levels appear to be increased and other stored metals such as copper are not correspondingly increased. Additionally, this interpretation would not account for the observed rise in aluminium in the PD nigra which is normally present in only minute concentrations.
(4) Infiltration of iron-rich microglia in response to regional cell degeneration could account for a local increase in iron.
(5) Breakdown in the blood–brain barrier. There is increasing interest in the possibility that an alteration in the blood–brain barrier could account for selective iron accumulation in regions undergoing neurodegeneration. Artificial disruption of the blood–brain barrier with neuroleptic agents causes an increase in cerebral iron-^{52} uptake on positron emission tomography (PET) studies in

monkeys (Leenders *et al.*, 1994) and a 200–300% increase in brain iron uptake in rodents (Ben-Shachar *et al.*, 1994). Increased iron uptake on Fe^{52} PET has also been noted in PD patients as well as in patients with Wilson's disease and Halervorden–Spatz's disease (Leenders *et al.*, 1993). In this regard it is noteworthy that neuroleptic treatment occasionally results in a persistent parkinsonian syndrome (Jimenez-Jimenez *et al.*, 1993). The mechanism responsible for the extrapyramidal syndrome that occurs in association with the use of neuroleptic agents is not known but could relate to iron accumulation and consequent cell damage due to a breakdown in the blood–brain barrier. We have recently shown that manganese intoxication is associated with a focal increase in iron and aluminium in the globus pallidus (Olanow *et al.*, 1995). These changes were most pronounced in endothelial cells and in a perivascular distribution consistent with a breakdown in the blood–brain barrier. The possibility that an inflammatory reaction could account for a focal breakdown in the blood–brain barrier in PD and AD patients has been suggested based on findings of proliferating reactive microglia (McGeer *et al.*, 1988) and increased interleukin-1 and -6 and tumour necrosis factor α in regions undergoing neurodegeneration (Youdim, 1995).

4.4 Neuroprotection

There exists a strong rationale and some experimental data to support the possibility that iron-induced oxidant stress contributes to cell death in PD and perhaps other neurodegenerative disorders. This raises the possibility that therapies designed to chelate iron and prevent it from participating in oxidation reactions may protect vulnerable neurones and slow the rate of disease progression in patients suffering from neurodegenerative disorders. Iron chelators have been demonstrated to effectively prevent iron-catalysed reactions from taking place and to correspondingly limit free radical formation and tissue injury (Gutteridge *et al.*, 1983; Aust and White, 1985). Iron chelators also limit the extent of cell degeneration in the 6-OHDA and iron infusion models of PD (Ben-Shachar *et al.*, 1991b; Arendash *et al.*, 1994).

There have only been a limited number of studies of iron chelators in human patients with neurodegenerative disorders. Crapper-McLachlan *et al.* (1991) reported that desferrioxamine slowed the rate of progression of patients with AD. The study has been criticized because of design flaws including lack of satisfactory blinding. Further, desferrioxamine does not normally cross the blood–brain barrier and is not known to reduce brain iron levels in adults. Accordingly, benefits observed in this clinical trial are not necessarily related to iron chelation, and, indeed, the authors attribute benefits to chelation of aluminium. There has been considerable interest in the 21 amino steroids or lazaroid group of drugs. Some of these agents inhibit lipid peroxidation possibly by chelating membrane-bound iron. Preliminary studies of lipid-soluble lazaroids thought to be capable of entering the brain have provided encouraging results in animal models of neurodegeneration (Hall, 1992). Clinical trials of lazaroids have been initiated for cerebral haemorrhage and spinal cord injury, but clear-cut benefits have not yet been documented. D-Penicillamine is an effective therapy for Wilson's disease, presumably because of its capacity to chelate and remove copper from the nervous system. However, iron concentrations are also increased in Wilson's disease. Although D-penicillamine preferentially chelates copper it also chelates iron, and it is possible that some of the benefits observed are related to this latter property. Two patients with Halervorden–Spatz's disease are known to have been treated with the iron chelator desferrioxamine. One experienced no benefit (Gallyas and Kornyey, 1968). The other had transient improvement with larger doses (Dooling *et al.*, 1974).

Thus, with the exception of Wilson's disease, there is no clear evidence at the present time of metal chelator-induced clinical benefit in the neurodegenerative diseases although they have not been adequately studied in these disorders. In considering the initiation of clinical trials of iron chelators in humans, a note of caution is warranted. Rodents treated with central iron chelators experience prolonged coma (Halliwell and Gutteridge, 1985). Further, patients that have received desferrioxamine in combination with neuroleptic agents that are thought to open the blood–brain barrier to desferrioxamine are reported to have experienced cerebral and ocular toxicity (Blake *et al.* 1985). Nonetheless, the development of an iron chelator that can penetrate the central nervous system and selectively bind iron without inducing significant toxicity is a rational goal of continued research efforts aimed at developing a neuroprotective therapy for patients with neurodegenerative disorders.

References

Aisen, P. (1992) *Ann. Neurol.* **32**, 62–68.

Arendash, G.W., Sengstock, G.J., Barone, S. & Olanow, C.W. (1994) In *Animal Models of Toxin-Induced Neurological Disorders* (eds Woodruff, M. & Nonneman, A.), pp. 176–212. Plenum Press, New York.

Aust, F.D. & White, B.C. (1985) *Adv. Free Rad. Biol. Med.* **1**, 1–17.

Ben-Schachar, D. & Youdim, M.B.H. (1991) *J. Neurochem.* **57**, 2133–2135.

Ben-Shachar, D., Yehuda, S., Finberg, J., Spanier, I. & Youdim, M.B.H. (1988) *J. Neurochem.* **56**, 1434–1438.

Ben-Shachar, D., Riederer, P. & Youdim, M.B.H. (1991a) *J. Neurochem.* **57**, 1609–1614.

Ben-Shachar, D., Eshel, G., Finberg, J.P.M. & Youdim, M.B.H. (1991b) *J. Neurochem.* **56**, 1441–1444.

Ben-Shachar, D., Levine, E., Spanier, I., Leenders, K.K. & Youdim, M.B.H. (1994) *J. Neurochem.* **62**, 1112–1119.

Bielski, B.H.J. (1992) *Ann. Neurol.* **32**, 28–32.

Blake, D.R., Winyard, P. & Lonec, J. (1985) *Quant. J. Med.* **219**, 345–355.

Connor, J., Snyder, B., Beard, J., Fine, R. & Mufson, E. (1992a) *J. Neurosci. Res.* **31**, 327–335.

Connor, J., Menzies, S., St Martin, S. & Mufson, E. (1992b) *J. Neurosci. Res.* **31**, 75–83.

Connor, J.R., Snyder, B.S., Arosio, P. *et al.* (1995) *J. Neurochem.* **65**, 717–724.

Crapper-McLachlan, D.R., Dalton, A.J., Kruck, T.P.A. *et al.* (1991) *Lancet* **337**, 1304–1308.

Dawson, V.L., Dawson, T.M., London, E.D., Bredt, D.S. & Snyder, S.H. (1991) *Proc. Natl Acad. Sci. USA* **88**, 6368–6371.
Dexter, D.T., Wells, F.R., Lees, A.J. *et al.* (1989) *J. Neurochem.* **52**, 1830–1836.
Dexter, D.T., Carayon, A., Vidaihet, M. *et al.* (1990) *J. Neurochem.* **55**, 16–20.
Dexter, D.T., Carayon, A., Javoy-Agid, F. (1991) *Brain* **114**, 1983–1975.
Dexter, D.T., Sian, J. & Rose, S. (1994) *Ann. Neurol.* **35**, 38–44.
Dooling, E.C., Schoene, W.C. & Richardson, E.P. (1974) *Arch. Neurol.* **30**, 70–83.
Drayer, B., Olanow, C.W., Burger, P., Johnson, A., Herfkins, R. & Riederer, S. (1986) *Radiology* **159**, 493–498.
Drysdale, J.W., Adelman, T.G., Arosio, P. *et al.* (1977) *Semin. Hematol.* **14**, 17–88.
Dwork, A.J., Lawler, G., Zybert, P. *et al.* (1990) *Brain Res.* **518**, 31–39.
Earle, K.M. (1966) *J. Neuropathol. Exp. Neurol.* **27**, 1–14.
Gallyas, F. & Kornyey, S. (1968) *Arch. Psychiatr. Nervenkr.* **212**, 33–45.
Good, P., Olanow, C.W. & Perl, D.P. (1992a) *Brain Res.* **593**, 3343–3346.
Good, P.F., Perl, D.P., Bierer, L.M. & Schmeidler, J. (1992b) *Ann. Neurol.* **31**, 286–292.
Gotz, M.E., Kunig, G., Riederer, P. & Youdim, M.B.H. (1994) *Pharm. Ther.* **63**, 37–122.
Gutteridge, J.M.C., Halliwell, B., Treffry, A. *et al.* (1983) *J. Biochem.* **209**, 557–560.
Gutteridge, J.M., Quinlan, G.J., Clark, I. & Halliwell, B. (1985) *Biochim. Biophys. Acta* **835**, 441–447.
Hall, E. (1992) *Ann. Neurol.* **32**, 137–142.
Hallgren, B. & Sourander, P. (1958) *J. Neurochem.* **3**, 41–51.
Halliwell, B. & Gutteridge, J. (1985) *Trends Neurol. Sci.* **8**, 22–29.
Halliwell, B. & Gutteridge, J.M.C. (1988) *ISI Atlas Sci. Biochem.* **I**, 48–52.
Hirsch, E.C., Brandel, J.P. & Galle, P. (1991) *J. Neurochem.* **56**, 446–451.
Jellinger, K., Kienzl, E., Rumpelmair, B. *et al.* (1992) *J. Neurochem.* **59**, 1168–1171.
Jenner, P., Schapira, A.H.V. & Marsden, C.D. (1992) *Neurology* **42**, 2241–2250.
Jimenez-Jimenez, F.J., Cabreravaldivia, F., Ayusoperalta, L. *et al.* (1993) *Mov. Disord.* **8**, 246–247.
Lauffer, R.B. (1992) *Iron and Human Diseases*. CRC Press, Boca Raton.
Leenders, K.K., Antonini, A., Schwarzbach, P. *et al.* (1993) In *Quantifications of Brain Function. Tracer Kinetics and Image Analysis in Brain PET* (ed. Uemurd, K.), pp. 145–146. Elsevier, Amsterdam.
Leenders, K.L., Antonini, A., Schwarzbach, R. *et al.* (1994) *J. Neural Transm.* **43** (Suppl.), 123–132.
Lewitt, P.A. & Brewer, G.J. (1994) In *Neurodegenerative Disorders* (ed. Calne, D.B.), pp. 667–683. W.B. Saunders, Philadelphia.
Liu, D., Yang, R., Yan X. & McAdoo, D. (1994) *J. Neurochem.* **62**, 37–44.
McGeer, P.L., Itagaki, S., Boyes, B.E. & McGeer, E.G. (1988) *Neurology* **38**, 1285–1291.
Mann, V.M., Cooper, J.M., Daniel, S.E. *et al.* (1994) *Ann. Neurol.* **3**, 876–881.
Mash, D.C., Pablo, J., Buck, B.E. *et al.* (1991) *Exp. Neurol.* **114**, 73–81.
Mena, I., Marin, O., Fuenzalida, S. & Cotzias, G.C. (1967) *Neurology* **17**, 128–136.
Monteiro, H. & Winterbourn, C. (1988) *Biochem. J.* **256**, 923–928.
Monteiro, H.P. & Winterbourn, C.C. (1989) *Biochem. Pharmacol.* **38**, 4177–4182.
Oestreicher, E., Sengstock, G.J., Riederer, P., Olanow, C.W., Dunn, A.J. & Arendash, G. (1994) *Brain Res.* **38**, 4177–4182.
Olanow, C.W. (1992) *Neurol. Clin. North Am.* 405–420.
Olanow, C.W. (1993a) *J. Neural Transm.* **91**, 161–180.
Olanow, C.W. (1993b) *TINS* **16**, 439–444.
Olanow, C.W. (1991) In *Neurodegenerative Disorders* (ed. Calne, D.B.), pp. 807–824. W.B. Saunders, Philadelphia.
Olanow, C.W. & Arendash, G. (1994) *Curr. Opin. Neurol.* **7**, 548–558.
Olanow, C.W. & Drayer, B. (1987) In *Recent Developments in Parkinson's Disease* (eds Fahn, S., Marsden, D., Calne, D. & Goldstein, M.), pp. 135–143. Macmillan, New York.

Olanow, C.W., Cohen, G., Perl, D.P. & Marsden, C.D. (1992) *Ann. Neurol.* **32**, 1–145.
Olanow, C.W., Calne, D.B., Perl, D. & Pate, B. (1993) *Mov. Disord.* **8**, 410.
Olanow, C.W., Good, P., Pate, B.D. *et al.* (1995) *Neurology* (in press).
Orrenius, S., Burkitt, M.J., Kass, G.E., Dypbukt, J.M. & Nicotera, P. (1992) *Ann. Neurol.* **32**, 33–42.
Ratan, R.R., Murphy, T.H. & Baraban, J.M. (1994) *J. Neurochem.* **62**, 376–379.
Riederer, P. & Wasserman, A. (1995) *J. Neural Transm.* (in press).
Reiderer, P., Rausch, W.D., Schmidt, B. & Youdim, M.B.H. (1988) *Mt Sinai J. Med.* **55**, 21–29.
Riederer, P., Sofic, E., Rausch, W.-D. *et al.* (1989) *J. Neurochem.* **52**, 515–520.
Richardson, J.S. (1993) *Ann. NY Acad. Sci.* **695**, 73–76.
Richter, C., Park, J.W. & Arnes, B.N. (1988) *Proc. Natl Acad. Sci. USA* **85**, 6465–6467.
Roberts, R., Sandra, A., Siek, G.C. *et al.* (1992) *Ann. Neurol.* **32**, 43–50.
Rosen, D.R., Siddique, T., Patterson, D. *et al.* (1993) *Nature* **362**, 59–62.
Sengstock, G., Olanow, C.W., Dunn, A.J. & Arendash, G.W. (1992) *Brain Res. Bull.* **28**, 645–649.
Sengstock, G.J., Olanow, C.W., Menzies, R.A., Dunn, A.J. & Arendash, G. (1993) *J. Neurosci. Res.* **35**, 67–82.
Sengstock, G., Olanow, C.W., Dunn, A.J. *et al.* (1994) *Exp. Neurol.* **130**, 82–94.
Sofic, E., Riederer, P., Heinsen, H. *et al.* (1988) *J. Neural Transm.* **74**, 199–205.
Temlett, J.A., Landsberg, J.P., Watt, F. & Grime, G.W. (1994) *J. Neurochem.* **62**, 134–146.
Wolff, S.P., Garner, A. & Dean, R.T. (1986) *Trends Biol. Sci.* **11**, 27–31.
Yasui, M., Ota, K. & Garrato, R.M. (1993) *Neurotoxicology* **14**, 445–450.
Yehuda, S. & Youdim, M.B.H. (1989) *Am. J. Clin. Nutr.* **50**, 618–630.
Youdim, M.B.H. (1988) *Brain Iron: Neurochemical and Behavioral Aspects*. Taylor and Francis, London.
Youdim, M.B.H. (1994) In *Neurodegenerative Disorders* (ed. Calne, D.B.), pp. 251–287. W.B. Saunders, Philadelphia.
Youdim, M.B.H., Ben-Shachar, D. & Riederer, P. (1989) *Am. J. Clin. Nutr.* **50**, 607–617.
Youdim, M.B.H., Ben-Shachar, D. & Riederer, P. (1994) *J. Neural Transm.* **43**.
Zaleski, S. (1887) *Arch. Exp. Pathol. Pharmakol.* **23**, 77–90.

CHAPTER 5

NITRONE RADICAL TRAPS PROTECT IN EXPERIMENTAL NEURODEGENERATIVE DISEASES

Robert A. Floyd and John M. Carney

Free Radical Biology & Aging Research, Oklahoma City, OK 73104, USA

Table of Contents

Neurodegeneration and Neuroprotection in Parkinson's Disease
ISBN 0-12-525445-8

5.1 Introduction

5.1.1 Overview of the protective potential of nitrone radical traps

Certain nitrone-based free radicals traps (NRTs) protect experimental animals from ischaemia/reperfusion injury (stroke), as well as 1-methyl-4-phenyl-1,2,3,6-tetrahydropyridine (MPTP)-induced Parkinson's disease. In addition, NRTs administered chronically reverse the age-associated increase in oxidatively damaged protein and the age-associated decrease in the activity of the oxidative-sensitive enzyme glutamine synthetase (GS) in the brain. Accompanying the NRT-mediated changes in oxidized protein and GS activity is a significant improvement in the performance of animals in behavioural tests measuring short-term spatial memory. These observations, along with several others detailed herein, indicate that NRTs may prove useful to treat age-related neurodegenerative diseases.

5.1.2 Oxygen free radicals in neurodegenerative diseases

Oxygen free radicals and oxidative damage are involved in brain ageing as well as in the age-related neurodegenerative diseases stroke, Parkinson's disease and Alzheimer's disease. This concept is strongly supported by recent results. It is highly likely that the primary sources of reactive oxygen species (ROSs) in the brain are produced mostly by mitochondria and that these organelles provide a constant oxidative stress which contributes to age-related alterations in brain.

5.1.3 Possible mechanism of NRT action

It is likely that NRTs react with and hence quench the primary ROSs and/or the subsequently produced secondary free radicals which arise and thus decrease the oxidative stress imposed. When there is a large oxidative stress (e.g. ischaemia/reperfusion insult), then the NRTs act to minimize the cascade of oxidative events which leads to tissue injury. As a part of this hypothesis we also suggest that during experimental Parkinson's disease (MPTP induced), mitochondria produce even higher levels of ROSs, which then trigger other oxidative events leading to tissue injury. Further, during brain ageing the tissue becomes more susceptible to primary ROSs and/or secondary oxidative events and that these age-associated changes, most likely due in large part to changes in mitochondria, predispose the tissue to oxidative damage.

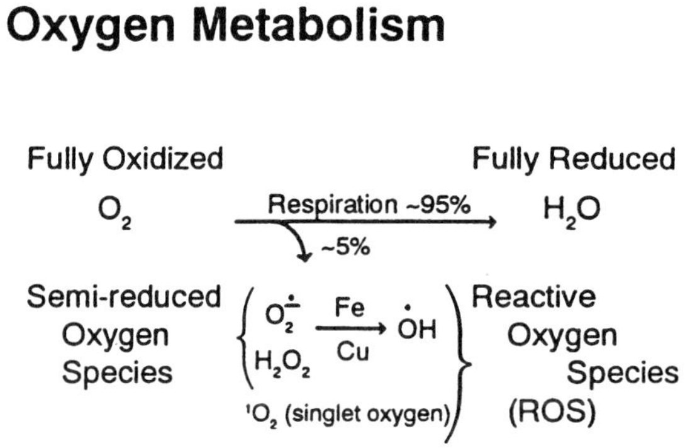

Oxygen Metabolism

Fully Oxidized — Fully Reduced
O_2 —Respiration ~95%→ H_2O
~5%
Semi-reduced Oxygen Species: $O_2^{\cdot-}$, H_2O_2 —Fe/Cu→ $\dot{O}H$, $^1O_2$ (singlet oxygen) — Reactive Oxygen Species (ROS)

Figure 1 The origin of ROSs which includes singlet oxygen as well as hydroxyl free radicals that are formed by the catalytic action of iron or copper in combination with the semi-reduced oxygen species superoxide and hydrogen peroxide, the last of which are by-products of respiratory activity whereby oxygen is fully reduced to form water.

5.2 Background

5.2.1 Oxidative stress in biological systems

Living systems which depend upon oxygen for life are constantly subjected to oxidative stress. The validity of this seemingly paradoxical statement is due to the fact that during the process of respiration, whereby molecular oxygen is reduced to water, a small amount, perhaps as much as 5% of the total oxygen consumed, of semi-reduced species of oxygen is produced. These semi-reduced species of oxygen are very reactive and thus initiate a series of oxidative reactions leading to oxidative stress. Superoxide ($O_2^{\cdot-}$) and hydrogen peroxide are semi-reduced species of oxygen present in biological systems which, in the presence of the trace metals iron and copper, react to form hydroxyl free radicals ($OH^{\cdot}$), another semi-reduced species of oxygen. Usually, singlet oxygen is also included; although, technically it is not a semi-reduced species of oxygen, and thus superoxide, hydrogen peroxide, hydroxyl free radicals and singlet oxygen are collectively known as ROSs. The production of semi-reduced and reactive oxygen species in relation to the respiratory use of oxygen is presented schematically in Figure 1.

Thus, aerobic living systems, simply because they utilize oxygen, are subjected to a continuous production of ROSs. The continuous presence of ROSs represents the endogenous oxidative stress under which biological systems exist. We term the oxidative stress to which biological systems are subjected as the oxidative damage potential, P_o. At times, depending upon the extent of the endogenous as well as the exogenous oxidative insults imposed, P_o may be significantly larger than at other times.

Nature protects herself from the imposed oxidative damage potential by utilizing various antioxidant systems which have evolved to combat oxidative stress, and these

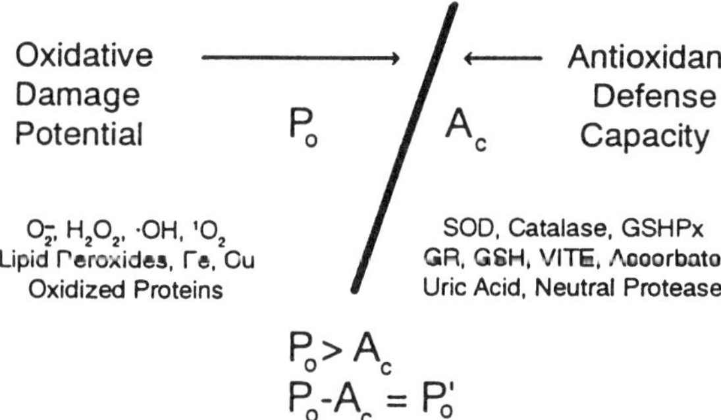

Figure 2 Aerobic biological systems experience oxidative stress whereby the oxidative damage potential P_o is imposed due to the presence of semi-reduced oxygen species, singlet oxygen, lipid peroxides, oxidized proteins and the redox active metals iron and copper. P_o is opposed by the antioxidant defence capacity of the system A_c, which is almost always smaller than P_o, thus resulting in a net flux of oxidatively damaging species, P_o'. GSH, reduced glutathione; GSHPx, glutathione peroxidase; GR, glutathione reductase; SOD, superoxide dismutase; VITE, vitamin E.

collectively are referred to as the antioxidant defence capacity, A_c, of the system. The oxidative damage potential is held in approximate balance by the antioxidant defence capacity of the system. This is represented schematically in Figure 2. P_o is normally slightly larger than A_c, such that there is a continuous low level flux, P_o', of ROSs.

5.2.2 Measurement of oxygen free radicals/oxidative damage *in vivo*

The amount of oxygen free radicals present at any specific times is considered to be extremely small, perhaps no more than 10^{-12}–10^{-11} M (Tyler, 1975) for superoxide and even less, perhaps no more than 10^{-15} M, for hydroxyl free radicals. The normal hydrogen peroxide content will depend upon many factors, including the type of tissue. Boveries *et al.* (1972) estimate free H_2O_2 production for liver to be at the rate of 90 nmol min^{-1} per gram of wet weight. The steady state level of H_2O_2 in tissue has been observed to be 10^{-7}–10^{-9} M (Chance *et al.*, 1979). Thus, it is quite apparent, since $O_2^{\cdot -}$, $OH^{\cdot}$ and H_2O_2 are in such low concentrations and since these species do not have unique and intense spectroscopic properties, their presence and flux in biological systems must be ascertained by indirect methods.

Basically, two general approaches have been used (Figure 3). These are: (1) the use of exogenous traps which react with oxygen free radicals to yield a stable and unique product that, with time, accumulates to a concentration such that it can be quantified and (2) the measurement of unique oxidation products produced when ROSs react with biological molecules. In the latter category, oxidized proteins and oxidized nucleic acid bases have been measured as an index of oxidative

Quantitation of Oxygen Free Radicals/Oxidative Damage

Approaches

A. Oxidative Damage to Biomolecules
   1. Measurement of oxidized nucleic acid bases
      8-oxo-guanine, thymine glycol, 5-hydroxymethyl uracil
      Note: Repair processes lower amounts of oxidized bases
   2. Measurement of oxidized protein, dinitrophenyl hydrazine reactive carbonyls, activity loss of sensitive enzymes, glutamine synthetase
      Note: Proteolytic enzymes catabolize oxidized proteins

B. Addition of Exogenous Traps to React with ROS
   1. Salicylate hydroxylation - index of $\dot{O}H$ flux
      Products are 2,3 and 2,5-dihydroxy benzoic acids (DHBAs)
   2. Nitrone-based free radical traps (NRTs)
      Example:
      α-phenyl-tert-butyl nitrone (PBN)
      $NRT + \dot{R} \longrightarrow \dot{R}\text{-NRT}$
      (radical) Trapped radical
      Characterize $\dot{R}$-NRT by electron paramagnetic resonance

Figure 3 *In vivo* quantification of oxygen free radicals and/or oxidative damage has depended largely upon two approaches: (1) the measurement of oxidatively damaged biomolecules and (2) the use of exogenous traps such as salicylate or NRTs that react to produce unique products which are quantified and characterized.

damage. Repair of oxidized nucleic acids and proteolysis of oxidized proteins are processes which certainly influence the level of oxidative products obtained. Products of oxidized lipids, such as malondialdehyde, have been very difficult to demonstrate in living tissues perhaps because malondialdehyde is a substrate for mitochondria.

A corollary to the measurement of oxidized protein is the loss of activity of an enzyme whose activity is very sensitive to oxidative damage, such as glutamine synthetase. Perhaps the best approach to assessing oxygen free radical flux/oxidative damage is the use of a combination of as many methods as possible. This has proven to be very useful in studying the mechanisms involved in the ischaemia/reperfusion insult (IRI) as well as the effects of chronic administration of NRTs on brain ageing in Mongolian gerbils (Cao *et al.*, 1988; Oliver *et al.*, 1990; Carney *et al.*, 1991). In these specific examples, salicylate hydroxylation (Cao *et al.*, 1988), protein oxidation and loss of glutamine synthetase activity (Oliver *et al.*, 1990; Carney *et al.*, 1991) were all successfully used on the same model. The data obtained from the combined methodologies were consistent and clearly implicated oxygen free radicals/oxidative damage as primary aetiological parameters in IRI-mediated brain damage and age-associated changes in brain (Floyd and Carney, 1991; Floyd, 1991).

5.2.3 Use of NRTs as analytical tools

The application of the concept that free radicals, $R^{\cdot}$, could be trapped by nitroso compounds to yield relatively stable nitroxide products (see equation (1)),

$$R'{-}N{=}O + R^{\cdot} \rightarrow R'{-}\overset{\overset{\displaystyle O}{|^{\cdot}}}{N}{-}R \tag{1}$$

which when characterized by electron paramagnetic resonance would yield information about R˙ seems to have arisen virtually simultaneously in two or three different locations (i.e. see Lagercrantz and Forshult, 1968; Chalfont and Perkins, 1968; Janzen, 1971). Most likely the first to report the use of a nitrone, α-phenyl-*t*-butyl nitrone (PBN), to trap a free radical was Iwamura and Inamoto (i.e. see Janzen, 1971).

Thus, the reaction of a nitrone with a free radical can be written as in equation (2),

$$X{-}\overset{\overset{\displaystyle H}{|}}{C}{=}\underset{+}{\overset{\overset{\displaystyle O+}{|}}{N}}{-}Y + R^{\cdot} \rightarrow X{-}\overset{\overset{\displaystyle H}{|}}{C}R{-}\overset{\overset{\displaystyle O}{|^{\cdot}}}{N}{-}Y \tag{2}$$

where, in the case of PBN, X=phenyl and Y=*t*-butyl. Janzen and Blackburn (1969) were the first to term this a spin-trapping reaction. The unreacted nitrone was thus termed the spin trap, and the product a spin adduct. In the present treatise, we refer to the nitrones which react with free radicals as nitrone-based free radical traps (NRTs) to more accurately reflect their action and chemical nature.

NRTs were at first used to investigate free radical intermediates in chemical reactions. Then in the mid-1970s they were first used in biochemical/biological systems by Harbour and Bolton (1975) as well as by McCay *et al.* (1976) and Poyer *et al.* (1978), in attempts to detect free radical intermediates. The first use of NRTs to trap free radicals in living animals was in 1979 by Lai *et al.* (1979) NRTs have proven very useful in some cases in elucidating free radical intermediates, such as in the metabolism of CCl_4 (Poyer *et al.*, 1978), yet the radical adducts of NRTs are in many cases subsequently metabolized, presumably reduced, to render the products diamagnetic and hence not 'visible' for characterization by electron paramagnetic resonance (i.e. see Mottley *et al.*, 1981; Floyd, 1983; Pou *et al.*, 1989; Kalyanaraman *et al.*, 1992).

5.2.4 Observations on mutagenic and carcinogenic processes

We showed very early that the commonly used NRTs were not mutagenic using the Ames Tester system (Hampton *et al.*, 1981). The tester strains used were TA98, TA100, TA1535 and TA1537, and the NRTs were tested at two concentrations, 50 and 100 μg per Petri plate.

There are very few studies on the effects of NRTs on carcinogenesis. Farber's group demonstrated that liver cell death caused by feeding a choline-devoid diet, which eventually leads to liver tumour formation, was prevented by administering nitroso *t* butane (Ghazarian *et al.*, 1989). Farber's group has also demonstrated that lipid peroxidation in intact liver, brought on by feeding a choline-devoid diet or after CCl_4 administration, was prevented by administering PBN at 100 mg kg^{-1}. Utilizing magnetic resonance imaging, Janzen *et al.* (1990) demonstrated that liver oedema caused by CCl_4 administration to rats was to a large extent prevented by PBN administration (125 mg kg^{-1}) 30 minutes prior to CCl_4 dosing.

In another pertinent study, Netto *et al.* (1992) demonstrated that α-4-pyridyl-*N*-oxide *N*-*t*-butylnitrone (POBN) given at the same time as the colon carcinogen 1,2-dimethylhydrazine significantly decreased the carcinogen-mediated methylation of DNA in target tissue. The DNA adducts N^7-methylguanine and O^6-methylguanine were significantly depressed by POBN, and this was attributed to POBN interference with P_{450}-mediated metabolism of the carcinogen; but of greater interest was the observation that C^6-methylguanine formation, which is considered to occur by free radical processes, was completely prevented by the co-administration of POBN.

5.2.5 Early observations on the protective action of NRTs

It is likely that the first report of a protective role for NRTs was by Novelli *et al.* in 1985 (1985a,b), who showed that PBN protected rats from lethality mediated by shock-induced by either superior mesenteric artery occlusion or by endotoxin injection, as well as by shock trauma caused by rapid rotation (58 rpm) of a drum in which the animals were confined. In all three cases, almost all (1 of 45 total) of the animals had died 24 hours after receiving the shock treatment. In contrast, almost all of the rats survived the treatments if they had been given PBN (150 mg kg^{-1}) prior to the treatment; and, in fact, PBN at 50 mg kg^{-1} prevented over half of the animals from treatment-induced lethality. PBN (150 mg kg^{-1}) was also effective in preventing death due to drum rotation trauma if given at least 60 minutes after treatment. Traumatic shock caused blood acidosis, excessive loss of bases and an increase in the haematocrit. PBN prevented the alteration in these parameters and, in fact, if given 30 or 60 minutes after traumatic shock, caused a time-dependent restitution to normal values. Novelli *et al.* (1985b) concluded that 'the brief time elapsing between the injection of the PBN and the progressive appearance of its effects . . . suggests that the drug diffuses and breaks the peroxidative chain reactors activated by trauma, endotoxin or ischemia.' The early observations of Novelli on the protective role of PBN from endotoxic shock were later independently confirmed by McKechnie *et al.* (1986), Hamburger and McCay (1989) and Pogrebniak *et al.* (1992).

Shortly after the observations of the protective role of PBN in shock, Hearse and Tosaki (1987) reported that PBN offered protection from ischaemia-caused reperfusion-induced ventricular fibrillation in the isolated Langendorff-perfused rat heart. Protection was maximal with 30–100 μM PBN in the perfusion fluid. PBN offered significant protection even if introduced shortly after reperfusion began. Following these observations, Bolli *et al.* (1988) reported that PBN present in the blood at a mean concentration of 1.85 mM before ischaemia helped to prevent the ischaemia-induced/reperfusion manifested loss of systolic thickening of *in vivo* 'stunned' dog myocardium.

Our research work on the neuroprotective action of NRTs began in early 1988 but was only reported in 1990 (Floyd, 1990). Table 1 summarizes some of the earlier observations regarding the protective role of NRTs in neurodegenerative diseases.

Table 1 Early demonstrations of the neuroprotective role of NRTs

Observations	Reference
1. PBN given before experimental stroke (IRI) prevented lethality in gerbils. Both old and young gerbils were protected, but old animals were more sensitive to IRI	Floyd (1990)
2. PBN (100 mg kg^{-1}) given before IRI in gerbils prevented IRI-induced increase in locomotor activity and significantly reduced IRI-induced loss in hippocampus CA_1 neurones	Phillis and Clough-Helfman (1990a,b)
3. PBN administered chronically (32 mg kg^{-1} twice daily) to old gerbils for 14 days caused a decrease back to young levels in the age-associated increase in brain oxidized protein levels, as well as an increase in the age-associated loss in brain glutamine synthetase activity and neutral protease activity. The age-associated increase in errors created in a radial arm maze was decreased back to young levels after chronic PBN administration	Carney *et al.* (1991)
4. PBN (100 mg kg^{-1}) given before or 30 minutes after an IRI (5 minute ischaemia) in gerbils prevented IRI-induced locomotor activity and significantly reduced damage to hippocampus CA_1 neurones observed 5 days postischaemia	Clough-Helfman and Phillis (1991)
5. PBN (75 or 150 mg kg^{-1}) administered before or immediately after IRI to gerbils helped to protect hippocampal CA_1 neurone loss. PBN was present in brain as confirmed by microdialysis. PBN (ECD_{50}=2.7 mM) protected cultured rat cerebellum neurones against 100 μM glutamate-induced toxicity	Yue *et al.* (1992)
6. NMDA induced neuronal deficits by direct injection into rat striatum was prevented by PBN (100 mg kg^{-1}) administered either pre-NMDA treatment or 30 minutes post-NMDA treatment	Li *et al.* (1992)

NMDA, *N*-methyl-D-aspartate.

Thus our early observations (Floyd, 1990) that PBN protected from IRI-induced lethality (Table 2) in gerbils was confirmed by Phillis and Clough-Helfman (1990a,b), who also showed that PBN was protective both pre- and post-treatment in mitigating IRI-mediated loss in hippocampal CA_1 neurones. This was further confirmed by Yue *et al.* (1992), who, in addition, showed that PBN protected against glutamate-induced

Table 2 Effect of PBN on the lethality of brain IRI (Floyd, 1990)

Gerbil age	Lethality of brain IRI	
	No PBN administered	Pretreated with PBN[c]
Young[a] (15 minute ischaemia)	~50% alive	All alive
Old[b] (10 minute ischaemia)	~15% alive	All alive

[a]'Young' refers to 3-month-old male gerbils.
[b]'Old' refers to 18-month-old retired breeder male gerbils.
[c]Animals were pretreated with 320 mg kg^{-1} of PBN 30 minutes prior to ischaemia.

toxicity in cultured cerebral neurones, and that PBN administered actually resulted in the presence of bona fide PBN in the brain extracellular fluid of gerbils. The protective role of PBN administered (either pre- or post-treatment) on *N*-methyl-D-aspartate (NMDA)-induced striated lesions in rats was demonstrated by Li *et al.* (1992).

Following our work on PBN protection in gerbil brain IRI, we showed that PBN when administered chronically at 32 mg kg^{-1} twice daily to old gerbils caused the age-associated increase in brain oxidized protein to decrease to that observed in young animals and to mediate the almost complete restoration of brain glutamine synthetase and neutral protease activity (Carney *et al.*, 1991). Concomitant with the PBN-induced brain chemical changes was a decrease in the age-associated increase in errors committed in a radial arm maze back to values obtained in young animals (Figure 4). The radial arm maze test in gerbils is an index of short-term spatial memory. Thus, it is quite clear from the earlier reports that PBN, representing only one NRT, has considerable potential as a neuroprotective agent in experimental animal models. The data presented later reinforce this conclusion.

5.2.6 Tissue distribution, pharmacokinetics of NRTs

Apart from a few studies published regarding the tissue distribution and pharmacokinetics of PBN in rats there is very little information known about this area. Chen *et al.* (1990a) first showed that if PBN was administered intraperitoneally in saline at 75 mg kg^{-1} then bona fide PBN appeared and peaked within all tissues examined (plasma, kidney, lung, liver, heart and brain) within about 20–30 minutes. The PBN levels reached were in general between 45 and 65 μg g^{-1} (i.e. nearly 60–87% of what would be expected if PBN was essentially equally distributed to all tissues of the body). The biological half-life of PBN was found to be 134 minutes. Very little bona fide PBN was found in the urine in the same time frame as the biological half-life, indicating that PBN probably was being rapidly metabolized to another compound. Later, Chen *et al.* (1990b) found that PBN was metabolized rapidly in the liver to essentially only one product, which was not characterized.

Utilizing *in situ* microdialysis implanted in brain (striatum) in gerbils, Yue *et al.*

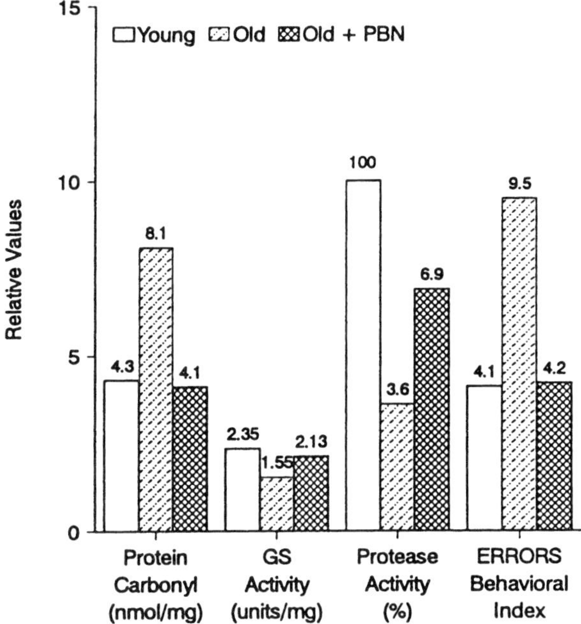

Figure 4 Data summarizing the results of Carney *et al.* (1991) and Floyd (1991) demonstrate that chronic administration of PBN to older gerbils caused a decrease in age-associated brain protein oxidation, a nearly complete recovery of glutamine synthetase activity and neutral protease activity and a concomitant decrease in the errors committed in a radical arm maze test which assesses short-term spatial memory.

(1992) demonstrated that the PBN concentration peaked in the brain 20 minutes after its intraperitoneal injection in saline. The concentration obtained in the extracellular microdialysate was about 250 and 450 μM of PBN after intraperitoneal injection of 75 and 150 mg kg^{-1}, respectively. If there had been completely equivalent distributions of PBN in all biological tissues, then the PBN concentration in the extracellular fluid supplied by the microdialysis device should have been 420 and 840 μM for the 75 and 150 mg kg^{-1} injections, respectively. Therefore, even though it is clear from this study that PBN rapidly reaches the brain, it apparently partitions within the brain.

In a more thorough study including not only PBN but also POBN, Cheng *et al.* (1993) showed that in hypertensive rats, PBN was concentrated in the brain tissue *per se* as compared to the brain microdialysate. That is, 100 minutes after administering 150 mg kg^{-1} intraperitoneally, PBN in the brain tissue was 587 μM, as against 331 μM for the microdialysate. POBN, which is less octanol-soluble than PBN, when administered intraperitoneally at an equivalent dose, was present in the brain tissue and brain microdialysate 100 minutes after injection but was at about one-half of the concentration of PBN. This was despite the fact that both PBN and POBN reached essentially equivalent (224 and 210 μM, respectively) steady state venous blood concentrations. The increased presence of PBN in the brain was considered to be due to the increased octanol/water solubility of PBN (16.59) versus POBN (0.18) (see values presented in Table 1 of Cheng *et al.* (1993)).

5.3 Recent experimental observations

The observations that NRTs protect from brain IRI-mediated lethality (Table 2) and if administered chronically mediate reversal of age-associated brain oxidative damage (Figure 4) clearly demonstrate the enormous potential these agents may have in treating neurodegenerative diseases. Recent results we have obtained and presented here underscore this conclusion.

5.3.1 Chronically administered NRTs protect from an IRI

Figure 5 shows the results of an experiment where old gerbils were given PBN (32 mg kg^{-1} intraperitoneally, twice daily) for 14 days, after which the animals were given bilateral carotid occlusion for 10 minutes at various times after ceasing PBN administration. After the ischaemic treatment, lethality was assessed at day 7. The data clearly show that 85–90% of the control animals (i.e. those not receiving PBN) were killed by 10 minutes of brain ischaemia. However, treatment for 14 days rendered the old gerbils essentially resistant to 10 minutes of brain ischaemia. The PBN-mediated protection remained for 3 days, and some protective activity was even noted 5 days after ceasing PBN dosing. It should be noted that the PBN half-life in rats has been shown to be 134 minutes; so, 3 days after ceasing PBN administration it is expected that, pharmokinetically, PBN would have experienced about 36 half-lives of washout, yet, as the data show, the protective effect is still largely intact. Our interpretation of these data is that the PBN administration has rendered the brain less susceptible to oxidative damage brought on by an IRI.

5.3.2 NRTs protect in brain ageing

The amount of oxidized protein in brain increases with age. The amount of oxidized protein in the brains of old gerbils is decreased by chronic dosing with PBN. As was noted previously, PBN given intraperitoneally twice daily at 32 mg kg^{-1} decreased oxidized protein levels back down to that level found in the young animals (see Figure 4). But, as the data in Figure 6 show, as little as 1 mg kg^{-1} of PBN administered twice daily for 14 days caused a significant decrease in the amount of oxidized protein. A near maximal decrease in oxidized protein content was achieved by giving PBN at 10 mg kg^{-1} twice daily. The 10 mg kg^{-1} dose represents 56 μmol kg^{-1}, and assuming that PBN essentially equally distributes to all organs, then, based on rat data (Chen *et al.*, 1990a,b), PBN in gerbil brain would be expected to achieve the ~40 μmol kg^{-1} level twice daily. Based on the PBN content of rat brain as assessed by microdialysis

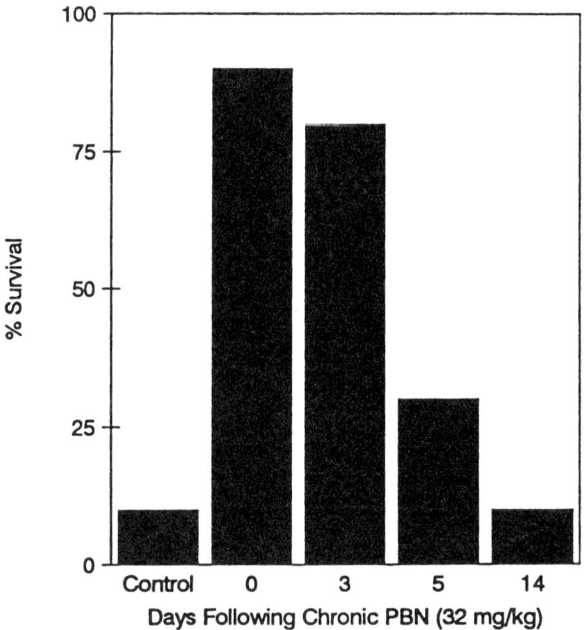

Figure 5 Illustration of the residual protective action of pretreatment with PBN. Older gerbils which had been pretreated with PBN (32 mg kg^{-1} intraperitoneally, twice daily) for 14 days were then given 10 minutes of bilateral carotid occlusion, and lethality was assessed 7 days after the treatment. Gerbils which had received the PBN pretreatment were protected 3 or even 5 days after ceasing PBN administration. The values presented are the average of 10 animals per group.

techniques, the intraperitoneal administration of 75 and 150 mg kg^{-1} caused the brain content to rise and remain relatively constant for up to 1 hour at levels of about 200 and 400 μmol PBN kg^{-1}, respectively. Thus, even though PBN in general is eliminated rather rapidly from rodents the brain content may lag behind during this general decrease. This is an area that needs more research effort, especially in view of the neurodegenerative protective action of NRTs when given in a chronic dosing pattern.

The vitamin E content of rodent brain is about 30 μmol kg^{-1}. Kogure *et al.* (1982) determined that the brain vitamin E content of 300–350 g male Wistar rats was 31.4 μmol kg^{-1}, and Yoshida *et al.* (1982), utilizing 240–320 g male Wistar rats, found the brain vitamin E content to be 34.8 μmol kg^{-1}. Zaspel and Csallany (1983) noted that the vitamin E content of brain from male Sprague–Dawley rats given a vitamin E-supplemented (300 mg kg^{-1}) diet was 62.2 μmol kg^{-1}. Thus, it is quite clear that PBN levels in brain are expected to achieve nearly equivalent amounts as vitamin E

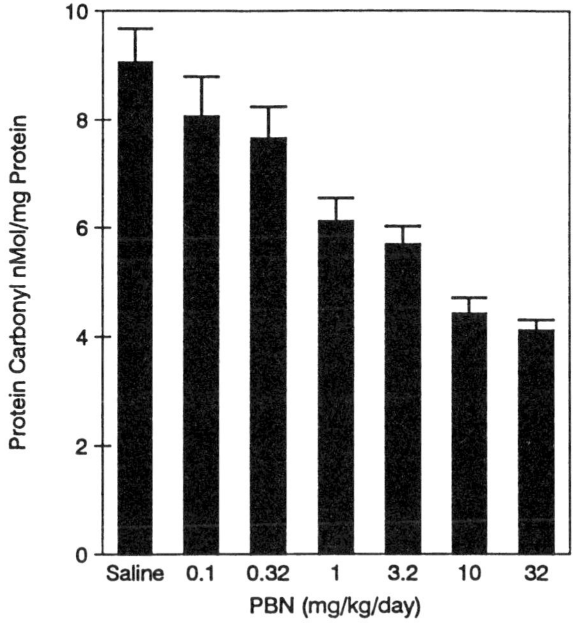

Figure 6 PBN administration mediated a decrease in oxidized protein in the brains of older gerbils. Protein oxidation was assessed after the animals had been treated with PBN (at the indicated dose) for 14 days.

when the PBN is administered at a dose of 10 mg kg^{-1}. Normally, vitamin E achieves a level of about one molecule per 1000 phospholipid molecules in natural membranes.

Figure 7 shows data obtained with old gerbils where PBN was administered intraperitoneally at various amounts 30 minutes prior to giving the animals a 10 minute ischaemic insult. The number of animals alive 7 days after the IRI was determined. The results clearly show that without PBN present about 85–90% of the animals were killed by 10 minutes of ischaemia. When PBN at 32 mg kg^{-1} was given, one-half of the old gerbils survived the 10 minutes of ischaemia. However, to essentially completely prevent IRI-mediated lethality, a dose of 320 mg kg^{-1} of PBN was required. Based on the work of Yue *et al.* (1992), who examined the PBN content of rat brain microdialysate, it is expected that an intraperitoneal dose of 320 mg kg^{-1} of PBN would yield a brain level of about 800–900 μmol kg^{-1}. In addition, it is expected that the brain levels would be sustained at these higher levels for over an hour. This level of PBN is nearly 30 times the expected vitamin E content of the brain.

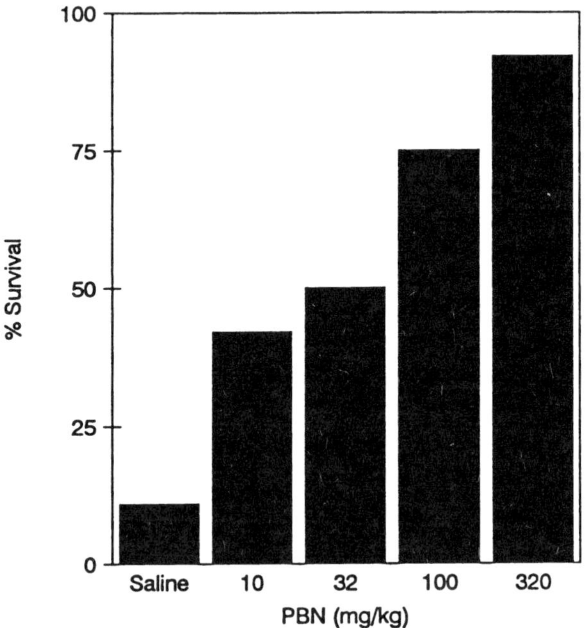

Figure 7 PBN-mediated protection from 10 minutes bilateral carotid ligation in aged gerbils (18–24 months old). Survival was assessed 7 days postischaemia. Each histogram is the mean of separate groups of 12 male gerbils. PBN was administered intraperitoneally 30 minutes prior to the ischaemia treatment.

5.3.3 NRTs protect in experimental Parkinson's disease

Langston and co-workers in 1983 discovered that a specific few South San Francisco drug users rapidly developed symptoms similar to Parkinson's disease (Lewin, 1984). They identified the responsible compound as MPTP, which was formed as a by-product in the home synthesis of so-called 'synthetic heroin'. Subsequent animal testing in mice and primates have confirmed that MPTP administration is a good method of inducing symptoms of Parkinson's disease in experimental animals.

We have utilized the C56BL/J6 mouse model to determine if MPTP-induced loss of dopamine is prevented by NRTs. The results clearly indicate that NRTs are very protective in preventing the MPTP-induced loss in brain dopamine (Carney, J.M. *et al.*, unpublished observations). The results presented in Tables 3–5 summarize the results obtained. Utilizing the C56BL/J6 male mouse model, we found that a single MPTP injection caused increased hydroxylation of salicylate in caudate and brain stem (Table 3). This clearly indicates increased hydroxyl free radical formation after MPTP administration.

Table 3 Data illustrating that MPTP administration to C57BL/J6 male mice caused an increase in salicylate hydroxylation (Carney *et al.*, 1994)

Tissue	MPTP	2,3-DHBA/salicylate ($\times 10^3$)
Brain stem	Yes	13.7±2.7
	No	3.9±0.4
Caudate	Yes	13.1±3.2
	No	6.7±3.0

2,3-DHBA, 2,3-dihydroxy-benzoic acid

Table 4 Data illustrating that the NRTs (PBN, POBN or DMPO), given at 10 mg kg^{-1} at the same time that MPTP was administered, prevented the loss in caudate dopamine in C57BL/J6 male mice (Carney *et al.*, 1994)

Treatment	Caudate dopamine (ng (mg protein)$^{-1}$)
1. No MPTP/no NRT	184±18
2. MPTP/no NRT	75±22
3. MPTP/PBN	204±25
4. MPTP/DMPO	180±23
5. MPTP/POBN	193±39

DMPO, 5,5-dimethyl-1-pyrroline *N*-oxide

Table 5 Data illustrating that PBN given at different levels at the same time that MPTP was administered prevented the loss in caudate dopamine in C57BL/J6 male mice (Carney *et al.*, 1994)

Treatment	Caudate dopamine (ng (mg protein)$^{-1}$)
1. No MPTP/no PBN	184±18
2. MPTP/no PBN	75±22
3. MPTP/PBN (0.1 mg kg^{-1})	93±21
4. MPTP/PBN (1 mg kg^{-1})	179±23
5. MPTP/PBN (3.2 mg kg^{-1})	160±5
6. MPTP/PBN (10 mg kg^{-1})	204±25
7. MPTP/PBN (32 mg kg^{-1})	178±22
8. MPTP/PBN (100 mg kg^{-1})	158±37

Table 4 presents data demonstrating that four doses of MPTP caused a significant decrease in the dopamine level in caudate and that NRTs when co-administered intraperitoneally with MPTP prevented the dopamine loss. Clearly not only PBN but also 5,5-dimethyl-1-pyrroline *N*-oxide (DMPO) and POBN were effective in preventing the MPTP-induced dopamine loss. Table 5 shows that PBN at 1 mg kg^{-1} when co-administered with MPTP was an effective dose in preventing the caudate dopamine loss.

5.4 Mechanisms of NRT protective action

The data presented in this report clearly show that NRTs are very protective in experimental models of neurodegenerative diseases. Why are NRTs protective in experimental models of neurodegenerative disease? What are the mechanisms involved? The simplest and most consistent mechanism is that NRTs are scavenging critical free radicals and thus preventing the chain of free radical/oxidative processes which, if unabated, leads to oxidative damage. This idea will be treated in a later section, but first it is necessary to consider, and in most cases rule out, other possibilities.

5.4.1 Do NRTs protect by producing nitric oxide?

Chamulitrat *et al.* (1993) suggested that the biological mechanism of PBN action could be due to its decomposition to form nitric oxide. They based their claims on *in vitro* data where they showed that: (1) a 2-week-old 150 mM PBN solution yielded 155 μM nitrite (the final nitric oxide product) and (2) ultraviolet photolysis, using a 1000 W source, for 10 minutes of a 150 mM PBN solution yielded 225 μM nitrite. Freshly prepared PBN did not yield any nitrite. We consider that nitric oxide production from PBN is not the mechanism of neurodegenerative protective action. Among the many reasons that can be marshalled against the nitric oxide formation idea is the consistent literature observations that PBN acts in an opposite fashion to that expected with nitric oxide.

Dawson *et al.* (1991) showed that nitric oxide enhanced the toxicity of glutamate in primary cortical neurone cultures. All of the evidence thus far indicates that glutamate build-up and its subsequent neurotoxic action in hippocampus appears to be the reason why CA_1 neurones die as a result of an ischaemia. We have found PBN to be very protective, and thus if PBN is producing nitric oxide, then, based on the results of Dawson *et al.* (1991), it is expected that it would not be protective. This general conclusion is supported by at least two other observations where nitric oxide production was found to enhance neurotoxicity. Oury *et al.* (1992) showed that the central nervous system damage brought on by breathing high oxygen levels in transgenic mice was enhanced by treating the animals with N^{ω}-nitro-L-arginine, a compound that inhibits nitric oxide production from arginine. This indicates nitric oxide production enhances central nervous system damage (i.e. opposite to the action of PBN).

Cazevieille *et al.* (1993) showed that nitric oxide also enhanced the superoxide-mediated damage to cultured neurones undergoing hypoxia/reoxygeneration injury. Thus, from all of the above reasons, plus the results of experiments we have conducted ourselves using N^{ω}-nitro-L-arginine, it is considered that the neuroprotective action of PBN does not involve the production of nitric oxide.

5.4.2 Do NRTs interact with guanylate cyclase?

It remains a possibility that NRTs may alter the action of nitric oxide, which binds to guanylate cyclase causing the accumulation of cGMP and mediates vasodilation. Neither Anderson *et al.* (1993) nor Sata *et al.* (1988) found any biological evidence for PBN interacting with guanylate cyclase and thus interfering with its action. Sata *et al.* (1988) did show that PBN at 1 mM blunted acetylcholine-induced increase in cGMP in rat thoracic aorta, but it had no effect on ATP or sodium nitroprusside-mediated increase in cGMP. They concluded that acetylcholine-mediated increase in cGMP mediated by free radicals which PBN trapped.

5.4.3 Do NRTs alter calcium shuttling?

Anderson *et al.* (1993) showed that NRTs caused the relaxation of preconstricted isolated rat pulmonary artery rings, but the concentrations required were quite high (ED_{50} values: PBN, 1.93 mM; POBN, 7.08 mM; DMPO, 24.5 mM). They also showed, using isolated pulmonary artery rings, that peak calcium influx, as well as steady state calcium levels, were significantly less when 3 mM PBN was present. The amount of NRTs required to alter calcium shuttling was much higher than that needed to show neuroprotection; therefore, we conclude that the primary action of NRTs is not by altering calcium shuttling.

5.4.4 Do NRTs influence prostaglandin synthesis?

The influence of NRTs on prostaglandin synthesis has not been researched extensively; however, from the limited amount of work available, it appears there is little influence. Thus, Anderson *et al.* (1993) found no evidence to suggest that NRTs altered prostaglandin synthesis in isolated rat pulmonary arteries. Smith *et al.* (1981) found that DMPO increased prostaglandin E_2 synthesis by 50% in ram seminal vesicles. It is known that the enzyme prostaglandin synthetase self-destructs and that the mechanism of its inactivation is probably by a free radical process, so the DMPO action may be by preventing the self-destruction of the enzyme. This is an area of research that needs more work.

5.4.5 Do NRTs alter mitochondrial respiration?

NRTs have been used to trap free radicals produced by mitochondria (Nohl *et al.*, 1981; Wong *et al.*, 1988). The question of whether NRTs influence mitochondria respiration has received only little attention. Wong *et al.* (1988) found that malate–glutamate respiration of isolated rat liver mitochondria was inhibited 15 and 24% by 24 mM DMPO and 4.7 mM PBN, respectively. In a more systematic study we found that PBN up to 500 μM had little if any effect on state 4, state 3 or ADP/O ratios of isolated pig liver mitochondria using either pyruvate, malate–glutamate or succinate as substrates. Therefore, it is highly unlikely that NRTs in the biologically meaningful range have any effect on mitochondrial respiration. It is possible, however, that NRTs do scavenge free radicals produced by mitochondria. This possibility is now being investigated.

5.4.6 Do NRTs alter P_{450}-mediated metabolism?

Very little research has been directed towards this research area, yet what has been done clearly indicates that NRTs do interact with P_{450}. Augusto *et al.* (1982) demonstrated that the P_{450} metabolism of certain dihydropyridine substrates was altered by NRTs. The P_{450} dissociation constants were 6100 μM for DMPO, 700 μM for POBN and 150 μM for PBN. PBN completely prevented the metabolism of certain dihydropyridine drugs. In another study, Netto *et al.* (1992) demonstrated that 1,2-dimethylhydrazine-mediated methylation of DNA in the target organ was completely prevented by the co-administration of POBN with the carcinogen. The NRT-mediated decrease in these DNA adducts was considered to be due to POBN prevention of P_{450}-mediated metabolism of the carcinogen.

5.4.7 NRTs act as potent antioxidants in oxidative stress

All of the work conducted thus far suggests that the action of NRTs must be looked at from a different perspective if on the one hand they are to be given chronically or on the other in a few large doses. Thus, we consider that if NRTs are to be given in a few large doses to blunt the oxidative stress imposed, their mechanism of action may be different than if they are given in a chronic fashion. We consider that the mechanism of action of NRTs given in a few large doses can best be rationalized as behaving as a trapping agent and thus acting in a mass action mechanism where the more present the more effective they are at preventing oxidative damage. Thus the demonstrations that PBN prevented IRI-mediated death, if given before the ischaemia or shortly thereafter, can best be viewed as acting as true scavengers of oxygen free radicals and/or the secondary free radicals produced.

On the other hand, if the NRTs are given at low doses in a chronic fashion, we consider that the presence of the NRTs alter the biological system such that it can more adequately withstand an oxidative stress. In the latter case, biological adaptation occurs, and hence the processes involved are much more complicated. The decrease in oxidized protein which occurs as a result of chronic administration of NRTs offers clues to the biological adaptation to the new oxidative damage state. The equilibrium between protein synthesis, protein oxidation and degradation appears to be important.

In a recent publication, Matts *et al.* (1993) demonstrated that denatured protein,

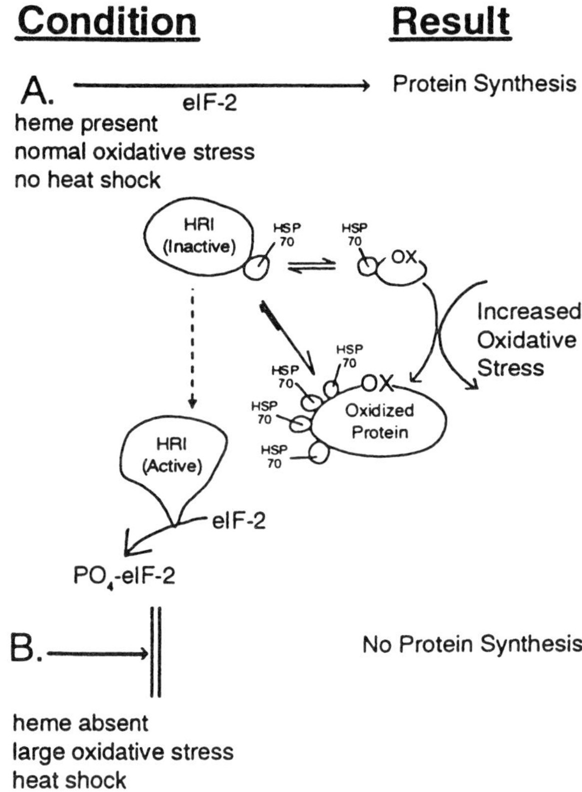

Figure 8 Model based on the data of Matts *et al.* (1993) illustrating that protein synthesis in a rabbit reticulocyte system was shut down when denatured protein was added to the system. Based on this model, increased levels of oxidized protein produced by increased oxidative stress is expected to act as a sink for heat shock 70 (HSP70) protein. The loss of HSP70 binding to an HRI activates this protein which is involved in mediating phosphorylation of eIF-2, which then inhibits protein synthesis. The important roles of HSP70 and oxidized protein are emphasized by this model.

when added to a protein synthesizing rabbit reticulocyte system, shut down protein synthesis. They demonstrated that the denatured protein acted as a sink for heat shock proteins. A small amount of heat shock protein binding to a haem-regulated inhibitor (HRI) of protein synthesis is necessary for protein synthesis to continue. Thus, addition of denatured proteins causing the competitive removal of heat shock proteins from HRI allowed phosphorylation of the HRI, which then shut down protein synthesis.

Thus, in this model system the denatured protein pool regulates protein synthesis. A generalized scheme to illustrate these results is shown in Figure 8. Based on these observations and our demonstrated results in mRNA (Carney *et al.*, 1993) of heat

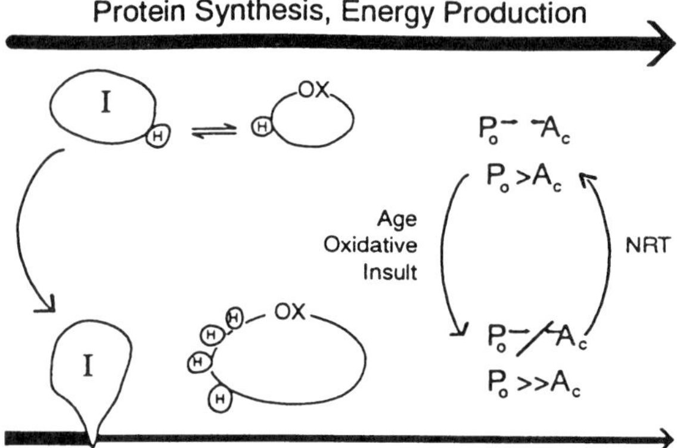

Figure 9 Generalized scheme illustrating that major cellular functions such as protein synthesis and energy production are normal when the normal levels of oxidized protein and effector molecules (H) are maintained, and hence inhibitors (I) remain inactive. Under increased oxidative stress, as is increasingly prevalent during ageing, the increased oxidative damage potential to the antioxidant defence capacity (P_o/A_c) ratio increases, resulting in increased levels of oxidized protein, which depletes H binding to I and thus activates I, resulting in decreased protein synthesis and energy production. Based on this highly generalized model, NRTs, administered chronically, would be expected to restore the normal oxidized protein levels, thereby allowing restoration of normal cellular functions and hence to restore a state which would be more resistant to oxidative stress.

shock protein 70 drastically increases after an IRI and that PBN prevents the IRI-mediated rise in heat shock protein mRNA combined with the fact that mRNA of heat shock protein 70 seems to be consistently elevated in older animals, allows us to generalize that NRTs, given chronically at low doses, act by lowering the oxidized protein pool, thus allowing protein synthesis and energy metabolism to occur normally, and hence the biological system re-adapts to where it can more readily withstand an oxidative stress. This mechanistic scheme is presented in Figure 9.

Acknowledgements

This research was supported by NIH grants AG09690 and NS 23307. Support is also appreciated from the Glen Foundation for Medical Research.

References

Anderson, D.E., Yuan, X.-J., Tseng, C.-M., Rubin, L.J., Rosen, G.M. & Tod, M.L. (1993) *Biochem. Biophys. Res. Commun.* **193**, 878–885.

Augusto, O., Beilan, H.S. & Ortiz de Montellano, P.R. (1982) *J. Biol. Chem.* **257**, 11288–11295.

Bolli, R., Patel, B.S., Jeroudi, M.O., Lai, E.K. & McCay, P.B. (1988) *J. Clin. Invest.* **82**, 476–485.

Boveris, A., Oshino, N. & Chance, B. (1972) *Biochem. J.* **128**, 617–630.
Cao, W., Carney, J.M., Duchon, A., Floyd, R.A. & Chevion, M. (1988) *Neurosci. Lett.* **88**, 233–238.
Carney, J.M. & Floyd, R.A. (1991) *J. Mol. Neurosci.* **3**, 47–57.
Carney, J.M., Starke-Reed, P.E., Oliver, C.N., Landrum, R.W., Chen, M.S., Wu, J.F. & Floyd, R.A. (1991) *Proc. Natl Acad. Sci. USA* **88**, 3633–3636.
Carney, J.M., Kindy, M.S., Smith, C.D., Wood, K., Tatsuno, T., Wu, J.F., Landrum, W.R. & Floyd, R.A. (1993) In *Cerebral Ischemia and Basic Mechanisms* (eds Hartman, A., Yatsu, F. & Kuschinsky, W.), pp. 301–311. Springer-Verlag, Berlin.
Cazevieille, C., Muller, A., Meynier, F. & Bonne, C. (1993) *Free Radic. Biol. Med.* **14**, 389–395.
Chalfont, G.R. & Perkins, M.J. (1968) *J. Am. Chem. Soc.* **90**, 7141–7142.
Chamulitrat, W., Jordan, S.J., Mason, R.P., Saito, K. & Cutler, R.G. (1993) *J. Biol. Chem.* **268**, 11520–11527.
Chance, B., Sies, H. & Boveris, A. (1979) *Physiol. Rev.* **59**, 527–605.
Chen, G., Bray, T.M., Janzen, E.G. & McCay, P.B. (1990a) *Free Radic. Res. Commun.* **9**, 317–323.
Chen, G., Griffin, M., Poyer, J.L. & McCay, P.B. (1990b) *Free Radic. Biol. Med.* **8**, 93–98.
Cheng, H.-Y., Liu, T., Feuerstein, G. & Barone, F.C. (1993) *Free Radic. Biol. Med.* **14**, 243–250.
Clough-Helfman, C. & Phillis, J.W. (1991) *Free Radic. Res. Commun.* **15**, 177–186.
Dawson, V.L., Dawson, T.M., London, E.D., Bred, D.S. & Snyder, S.H. (1991) *Proc. Natl Acad. Sci. USA* **88**, 6368–6371.
Floyd, R.A. (1983) *Biochim. Biophys. Acta* **756**, 204–216.
Floyd, R.A. (1990) *FASEB J.* **4**, 2587–2597.
Floyd, R.A. (1991) *Science* **254**, 1597.
Floyd, R.A. & Carney, J.M. (1991) *Arch. Gerontol. Geriatr.* **12**, 155–177.
Ghazarian, D.M., Bubnic, S., Farber, E. & Ghoshal, A.K. (1989) *Proc. Am. Assoc. Cancer Res.* **30**, 202 (abstract).
Hamburger, S.A. & McCay, P.B. (1989) *Circ. Shock* **29**, 329–334.
Hampton, M.J., Floyd, R.A. & Janzen, E.G. (1981) *Mutat. Res.* **91**, 179–283.
Harbour, J.R. & Bolton, J.R. (1975) *Biophys. Res. Commun.* **64**(3), 803–807.
Hearse, D.J. & Tosaki, A. (1987) *Circ. Res.* **60**, 375–383.
Janzen, E.G. (1971) *Acc. Chem. Res.* **4**, 31–40.
Janzen, E.G. & Blackburn, B.J. (1969) *J. Am. Chem. Soc.* **91**, 4481–4490.
Janzen, E.G., Towner, R.A. & Yamashiro, S. (1990) *Free Radic. Res. Commun.* **9**, 325–335.
Kalyanaraman, B., Perez-Reyes, E. & Mason, R.P. (1992) *Tetrahedron Lett.* **50**, 4809–4812.
Kogure, K., Watson, B.D., Busto, R. & Abe, K. (1982) *Neurochem. Res.* **7**(4), 437–454.
Lagercrantz, C. & Forshult, S. (1968) *Nature* **218**, 1247–1248.
Lai, E.K., McCay, P.B., Noguchi, T. & Fong, K.-L. (1979) *Biochem. Pharmacol.* **28**, 2231–2235.
Lewin, R. (1984) *Science* **225**, 1460–1462.
Li, M., Szczepanik, A.M., Brooks, K.M. & Wilmot, C.A. (1992) *Soc. Neurosci. Abstr.* **18**, 645.
McCay, P.B., Poyer, J.L., Floyd, R.A., Fong, K.-L. & Lai, E.K. (1976) *Fed. Proc.* **35**, 421 (abstract).
McKechnie, K., Furman, B.L. & Parratt, J.R. (1986) *Circ. Shock* **19**, 429–439.
Matts, R.L., Hurst, R. & Xu, Z. (1993) *Biochem.* **32**, 7323–7328.
Mottley, C., Kalyanaraman, B. & Mason, R.P. (1981) *FEBS Lett.* **130**, 12–14.
Netto, L.E.S., Ramakrishna, N.V.S., Kolar, C., Cavalieri, E.L., Rogan, E.G., Lawson, T.A. & Augusto, O. (1992) *J. Biol. Chem.* **267**, 21524–21527.
Nohl, H., Jordan, W. & Hegner, D. (1981) *FEBS Lett.* **123**, 241–244.
Novelli, G., Angiolini, P., Cansales, G., Lippi, R. & Tani, R. (1985a) In *Oxygen Free Radicals in Shock* (eds Novelli, G.P. & Ursini, F.), pp. 119–124. Karger, Basel.
Novelli, G.P., Anglolini, P. & Tani, R. (1985b) In *Free Radicals In Liver Injury* (eds Poli, G., Cheeseman, K.H., Dianzani, M.U. & Slater, T.F.), pp. 225–228, IRL Press, Oxford.
Oliver, C.N., Starke-Reed, P.E., Stadtman, E.R., Liu, G.J., Carney, J.M. & Floyd, R.A. (1990) *Proc. Natl Acad. Sci. USA* **87**, 5144–5147.

Oury, T.D., Ho, Y.-S., Piantadosi, C.A. & Crapo, J.D. (1992) *Proc. Natl Acad. Sci. USA* **89**, 9715–9719.
Phillis, J.W. & Clough-Helfman, C. (1990a) *Med. Sci. Res.* **18**, 403–404.
Phillis, J.W. & Clough-Helfman, C. (1990b) *Neurosci. Lett.* **116**, 315–319.
Pogrebniak, H.W., Merino, M.J., Hahn, S.M., Mitchell, J.B. & Pass, H.I. (1992) *Surgery* **112**, 130–139.
Pou, S., Hassett, D.J., Britigan, B.E., Cohen, M.S. & Rosen, G.M. (1989) *Anal. Biochem.* **177**, 1–6.
Poyer, J.L., Floyd, R.A., McCay, P.B., Janzen, E.G. & Davis, E.R. (1978) *Biochem. Biophys. Acta* **539**, 402–409.
Sata, T., Kubota, E. & Yoshitake, J. (1988) *Fukuoka Acta Med.* **79**, 674–682.
Smith, F.L., Floyd, R.A. & Carpenter, M.P. (1981) In *Proc. Int. Conf. Oxygen and Oxy-Radicals in Chemistry and Biology, Austin, TX, May 1980* (eds Rodgers, M.A.J. & Powers, E.L.), pp. 743–745, Academic Press, New York.
Tyler, D.D. (1975) *Biochem. J.* **147**, 493–504.
Wong, P.K., Poyer, J.L., DuBose, C.M. & Floyd, R.A. (1988) *J. Biol. Chem.* **263**, 11296–11301.
Yoshida, S., Abe, K., Busto, R., Watson, B.D., Kogure, K. & Ginsberg, M.D. (1982) *Brain Res.* **245**, 307–316.
Yue, T.-L., Gu, J.-L., Lysko, P.G., Cheng, H.-Y., Barone, F.C. & Feuerstein, G. (1992) *Brain Res.* **574**, 193–197.
Zaspel, B.J. & Csallany, A.S. (1983) *Anal. Biochem.* **130**, 146–150.

CHAPTER 6

THERAPEUTIC EFFECTS OF NITRIC OXIDE SYNTHASE INHIBITION IN NEURONAL INJURY

M. Flint Beal

Neurology Service,
Massachusetts General Hospital,
Boston, MA 02114, USA

Table of Contents

6.1 Introduction

Nitric oxide (NO) is a unique biological messenger molecule in the central nervous system (Dawson and Dawson, 1994). It is an unusual transmitter since it is a free radical gas which is not stored in synaptic vesicles. NO is synthesized by nitric oxide synthase (NOS) from L-arginine, and it simply diffuses from nerve terminals.

Early studies led to the discovery that NO is the primary endogenous vasodilator released from vascular endothelium. A role in macrophage-induced cytotoxicity was established by the ability of NOS inhibitors to block cytotoxic effects. Initial evidence for a transmitter role in the nervous system was the observation that cultured cerebellar neurones release NO following exposure to glutamate (Garthwaite *et al.*, 1988). Subsequent work showed that NOS inhibitors block cGMP synthesis produced in brain slices following activation of *N*-methyl-D-aspartate (NMDA) receptors (Garthwaite *et al.*, 1988; Bredt and Snyder, 1989).

Neurodegeneration and Neuroprotection
in Parkinson's Disease
ISBN 0-12-525445-8

NO is formed from the guanidino nitrogen of L-arginine, which is converted to L-citrulline by an oxidative–reductive pathway that consumes five electrons. The NOS enzyme is calcium/calmodulin-dependent (Bredt and Snyder, 1989). It requires NADPH, as well as flavin mononucleotide (FMN) and flavin adenine dinucleotide (FAD), which bind NOS stoichiometrically. NOS also uses tetrahydrobiotin as an electron-transferring cofactor. The structure and function of NOS have been clarified by the molecular cloning of the cDNAs for the brain, endothelial, macrophage and non-macrophage-inducible NOSs (Dawson and Snyder, 1994). Both neuronal and endothelial NOSs are constitutive in the sense that their activation does not require new enzyme protein synthesis. The neuronal isoform is activated by glutamate neurotransmission acting at NMDA receptors, which leads to increases in intracellular calcium which then activate NOS by calmodulin. The endothelial form of NOS is also activated by increases in intracellular calcium levels. Agonists such as acetylcholine or bradykinin act at receptors to activate the phosphoinoside cycle to generate calcium.

Macrophage NOS is considered inducible since macrophages normally contain no detectable NOS protein. In response to stimulation with cytokines such as γ-interferon and lipopolysaccharide, macrophages produce large quantities of NO, which mediates destruction of micro-organisms and tumours and may also lead to pathological tissue damage. Inducible NOS is not stimulated by calcium, which appears to be due to tight binding of calmodulin to inducible NOS, such that its binding is unaffected by calcium concentrations.

Molecular cloning studies showed consensus sites for phosphorylation in the various isoforms of NOS. Neuronal NOS can be phosphorylated by calcium/calmodulin-dependent protein kinase, cAMP-dependent protein kinase, cGMP-dependent protein kinase and protein kinase C. Phosphorylation of the enzyme decreases its catalytic activity, whereas dephosphorylation by calcineurin increases catalytic activity (T.M. Dawson *et al.*, 1993).

NOS neurones are widely distributed in the nervous system, where they constitute 1–2% of neurones in most structures. They colocalize with neurones identifiable with the histochemical stain NADPH-diaphorase (T.M. Dawson *et al.*, 1991; Hope *et al.*, 1991). In cerebral cortex they are aspiny neurones which colocalize with γ-aminobutyrate, somatostatin and neuropeptide Y. In the striatum they are medium-sized aspiny neurones which colocalize with somatostatin and neuropeptide Y. NOS neurones in the pedunculopontine nucleus colocalize with choline acetyltransferase. In the cerebellum NOS is present in granule cells and basket cells but not in Purkinje cells. It is thought that NMDA receptor stimulation activates granule cell NOS which then produces NO, which can stimulate guanylylcyclase in Purkinje cells to produce cGMP. NO binds the haem of guanylyl to alter the conformation of the enzyme, leading to its activation.

6.2 Toxic effects of NO

Excitotoxicity mediated by glutamate receptors has been implicated in neuronal death accompanying ischaemia, brain and spinal cord trauma, hypoglycaemia and epilepsy. Excitotoxic mechanisms have also been implicated in neurodegenerative diseases such as Alzheimer's disease (AD), Huntington's disease (HD), Parkinson's disease (PD) and amyotrophic lateral sclerosis (ALS). In these diseases, weak or slow excitotoxicity may occur either due to abnormalities in excitatory amino acid receptors or as a consequence of defects in energy metabolism (Beal, 1992; Albin and Greenamyre, 1992). Excitotoxic neuronal cell death is accompanied by increases in intracellular calcium, which then lead to a number of deleterious consequences including activation of intracellular proteases, lipases, nucleases, free radical generation and activation of NOS.

The role of NO in excitotoxicity was originally reported by Dawson and colleagues (V.L. Dawson *et al.*, 1991). They showed that following brief exposure of cultured cortical neurones to NMDA, inhibitors of NOS, calmodulin antagonists, flavoprotein inhibitors and reduced haemoglobin, which scavenges NO, markedly attenuated neurotoxicity. Removal of L-arginine from the media also blocked toxicity. Subsequent work showed that NOS inhibitors also blocked glutamate neurotoxicity in cultured striatal and hippocampal neurones (V.L. Dawson *et al.*, 1993). Furthermore, pretreatment of the cultures with quisqualate, which preferentially kills NOS neurones, blocked glutamate neurotoxicity in cultured cortical and striatal neurones. The development of NMDA neurotoxicity over time in culture was coincident with expression of NOS. The NOS inhibitors and haemoglobin had no effect on kainate neurotoxicity. A small protection was produced against quisqualate neurotoxicity, but it was thought that this might be related to activation of NMDA receptors (V.L. Dawson *et al.*, 1991). Other studies also showed no effect of NOS inhibitors on kainate and α-amino-3-hydroxy-5-methyl-4-isoxazoleproprionate (AMPA) neurotoxicity *in vitro*. Sodium nitroprusside, which releases NO, was shown to produce neurotoxicity which showed a similar concentration effect and time relationship to neurotoxicity produced by NMDA (V.L. Dawson *et al.*, 1993). The immunosuppressant FK506 which inhibits calcineurin-mediated dephosphorylation of NOS, and thereby diminishes NOS catalytic activity, inhibited NMDA neurotoxicity in cultured cortical neurones (T.M. Dawson *et al.*, 1993), but had no effect on kainate- and quisqualate-mediated neurotoxicity. Vigé and colleagues found that inhibition of NOS for 24 hours after a 5 minute glutamate exposure produced a 74% protection (Vigé *et al.*, 1993). The NOS inhibitor L-nitroarginine protected NMDA but not kainate neurotoxicity to striatal neurones in brain slices *in vitro* (Kollegger *et al.*, 1993). An *in vivo* study showed that inhibition of NOS decreased NMDA-induced loss of hippocampal neurones (Moncada *et al.*, 1992), and NMDA striatal injury (Buisson *et al.*, 1993).

In contrast to these results, however, several other studies show no effects of NOS inhibitors on glutamate neurotoxicity *in vitro* (Regan *et al.*, 1993; Hewett *et al.*, 1993; Demerle-Pallardy *et al.*, 1991; Pauwels and Leysen, 1992; Puttfarcken *et al.*, 1992). Similarly, chronic inhibition of NOS failed to protect hippocampal neurones from

NMDA toxicity in an *in vivo* study (Lerner-Natoli *et al.*, 1992). Lipton and colleagues attempted to reconcile these discrepant results by demonstrating that NO may have differing effects depending on the redox milieu of the cell (Lipton *et al.*, 1993). They demonstrated that the toxic effects of NO were likely due to the generation of peroxynitrite ($ONOO^-$), which is formed by the reaction of $O_2^{\cdot -}$ with NO. They showed that superoxide dismutase blocked the toxicity of an NO donor, but had no effect on $ONOO^-$ toxicity. Dawson and colleagues also showed a protective effect of superoxide dismutase against NMDA neurotoxicity consistent with a role of $ONOO^-$ (V.L. Dawson *et al.*, 1993). The studies of Lipton and colleagues showed that NO in the alternative redox state NO^+ (nitrosonium ion) could prevent NMDA neurotoxicity by down-regulation of the NMDA receptor by *S*-nitrosylation of the critical thiols of the NMDA receptor redox modulatory site (Lipton *et al.*, 1993). In a reducing environment NO was more likely to be toxic, whereas in an oxidizing environment it was more likely to inhibit the NMDA receptor. Other authors also showed an inhibitory effect of NO on NMDA receptors (Manzoni *et al.*, 1992; Manzoni and Bockaert, 1993). Recent strong evidence favouring a role of neuronal NO in NMDA excitotoxicity is the observation that cortical neurones cultured from mutant mice deficient in neuronal NOS are resistant to NMDA neurotoxicity (Dawson and Dawson, 1994).

6.3 NO in ischaemia

During ischaemia there is a massive release of excitatory amino acids into the extracellular space which stimulates NMDA receptors and leads to increases in NO production. Increases in NO to as high as 5 μM were shown during focal ischaemia using an NO-sensitive electrode (Malinski *et al.*, 1993). Other studies used electron paramagnetic resonance to show an increase in NO during both focal ischaemia and global forebrain ischaemia (Sato *et al.*, 1994; Tominaga *et al.*, 1994). An increase in nitrosylhaemoglobin and in plasma NO end-products was demonstrated during focal cerebral ischaemia and reperfusion in rats (Kumura *et al.*, 1994a,b). Both NOS activity and cGMP levels increase during focal cerebral ischaemia in rats (Kader *et al.*, 1993).

Studies of the effects of NOS inhibitors in models of cerebral ischaemia have been controversial. An initial report showed that L-nitroarginine at 1 mg kg^{-1} reduced infarct volume in a mouse model of focal cerebral ischaemia (Nowicki *et al.*, 1991). The same dosing regimen was effective in reducing size following middle cerebral artery occlusion in rats (Nagafuji *et al.*, 1992). Other studies showed that NOS inhibition reduced caudate injury in cats following focal ischaemia, although there was no effect on the cortical infarct volume (Nishikawa *et al.*, 1994). In contrast, several studies showed that inhibition of NOS increased the infarct volume following focal ischaemia in rats (Dawson *et al.*, 1992; Kuluz *et al.*, 1993; Yamamoto *et al.*, 1992). These studies generally used higher doses of NOS inhibitors, which will inhibit endothelial NOS, leading to vasoconstriction and reduced cerebral perfusion. Consistent with this effect, NO donors increase blood flow and reduce brain damage in focal ischaemia (Zhang *et al.*, 1994). These findings have been clarified by the observation that low doses of NOS inhibitors are neuroprotective, whereas higher doses are ineffective in the mouse focal ischaemic model, consistent with adverse vascular effects at higher dose levels (Carreau *et al.*, 1994). Similar controversy exists concerning neuroprotective effects of NOS inhibition on survival of hippocampal CA_1 neurones following global ischaemia (Buchan *et al.*, 1994; Caldwell *et al.*, 1994; Sancesario *et al.*, 1994), with more consistent protection with lower dosage of NOS inhibitors (Shapira *et al.*, 1994).

The most convincing evidence for a role of NO in cerebral ischaemia has come from studies of mutant mice deficient in neuronal NOS activity. In these mice, infarct volumes were significantly reduced at 24 and 72 hours after middle cerebral artery occlusion (Huang *et al.*, 1994). There was no change in blood flow nor any alteration in vascular anatomy in these mice. Following administration of nitro-L-arginine, which inhibits endothelial NOS, the infarct volume in the mutant mice was increased. These data therefore show that neuronal NO production exacerbates ischaemic injury, whereas endothelial NO protects after middle cerebral artery occlusion.

6.4 Neuronal NOS inhibitors

Improved inhibitors of NOS have recently been described. One of these is 7-nitroindazole (7-NI), which has a high degree of specificity for the neuronal isoform of NOS *in vivo*. Although *in vitro* studies suggest that it inhibits endothelial, macrophage and neuronal NOS (Wolff *et al.*, 1994), *in vivo* studies showed no effects on blood pressure, on endothelium-dependent blood vessel relaxation and on acetylcholine-induced blood vessel relaxation (Wolff *et al.*, 1994; Babbedge *et al.*, 1993; Moore *et al.*, 1993; Yoshida *et al.*, 1994). It has antinociceptive effects and it inhibits brain NOS activity. It acts as an inhibitor by competing with L-arginine for the prosthetic haem group of NOS, and it additionally affects the pteridine site of the enzyme (Mayer *et al.*, 1994).

Yoshida and colleagues showed that 7-NI at 25 or 50 mg kg^{-1} significantly reduced the infarct volume following middle cerebral artery occlusion in rats (Yoshida *et al.*, 1994). The protective effect was reversed by co-administration of L-arginine. We studied the neuroprotective effects of 7-NI against striatal lesions produced by excitotoxins *in vivo* (Schulz *et al.*, 1994). 7-NI significantly attenutated lesions produced by NMDA, but it had no effect on lesions produced by either kainate or AMPA.

We also investigated the effects of 7-NI in animal models of neurodegenerative diseases. 1-Methyl-4-phenyl-1,2,3,6-tetrahydropyridine (MPTP) produces clinical, biochemical and neuropathological changes which are analogous to those observed in idiopathic PD. The toxic effects of MPTP are mediated by its metabolite 1-methyl-4-phenylpyridinium (MPP+), which is a selective inhibitor of complex I of the electron transport chain (Tipton and Singer, 1993). The interruption of oxidative phosphorylation results in decreased levels of ATP, which may lead to partial neuronal depolarization and secondary activation of voltage-dependent NMDA receptors, resulting in excitotoxic neuronal cell death (Beal, 1992; Chan *et al.*, 1991). We therefore examined whether 7-NI could block dopaminergic neurotoxicity in mice after systemic administration of MPTP. 7-NI dose dependently protected against MPTP-induced dopamine depletion using two different dosing regimens of MPTP which produced varying degrees of dopamine depletion (Schulz *et al.*, 1995). At a dose of 50 mg kg^{-1}, 7-NI produced almost complete protection in both dosing regimens. As noted above, NO may interact with $O_2^{\cdot -}$ to generate $ONOO^-$ (Beckman *et al.*, 1990). A consequence of $ONOO^-$ is oxidative damage to lipids, proteins and DNA. Peroxynitrite can be protonated to form ONOOH, which may then decompose to $OH^{\cdot}$ (Vigé *et al.*, 1993; Crow *et al.*, 1994). In the presence of superoxide dismutase it can also generate nitronium ions, which nitrate tyrosine residues (Ischiropoulos *et al.*, 1992). Consistent with this we found that MPTP neurotoxicity in mice was accompanied by significant increases in 3-nitrotyrosine which were blocked by 7-NI.

We also examined the effects of 7-NI on neurotoxicity produced by striatal injections of the reversible succinate dehydrogenase inhibitor malonate or systemic administration of the irreversible succinate dehydrogenase inhibitor 3-nitropropionic acid (Schulz *et al.*, 1994). These compounds produce behavioural, neurochemical and histopathological changes which closely resemble those of HD. We found that 7-NI attenuated striatal malonate lesions and that the protection was reversed by L-arginine but not by D-arginine. Another report also recently showed that 3 mg kg^{-1} of the NOS inhibitor *N*-nitro-L-arginine significantly attenuated striatal malonate lesions (Maragos and Silverstein, 1995). We also found that striatal malonate lesions were attenuated in mutant mice deficient in neuronal NOS (J.B. Schulz *et al.*, unpublished findings). 7-NI protected against malonate induced decreases in ATP, and increases in striatal lactate as assessed *in vivo* using chemical shift magnetic resonance spectroscopy (Schulz *et al.*, 1994). 7-NI had no effects on spontaneous electrophysiological activity in the striatum, arguing that its neuroprotective effects were not mediated by an interaction with excitatory amino acid receptors.

We co-administered 7-NI with 3-nitropropionic acid for 5 days. Under these conditions all of the control animals showed striatal lesions, whereas there was complete protection in the 7-NI treated animals. Furthermore, 3-nitropropionic acid administration produced increases in hydroxyl radical and 3-nitrotyrosine levels in the striatum, which were significantly attenuated by 7-NI treatment. These findings argue for a role of $ONOO^-$ in 3-nitropropionic acid neurotoxicity which is blocked by inhibiting neuronal NOS. Further evidence is from observations in transgenic mice overexpressing the enzyme superoxide dismutase (Beal *et al.*, 1995). Increased scavenging of superoxide will block its interaction with NO to produce $ONOO^-$. We found that 3-nitropropionic acid induced lesions, 3-nitropropionic acid induced increases in hydroxyl radical generation and increases in 3-nitrotyrosine were significantly attenuated in these mice (Beal *et al.*, 1995).

Neuronal NOS may also play a role in neuronal injury in the spinal cord. NMDA causes release of NO from rat spinal cord *in vitro* (Li *et al.*, 1994). Following ventral root avulsion, spinal motoneurones express NOS, and the lesion-induced NOS is coincident with death of the injured neurones (Wu, 1993). Inhibition of NOS significantly reduced the death of motoneurones due to spinal root avulsion (Wu and Li, 1993).

6.5 Cellular mechanisms of NO toxicity

NO has a number of cellular effects which may contribute to its toxicity. NO can bind to enzymes containing iron–sulfur complexes and inactivate them. Cytotoxic macrophages were shown to injure tumour cells by NO-mediated inhibition of complexes I and II of the electron transport chain and aconitase (Hibbs *et al.*, 1988; Stadler *et al.*, 1991). In rat skeletal muscle mitochondria, NO rapidly and reversibly inactivates cytochrome-*c* oxidase (Cleeter *et al.*, 1994). Similarly in cultured astrocytes, NO inhibits cytochrome oxidase with a lesser inhibition of succinate–cytochrome-*c* reductase activity, but has no effect on complex I activity (Bolanos *et al.*, 1994). A vitamin E analogue protects against the inhibition of cytochrome oxidase (Heales *et al.*, 1994). NO also blocks cytochrome oxidase in brain synaptosomes (Brown and Cooper, 1994). NO can reversibly deenergize mitochondria and kill hepatocytes by mobilizing mitochondrial calcium (Richter *et al.*, 1994).

NO also is a signal for ADP-ribosylation of proteins. One protein which undergoes ADP-ribosylation is glyceraldehyde-3-phosphate dehydrogenase (Zhang and Snyder, 1992). This ribosylation involves a cysteine which is involved in enzyme activity. The ribosylation may therefore lead to a decrease in ATP produced by glycolysis. NO also inhibits ribonucleotide reductase, the enzyme which converts ribonucleotide to deoxynucleotides for DNA synthesis, which may inhibit DNA repair (Lepoivre *et al.*, 1990).

As discussed above, the interaction of NO with $O_2^{\cdot -}$ leads to the formation of $ONOO^-$ (Beckman *et al.*, 1990). This reaction occurs at an extremely fast rate of $6.7 \times 10^9\ \text{M}^{-1}\ \text{s}^{-1}$, which is essentially diffusion limited and does not require transition metals. It therefore depends on the concentrations of superoxide and nitric oxide in the cell. Both superoxide and NO can be produced by NOS (Heinzel *et al.*, 1992; Pou *et al.*, 1992). Peroxynitrite can be protonated to produce ONOOH, which can then decompose to OH^- (Beckman *et al.*, 1990). Peroxynitrite itself is a highly oxidizing agent which can cause tissue damage (Kooy *et al.*, 1994). At physiological pH, $ONOO^-$ may be able to diffuse over several cell diameters to produce cell damage by oxidizing lipids, proteins and DNA.

Peroxynitrite can inhibit mitochondrial function. In trypanosomes it inhibits both succinate dehydrogenase and fumarate reductase (Rubbo *et al.*, 1994). In heart mitochondria it inhibits both NADH dehydrogenase and succinate dehydrogenase (Radi *et al.*, 1994). Recent studies show that ONOO inactivates aconitases but that NO does not (Castro *et al.*, 1994; Hausladen and Fridovich, 1994). Peroxynitrite is a more potent inhibitor of mitochondrial respiration than NO produced by immunostimulated macrophages (Szabo and Salzman, 1995). In cultured neurones $ONOO^-$ produces dose-dependent decreases in succinate–cytochrome-*c* reductase and cytochrome oxidase activities (Bolanos *et al.*, 1995). Peroxynitrite dose dependently damages neurones but spares astrocytes. Peroxynitrite also produces calcium efflux from mitochondria (Packer and Murphy, 1994). Comparative studies of the toxicity of NO and $ONOO^-$ to *Escherichia coli* showed that 1 mM NO had no toxicity while 1 mM $ONOO^-$ was completely bactericidal after 5 seconds (Brunelli *et al.*, 1995).

A critical site for the effects of either NO or $ONOO^-$ may be damage to DNA. NO produces deamination damage and strand breaks in DNA (Wink *et al.*, 1991; Nguyen *et al.*, 1992). DNA strand breaks are also produced by $ONOO^-$ (King *et al.*, 1992). NO-treated DNA stimulates poly(ADP-ribose) polymerase activity in rat brain extracts (Zhang *et al.*, 1994). In the presence of DNA strand breaks this chromatin-bound enzyme transfers ADP-ribose from NAD to nuclear proteins and to the ADP-ribose polymer itself, with concomitant release of nicotinamide (Schraufstatter *et al.*, 1986; Althaus, 1992). For each mole of ADP-ribose transferred, 1 mol of NAD is consumed and the regeneration of NAD consumes 4 mol of ATP. Cell death may then occur as a consequence of both NAD and ATP depleton. In cultures of rat cerebral cortex, inhibitors of poly(ADP-ribose) polymerase blocked neurotoxicity produced by either NMDA or NO (Zhang *et al.*, 1994). Inhibition of neurotoxicity correlated with the potencies of various inhibitors as blockers of poly(ADP-ribose) activity. Inhibitors of poly(ADP-ribose) polymerase also blocked NO neurotoxicity in hippocampal slices (Wallis *et al.*, 1991).

6.6 Inducible NOS inhibitors

The role of inducible NOS in neuronal injury is receiving increasing attention. Inducible NOS has been localized to both astrocytes and meningeal fibroblasts (Fujisawa *et al.*, 1995; Skaper *et al.*, 1995). Studies of mixed glial–neuronal cultures exposed to γ-interferon or lipopolysaccharide, which induce NOS activity in glia, show that induction of NOS is associated with neurotoxicity which is blocked by NOS inhibitors (Skaper *et al.*, 1995; Boje and Arora, 1992; Dawson *et al.*, 1994). The neurotoxicity is attenuated by superoxide dismutase, indicating that peroxynitrite may play a role (Skaper *et al.*, 1995). In one study, activation of inducible NOS potentiated NMDA but not kainate neurotoxicity (Hewett *et al.*, 1994). Inducible NOS is activated after focal ischaemia, and treatment with aminoguanidine, which blocks inducible NOS, attenuates neuronal injury after focal ischaemia (Iadecola *et al.*, 1995a,b). Inducible NOS may also play a role in neuronal injury associated with the acquired immune deficiency syndrome (AIDS) and injury to oligodendrocytes in multiple sclerosis (Dawson and Dawson, 1994). Inhibition of inducible NOS attenuates experimental allergic encephalitis (Koprowski *et al.*, 1993). A potential role of inducible NOS in AD is also suggested by the finding that β-amyloid protein activates microglial cells and triggers the production of reactive nitrogen intermediates (Meda *et al.*, 1995). NOS activity is also increased in brain microvessels in AD (Dorheim *et al.*, 1994).

6.7 Conclusions

A role of neuronal NOS in both cerebral ischaemia and in excitotoxic induced neuronal injury is now well established. The most convincing evidence comes from mutant mice deficient in neuronal NOS activity. This suggests that the development of selective neuronal NOS inhibitors may prove efficacious in the treatment of both acute and chronic neurological illnesses in which excitotoxicity has been implicated. Recent evidence in experimental models of stroke, PD and HD has shown that the relatively selective neuronal NOS inhibitor 7-NI is neuroprotective. A role of inducible NOS has been implicated in other neurological diseases which may involve activation of the immune system, such as AIDS, multiple sclerosis, and microglia surrounding amyloid plaques in AD. It is therefore possible that selective inhibitors of inducible NOS might prove useful in these disorders.

References

Albin, R.L. & Greenamyre, J.T. (1992) *Neurology* **42**,733–738.

Althaus, F.R. (1992) *J. Cell Sci.* **102**, 663–670.

Babbedge, R.C., Bland-Ward, P.A., Hart, S.L. & Moore, P.K. (1993) *Br. J. Pharmacol.* **110**, 225–228.

Beal, M.F. (1992) *Ann. Neurol.* **31**, 119–130.

Beal, M.F., Ferrante, R.J., Henshaw, R., Matthews, R.T., Chan, P.H., Kowall, N.W., Epstein, C.J. & Schulz, J.F. (1995) *J. Neurochem.* **65**, 919–922.

Beckman, J.S., Beckman, T.W., Chen, J., Marshall, P.A. & Freeman, B.A. (1990) *Proc. Natl Acad. Sci. USA* **87**, 1620–1624.

Boje, K.M. & Arora, P.K. (1992) *Brain Res.* **587**, 250–256.

Bolanos, J.P., Peuchen, S., Heales, S.J.R. *et al.* (1994) *J. Neurochem.* **63**, 910–916.

Bolanos, J.P., Heales, S.J.R., Land, J.M. & Clark, J.B. (1995) *J. Neurochem.* **64**, 1965–1972.

Bredt, D.S. & Snyder, S.H. (1989) *Proc. Natl Acad. Sci. USA* **86**, 9030–9033.

Brown, G.C. & Cooper, C.E. (1994) *FEBS Lett.* **356**, 295–298.

Brunelli, L., Crow, J.P. & Beckman, J.S. (1995) *Arch. Biochem. Biophys.* **316**, 327–334.

Buchan, A.M., Gertler, S.Z., Huang, Z.-G., Li, H., Chaundy, K.E. & Xue, D. (1994) *Neuroscience.* **61**, 1–11.

Buisson, A., Margaill, I., Callebert, J., Plotkine, M. & Boulu, R.G. (1993) *J. Neurochem.* **61**, 690–696.

Caldwell, M., O'Neill, M., Earley, B. & Leonard, B. (1994) *Eur. J. Pharmacol.* **260**, 191–200.

Carreau, A., Duval, D., Poignet, H., Scatton, B., Vige, X. & Nowicki, J.-P. (1994) *Eur. J. Pharmacol.* **256**, 241–249.

Castro, L., Rodriguez, M. & Radi, R. (1994) *J. Biol. Chem.* **269**, 29409–29415.

Chan, P., DeLanney, L.E., Irwin, I., Langston, J.W. & Di Monte, D. (1991) *J. Neurochem.* **57**, 348–351.

Cleeter, M.W.J., Cooper, J.M., Darley-Usmar, V.M., Moncada, S. & Schapira, A.H.V. (1994) *FEBS Lett.* **345**, 50–54.

Crow, J.P., Spruell, C., Chen, J., Gunn, C., Ischiropoulos, H., Tsai, M., Smith, C.D., Radi, R., Koppenol, W.H. & Beckman, J.S. (1994) *Free Radic. Biol. Med.* **16**, 331–338.

Dawson, D.A., Kusumoto, K., Graham, D.I., McCulloch, J. & Macrae I.M. (1992) *Neurosci. Lett.* **142**, 151–154.

Dawson, T.M. & Dawson, V.L. (1994) *Neuroscientist*, 9–20.

Dawson, T.M. & Snyder, S.H. (1994) *J. Neurosci.* **14**, 5147–5159.

Dawson, T.M., Bredt, D.S., Fotuhi, M., Hwang, P.M. & Snyder, S.H. (1991) *Proc. Natl Acad. Sci. USA* **88**, 7797–7801.

Dawson, T.M., Steiner, J.P., Dawson, V.L., Dinerman, J.L., Uhl, G.R. & Snyder, S.H. (1993) *Proc. Natl Acad. Sci. USA* **90**, 9808–9812.
Dawson, V.L., Dawson, T.M., London, E.D., Bredt, D.S. & Snyder, S.H. (1991) *Proc. Natl Acad. Sci. USA* **88**, 6368–6371.
Dawson, V.L., Dawson, T.M. Bartley, D.A., Uhl, G.R. & Snyder, S.H. (1993) *J. Neurosci.* **13**, 2651–2661.
Dawson, V.L., Brahmbhatt, H.P., Mong, J.A. & Dawson, T.M. (1994) *Neuropharmacology* **33**, 1425–1430.
Demerle-Pallardy, Lonchampt, M.-O., Chabrier, P.-E. & Braquet, P. (1991) *Biochem. Biophys Res. Commun.* **181**, 456–464.
Dorheim, M.A., Tracey, W.R., Pollock, J.S. & Grammas, P. (1994) *Biochem. Biophys. Res. Commun.* **205**, 659–665.
Fujisawa, H., Ogura, T., Hokari, A., Weisz, A., Yamashita, J. & Esumi, H. (1995) *J. Neurochem.* **64**, 85–91.
Garthwaite, J., Charles, S.L. & Chess-Williams, R. (1988) *Nature* **336**, 385–388.
Hausladen, A. & Fridovich, I. (1994) *J. Biol. Chem.* **269**, 29405–29408.
Heales, S.J.R., Bolanos, J.P., Land, J.M. & Clark, J.B. (1994) *Brain Res.* **668**, 243–245.
Heinzel, B., John, M., Klatt, P., Bohme, E. & Mayer, B. (1992) *Biochem. J.* **281**, 627–630.
Hewett, S.J., Corbett, J.A., McDaniel, M.L. & Choi, D.W. (1993) *Brain Res.* **625**, 337–341.
Hewett, S.J., Csernansky, C.A. & Choi, D.W. (1994) *Neuron* **13**, 487–494.
Hibbs, J.B. Jr., Taintor, R.R., Vavrin, Z. & Rachlin, E.M. (1988) *Biochem. Biophys. Res. Commun.* **157**, 87–94.
Hope, B.T., Michael, G.J., Knigge, K.M. & Vincent, S.R. (1991) *Proc. Natl Acad. Sci. USA* **88**, 2811–2814.
Huang, Z., Huang, P.L., Panahian, N., Dalkara, T., Fishman, M.C. & Moskowitz, M.A. (1994) *Science* **265**, 1883–1885.
Iadecola, C., Xu, X., Zhang, F., El-Fakahany, E.E. & Ross, M.E. (1995a) *J. Cereb. Blood Flow Metab.* **15**, 52–59.
Iadecola, C., Zhang, F. & Xu, X. (1995b) *Am. J. Physiol.* **268**, R286–R292.
Ischiropoulos, H., Zhu, L., Chen, J., Tsai, M., Martin, J.C., Smith, C.D. & Beckman, J.S. (1992) *Arch. Biochem. Biophys.* **298**, 431–437.
Kader, A., Frazzini, V.I., Solomon, R.A. & Trifletti, R.R. (1993) *Stroke* **24**, 1709–1716.
King, P.A., Anderson, V.E., Edwards, J.O., Gustafson, G., Plumb, R.C. & Suggs, J.W. (1992) *J. Am. Chem. Soc.* **114**, 5430–5432.
Kollegger, H., McBean, G.J. & Tipton, K.F. (1993) *Biochem. Pharmacol.* **45**, 260–264.
Kooy, N.W., Royall, J.A., Ischiropoulos, H. & Beckman, J.M. (1994) *Free Radic. Biol. Med.* **16**, 149–156.
Koprowski, H., Zheng, Y.M., Heber-Katz, E., Fraser, N., Rorke, L., Fu Z.F., Hanlon, C. & Dietzschold, B. (1993) *Proc. Natl Acad. Sci. USA* **90**, 3024–3027.
Kuluz, J.W., Prado, R.J., Dietrich, W.D., Schleien, C.L. & Watson, B.D. (1993) *Stroke* **24**, 2023–2029.
Kumura, E., Kosaka, H., Shiga, T., Yoshimine, T. & Hayakawa, T. (1994a) *J. Cereb. Blood Flow Metab.* **14**, 487–491.
Kumura, E., Yoshimine, T., Tanaka, S., Hayakawa, T., Shiga, T. & Kosaka, H. (1994b) *Neurosci. Lett.* **177**, 165–167.
Lepoivre, M., Chenais, B., Yapo, A., Lemaire, G., Thelander, L. & Tenu, J.-P. (1990) *J. Biol. Chem.* **265**, 14143–14149.
Lerner-Natoli, M., Rondouin, G., de Block, F. & Bockaert, J. (1992) *NeuroReport* **3**, 1109–1112.
Li, P., Tong, C., Eisenach, J.C. & Figueroa, J.P. (1994) *Brain Res.* **637**, 287–291.
Lipton, S.A., Choi, Y.-B., Pan, Z.-H., Lei, S.Z., Chen, H.V., Sucher, N.J., Loscarzo, J., Singel, D.J., & Stamler, J.S. (1993) *Nature* **364**, 626–632.
Malinski, T., Bailey, F., Zhang, Z.G. & Chopp, M. (1993) *J. Cereb. Blood Flow Metab.* **13**, 355–358.
Manzoni, O. & Bockaert, J. (1993) *J. Neurochem.* **61**, 368–370.

Manzoni, O., Prezeau, L., Marin, P., Deshager, S., Bockaert, J. & Fagni, L. (1992) *Neuron* **8**, 653–662.
Maragos, W.F. & Silverstein, F.S. (1995) *J. Neurochem.* **64**, 2362–2365.
Mayer, B., Klatt, P., Werner, E.R. & Schmidt, K. (1994) *Neuropharmacology* **33**, 1253–1259.
Meda, L., Cassatella, M.A., Szendrel, G.I., Otvos, Jr L., Baron, P., Villalba, M., Ferrari, D. & Rossi, F. (1995) *Nature* **374**, 647–650.
Moncada, C., Lekieffre, D., Arvin, B. *et al.* (1992) *NeuroReport* **3**, 530–532.
Moore, P.K., Wallace, P., Gaffen, Z., Hart, S.L. & Babbedge, R.C. (1993) *Br. J. Pharmacol.* **110**, 219–224.
Nagafuji, T., Matsuii, T., Koide, T. & Asano, T. (1992) *Neurosci. Lett.* **147**, 159–162.
Nguyen, T., Brunson, D., Crespi, C.L., Pennman, B.W., Wishok, J.S. & Tannenbaum, S.R. (1992) *Proc. Natl Acad. Sci. USA* **89**, 3030–3034.
Nishikawa, T., Kirsch, J.R., Koehler, R.C., Miyabe, M. & Traystman, R.T. (1994) *Stroke* **25**, 877–885.
Nowicki, J.P., Duval, D., Poignet, D.H. & Scatton, B. (1991) *Eur. J. Pharmacol.* **204**, 339–340.
Packer, M.A. & Murphy, M.P. (1994) *FEBS Lett.* **345**, 237–240.
Pauwels, P. & Leysen, J.E. (1992) *Neurosci. Lett.* **143**, 27–30.
Pou, S., Pou, W.S., Bredt, D.S., Snyder, S.H. & Rosen, G.M. (1992) *J. Biol. Chem.* **267**, 24173–24176.
Putfarcken, P.S., Lyons, W.E. & Coyle, J.T. (1992) *Neuropharmacology* **31**, 565–575.
Radi, R., Rodriguez, M., Castro, L. & Telleri, R. (1994) *Arch. Biochem. Biophys.* **308**, 89–95.
Regan, R.F., Renn, K.E. & Panter, S.S. (1993) *Neurosci. Lett.* **153**, 53–56.
Richter, C., Gogvadze, V., Schlapbach, R., Schweizer, M. & Schlegel, J. (1994) *Biochem. Biophys. Res. Commun.* **205**, 1143–1150.
Rubbo, H., Denicola, A. & Radi, R. (1994) *Arch. Biochem. Biophys.* **308**, 96–102.
Sancesario, G., Iannone, M., Morello, M., Nistico, G. & Bernardi, G. (1994) *Stroke* **25**, 436–444.
Sato, S., Tominaga, T., Ohnishi, T. & Ohnishi, S.T. (1994) *Brain Res.* **647**, 91–96.
Schraufstatter, I.U., Hyslop, P.A., Hinshaw, D.B., Spragg, R.G., Sklar, L.A. & Cochrane, C.G. (1986) *Proc. Natl Acad. Sci. USA* **83**, 4908–4912.
Schulz, J.B., Matthews, R.T., Henshaw, D.R. & Beal, M.F. (1994) *Soc. Neurosci. Abstr.* **20**, 1661.
Schulz, J.B., Matthews, R.T., Muqit, M.M.K., Browne, S.E. & Beal, M.F. (1995) *J. Neurochem.* **64**, 936–939.
Shapira, S., Kadar, T. & Weissman, B.A. (1994) *Brain Res.* **668**, 80–84.
Skaper, S.T., Facci, L. & Leon, A. (1995) *J. Neurochem.* **64**, 266–276.
Stadler, J., Billiar, T.R., Curran, R.D., Stuehr, D.J., Ochoa, J.B. & Simmons, R.L. (1991) *Am. J. Physiol.* **260**, C910–C916.
Szabo, C. & Salzman, A.L. (1995) *Biochem. Biophys. Res. Commun.* **209**, 739–743.
Tipton, K.F. & Singer, T.P. (1993) *J. Neurochem.* **61**, 1191–1206.
Tominaga, T., Sato, S., Ohnishi, T. & Ohnishi, S.T. (1994) *J. Cereb. Blood Flow Metab.* **14**, 715–722.
Vigé, X., Carreau, A., Scatton, B. & Nowicki, J.P. (1993) *Neuroscience* **55**, 893–901.
Wallis, R.A., Panizzon, K.L., Henry, D. & Wasterlain, C.G. (1991) *J. Neurochem.* **57**, 1288–1295.
Wink, D.A., Kasprzak, K.S., Maragos, C.M., Elespura, R.K., Misra, M., Dunhams, T.M., Cebula, T.A., Koch, W.H., Andrews, A.H., Allen, J.S. & Keffer L.K. (1991) *Science* **254**, 1001–1003.
Wolff, D.J., Lubeskie, A. & Umansky, S. (1994) *Arch. Biochem. Biophys.* **314**, 360–366.
Wu, W. (1993) *Exp. Neurol.* **120**, 153–159.
Wu, W. & Li, L. (1993) *Neurosci. Lett.* **153**, 121–124.
Yamamoto, S., Golanov, E.V., Berger, S.B. & Reis, D.J. (1992) *J. Cereb. Blood Flow Metab.* **12**, 717–726.
Yoshida, T., Limmroth, V., Irikura, K. & Moskowitz, M.A. (1994) *J. Cereb. Blood Flow Metab.* **14**, 924–929.
Zhang, F., White, J.G. & Iadecola, C. (1994) *J. Cereb. Blood Flow Metab.* **14**, 217–226.
Zhang, J. & Snyder, S.H. (1992) *Proc. Natl Acad. Sci. USA* **89**, 9382–9385.
Zhang, J., Dawson, V.L., Dawson, T.M. & Snyder, S.H. (1994) *Science* **263**, 687–689.

CHAPTER 7

MITOCHONDRIAL DYSFUNCTION IN NEURODEGENERATION: PROSPECTS FOR NEUROPROTECTION

A.H.V. Schapira

Department of Clinical Neurosciences,
Royal Free Hospital School of Medicine,
Rowland Hill Street, London NW3 2PF, UK

Table of Contents

Neurodegeneration and Neuroprotection
in Parkinson's Disease
ISBN 0-12-525445-8

7.1 Introduction

Mitochondria play a central role in the provision of energy for cellular metabolism. Dysfunction of the mitochondrial pathways for ATP synthesis might be expected to have a profound effect upon the function of oxidatively dependent tissues, of which brain and skeletal muscle are prime examples. Thus, abnormalities of energy metabolism might logically be considered as a cause of neurodegeneration. Recognition that mitochondrial defects may result in human disease is relatively recent, although our knowledge of this area has grown exponentially over the last 30 years. Only in the last 5 years, however, has the subject of the direct role of the mitochondrion in neurodegeneration received significant attention. The purpose of this chapter is to review the evidence for the contribution of mitochondrial dysfunction in neurodegenerative disorders and to explore potential mechanisms to circumvent the defects responsible. An understanding of the potential contribution of mitochondria to the pathogenesis of neurodegenerative disorders must be based upon a clear appreciation of the structure and function of the mitochondrial respiratory chain and mitochondrial DNA (mtDNA).

7.2 The mitochondrial respiratory chain and oxidative phosphorylation system

7.2.1 Function

The respiratory chain and oxidative phosphorylation system comprise five multi-subunit protein complexes (complexes I–V) and two mobile electron carriers (ubiquinone and cytochrome *c*) located on the mitochondrial inner membrane (Figure 1 and Table 1). The purpose of the respiratory chain is to harness the redox energy of NADH and $FADH_2$, which in turn are generated by the tricarboxylic acid (TCA) cycle and fatty acid β-oxidation. NADH is oxidized by complex I. $FADH_2$ produced by the TCA cycle is oxidized by complex II, whilst $FADH_2$ from β-oxidation is oxidized by electron-transferring protein. The reducing equivalents produced from the oxidation of NADH and $FADH_2$ are transported via electron carriers to ubiquinone (Figure 2). From ubiquinone, electrons are passed in turn to complex III, cytochrome

Table 1 Enzymes of the mitochondrial respiratory chain

Complex I	NADH:ubiquinone (CoQ) reductase
Complex II	succinate:ubiquinone reductase
Complex III	ubiquinol–cytochrome–*c* reductase
Complex IV	cytochrome-*c* oxidase
Complex V	ATPase

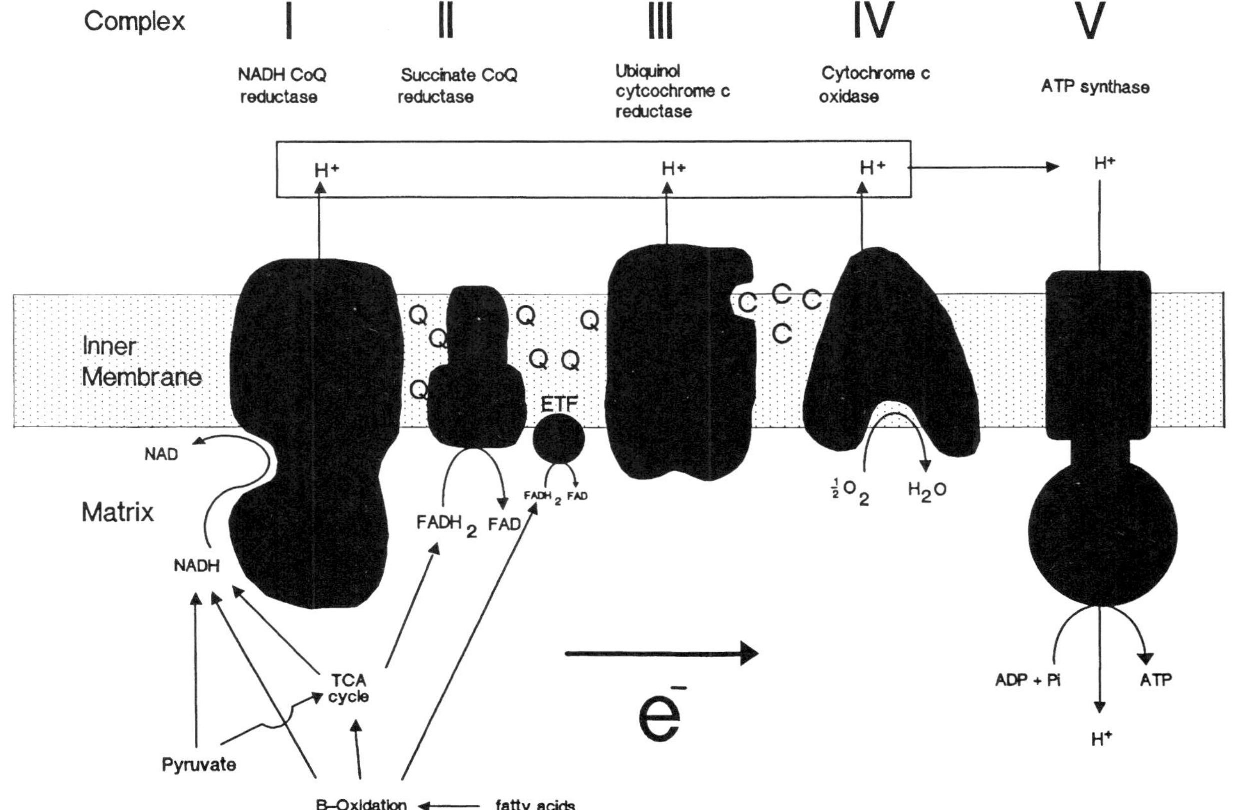

Figure 1 The mitochondrial respiratory chain. Q, ubiquinone; C, cytochrome *c*; ETF, electron-transferring flavoprotein. (Reproduced by permission of Butterworth/Heinemann.)

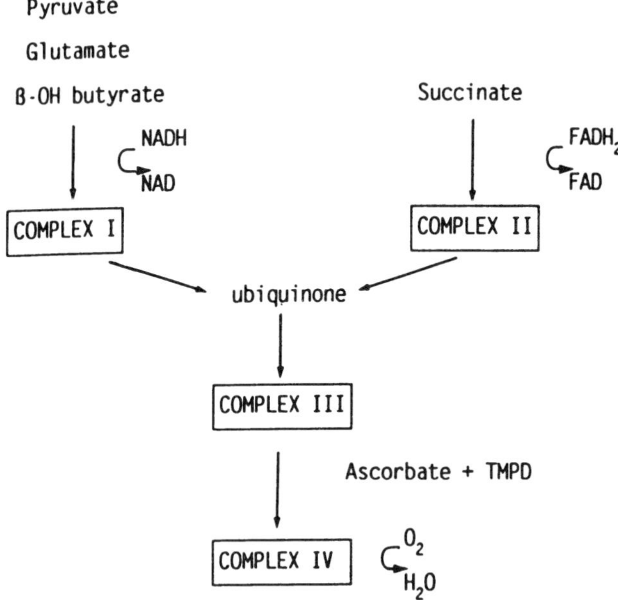

Figure 2 The mitochondrial respiratory chain. TMPD, *N,N,N,N*-tetramethyl-*p*-phenylenediamine. (Reproduced permission of Raven Press.)

c and complex IV to oxygen, which is reduced to water. The passage of electrons down the chain is associated with the vectoral movement of proteins from the matrix across the inner membrane and the generation of a protein motive which is used by complex V to generate ATP.

7.2.2 Structure–mtDNA

MtDNA is a small G-C-rich, circular, double-stranded molecule 16.5 kb long (Figure 3). Each mitochondrion has 2–10 molecules of mtDNA. MtDNA has two strands, heavy (H) and light (L), which have separate origins of replication, that of the H strand being contained within the displacement (D) loop, whilst that of the L strand is located 10 kb away. The mtDNA molecule has two separate transcriptional promoters located between the 16S gene and the D loop. MtDNA is highly efficient with virtually no introns. It encodes 24 RNA genes and 13 polypeptides, all the latter being part of the respiratory chain (Table 2). All genes are located on the H strand with the exception of eight tRNAs and the ND1 gene, which are encoded by the L strand. Transcription of mtDNA genes is not uniform, the 12S and 16S rRNA genes being transcribed 15–20 times that of the remaining molecule. This is possible because of a separate transcription termination codon located within the $tRNA^{Leu(UUR)}$ gene. MtDNA has no protective histone coat and lacks the repertoire of repair enzymes normally associated with nuclear DNA. The implications for the vulnerability of mtDNA to damage are discussed below.

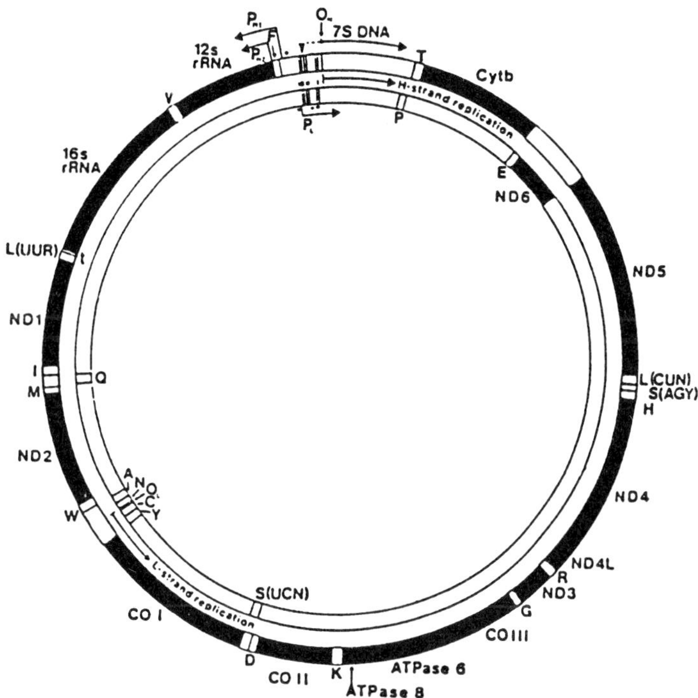

Figure 3 MtDNA. Heavy (H) and light (L) strands of DNA. Dark shaded areas encode rRNA or polypeptides. O_H and O_L are the origins of H and L strand replication, respectively. P_H and P_L are H and L strand promoters, respectively. (Reproduced by permission of INSERM.)

Table 2 RNA and polypeptides encoded by mtDNA

12S and 16S ribosomal RNA
22 transfer RNAs
Complex I subunits, ND1, ND2, ND3, ND4, ND4L, ND5, ND6
Complex II subunit, cytochrome *b*
Complex IV subunits, COI, COII, COIII
Complex V subunits, A6, A6L

7.2.3 Structure–complexes I–V

Complexes I, III, IV and V are unique in that they are the product of two separate genomes. Clearly both nuclear and mitochondrial DNA must act in a coordinated fashion in order to produce the holoenzymes. While the mtDNA gene products must traverse only the mitochondrial matrix to be incorporated into the inner membrane, nuclear encoded products are translated on cytoribosomes and must be transported across the cytoplasm and the outer mitochondrial membrane, or in some cases both the inner and outer membranes. The transport of nuclear encoded respiratory chain proteins involves, in the majority of cases, an N-terminal amino pre-sequence which targets outer mitochondrial membrane receptors. The pre-protein is transported across one or both membranes, and the pre-sequence is cleaved before insertion into the holoenzyme (for a review, see Voos *et al.*, 1994).

Complex I is the target of all the respiratory chain proteins, comprising 41 subunits, 6–8 of which are iron–sulfur (FeS) centres. Complex II has only four subunits, which are all encoded by nuclear DNA. The two largest subunits are associated with a succinate dehydrogenase (SDH) activity of the TCA cycle. Complex III has 11 subunits, one of which, cytochrome *b*, is mitochondrially encoded. Complex IV has 13 units and contains copper and cytochromes *a* and a_3 which are all associated with the three mitochondrially encoded subunits. Complex V has 14 subunits, A6 and A6L being encoded by mtDNA.

7.2.4 Respiratory chain inhibitors

Respiratory chain inhibitors may provide valuable models for putative toxins that could be responsible for neuronal degeneration alone or in concert with genetic or other endogenous factors. In addition, their relevance is highlighted by the ability of some of these toxins to induce animal models of neurodegenerative disease.

7.2.4.1 Complex I inhibitors

Rotenone, piericidin A, amytal, diphenylene iodonium and 1-methyl-4-phenylpyridinium (MPP^+) are all complex I inhibitors. Rotenone is particularly potent and inhibits complex I by preventing the reduction of ubiquinone but not ferricyanide – suggesting its site of action is between the highest potential FeS centre and ubiquinone. The binding of tritiated dihydrorotenone to complex I produced a 33 kDa subunit as the main photolabelled polypeptide (Earley *et al.*, 1987). This subunit was identified as the ND1 product of mtDNA. Stereotactic injection of rotenone into rat striatum produced selective loss of dopaminergic neurones (Heikkila *et al.*, 1985), suggesting a particular sensitivity of these neurones to respiratory chain inhibition.

MPP^+, the metabolite of 1-methyl-4-phenyl-1,2,3,6-tetrahydropyridine (MPTP) is probably the most widely studied complex I inhibitor. Although first recognized to cause parkinsonism in humans in 1979 (Davis *et al.*, 1979), it did not receive wide attention as a neurotoxin until 1983 (Langston *et al.*, 1983). MPTP is metabolized to MPP^+ mainly by monoamine oxidase B (MAO-B), which confines its site of action to the central nervous system. MPP^+ is a substrate for the dopaminergic reuptake pathway (Salach *et al.*, 1984) and so is selectively accumulated into these neurones. An energy-dependent mechanism for lipophilic cations then concentrates MPP^+ into mitochondria (Ramsay *et al.*, 1986), where it is a specific inhibitor of complex I (Nicklas *et al.*, 1985). The binding site of MPP^+ to complex I is at or near that of rotenone (Ramsay *et al.*, 1991) and may in fact involve two sites, one accessible to hydrophilic compounds and the other buried within the hydrophobic region of the complex (Gluck *et al.*, 1994). The suggestion that MPP^+ binds to a hydrophobic region within complex I is supported by the observation that increasing the length of an *n*-alkyl substitution enhances the potency of MPP^+ derivatives (Ramsay *et al.*, 1991) and that tetraphenylboron increases MPP^+ inhibition of complex I even in submitochondrial particles, i.e. it does not simply potentiate MPP^+ penetration of mitochondria (Ramsay *et al.*, 1989).

MPTP/MPP^+ are capable of inducing parkinsonism in primates (Burns *et al.*, 1983) and rats (Heikkila *et al.*, 1984). Their ability to do so is thought to be related to the fall in ATP synthesis consequent upon their inhibition of complex I. Indeed, there is evidence of a good correlation between MPP^+ concentration, decline in ATP synthesis and cell death in hepatoma cells (Di Monte and Smith, 1988). However, recent experiments have also shown that the interaction of MPP^+ with complex I may also induce superoxide formation (Hasegawa *et al.*, 1990; Cleeter *et al.*, 1992). The initial binding of MPP^+ with complex I is weak, and can be totally reversed by washing. Incubations longer than 15 minutes in the presence of reduced complex IV results in a severe, progressive and irreversible inhibition of complex I (Cleeter *et al.*, 1992). This irreversible component of complex I inhibition can be prevented with free radical scavengers, implicating oxidative damage as responsible for, or at least participating in, this phase of complex I inhibition. Interestingly, complexes II/III were not affected. Complex IV activity could not be assayed as it was kept in a reduced state. These experiments suggest that MPP^+ might bind to complex I and, after a preliminary reversible phase, initiate the generation of free radicals which subsequently appear to target complex I selectively. This in turn might suggest the development of a self-precipitating and self-amlifying cycle of complex I inhibition, free radical formation and oxidative damage. The requirement for complex IV to be reduced might suggest that the flow of electrons through the end of the chain must be impeded to initiate this cycle.

The proposition that MPTP might produce oxidative damage was partly based on the structural similarity of MPP^+ to paraquat. Paraquat causes cell death through oxidative damage (Bus and Gibson, 1984), and it was suggested that MPTP might be able to redox cycle through MPTP and MPP^+ (Johannessen *et al.*, 1986), although the redox potential of MPTP is considered too high to generate free radicals (Frank *et al.*, 1987). However, the observation that transgenic mice with enhanced superoxide dismutase activity are relatively resistant to MPP^+ (Przedborski *et al.*, 1992) and that ascorbate can protect human neuroblastoma cell lines against MPP^+ toxicity (Pardo *et al.*, 1993) both provide indirect evidence at least that oxidative damage is important in mediating the toxicity of MPP^+.

The tetrahydroisoquinolines (TIQs) are structurally related to MPTP and can be formed in the brain spontaneously by Pictet–Spengler cyclization of tryptamines and catecholamines. *N*-methyl-TIQ is a selective inhibitor of complex I and may be concentrated by mitochondria in a similar way to MPP^+ (Suzuki *et al.*, 1989). Long-term administration of TIQs may decrease dopamine content and induce parkinsonism in primates (Nagatsu and Yoshida, 1988). The concentration of TIQs has been found to be increased in the frontal cortex of one patient with Parkinson's disease (PD) (Niwa *et al.*, 1987).

7.2.4.2 Inhibition of other complexes

Complex II is irreversibly inhibited by 3-nitroproprionate (3-NP), an analogue of succinate which forms a covalent adduct with flavin active site (Alston *et al.*, 1977). Interestingly, 3-NP is also used to produce an animal model of Huntington's disease.

Antimycin, myxothiazol and stigmatellin are all complex III inhibitors, but none is currently associated with human disease or neurotoxic models.

Cyanide, carbon monoxide, sulfite and azide inhibit complex IV by interacting with cytochrome a_3. Interestingly, partial carbon monoxide poisoning is known to result in an akinetic rigid syndrome resembling PD.

Nitric oxide (NO) may be produced from L-arginine by NO synthase and is known to be produced by a variety of cells including glia and activated macrophages. NO synthase exists in inducible and constitutive forms, the latter being present in neurones and glia. NO has an affinity for iron-containing proteins and has been reported to inhibit complexes I and II in cell culture studies (Drapier and Hibbs, 1988; Stuehr and Nathan, 1989). There is also evidence that NO may reversibly inhibit complex IV in isolated mitochondrial preparations (M.J.W. Cleeter *et al.*, personal communication).

7.3 Diseases of the mitochondrial respiratory chain and mutations of mtDNA

Inborn metabolic deficiencies of the respiratory chain and mutations of mtDNA have now been associated with a large number of human diseases. The archetypal respiratory chain defects are represented clinically by the mitochondrial myopathies and encephalopathies. These include a variety of phenotypes including myopathy, cardiomyopathy and other presentations known by their acronyms: CPEO (chronic progressive external ophthalmoplegia), KSS (Kearn–Sayre syndrome), MELAS (mitochondrial myopathy, encephalopathy, lactic acidosis and stroke-like episodes), MERRF (myoclonic epilepsy and ragged red fibres), NARP (neurogenic weakness, ataxia, retinitis pigmentosa) and MNGIE (myoneurogastrointestinal encephalopathy). To this list must be added Pearson's syndrome (sideroblastic anaemia and pancreatic insufficiency), Leber's hereditary optic neuropathy (LHON), diabetes mellitus and maternally inherited deafness. This classification is only approximate and there is often considerable overlap between groups.

The morphological hallmark of the mitochondrial myopathies is the ragged red fibre, as seen by the modified Gomori trichrome stain, or the dense peripheral subsarcolemmal accumulation of reactive substrate as seen with the SDH or cytochrome oxidase stains. Electron microscopy may reveal paracrystalline inclusions in some fibres of a proportion of patients.

Lactic acidosis is an invariable accompaniment of a significant respiratory chain defect. This is the result of a block in oxidative phosphorylation producing an accumulation of the pyruvate formed by glycolysis being transferred into lactate. The lactate level may be normal or raised at rest but is invariably elevated by increased oxidative demand such as during exercise.

These disorders may be associated with respiratory chain defects of varying severity and pattern. There is no correlation between phenotype and the site of the respiratory chain deficiency. In addition, a wide variety of mtDNA mutations have been described in association with these disorders. These include deletions of varying size, duplications, and point mutations in tRNA, rRNA or protein-coding genes. There is a reasonable, but not absolute, correlation between some of these mutations and clinical presentation. For instance, deletions are often associated with CPEO, the $tRNA^{Leu(UUR)}$ mutation with MELAS and the $tRNA^{Lys}$ with MERRF. Point mutations in complex I-coding genes are seen in LHON, and in ATPase in NARP. It is interesting that mtDNA mutations involving tRNAs are associated with ragged red fibres, whilst those involving coding genes are not.

Caution must be exercised in the analysis of mtDNA mutations, and over-interpretation of their potential pathogenicity avoided. MtDNA is highly polymorphic and some published 'mutations' have subsequently been identified as polymorphisms. Heteroplasmy, the co-existence of mutant and wild-type mtDNA genomes, has been taken as a hallmark of pathogenicity. However, some mutations are homoplasmic (i.e. present only in the mutant form), and some apparent polymorphisms have recently been found to exist in the heteroplasmic form (Gill *et al.*, 1994). Finally, a mutation was thought to be pathogenic if it involved a conserved sequence and if it resulted in a significant change in amino acid residues. However, much of mtDNA can be considered 'conserved', and polymorphisms themselves may result in an alteration of amino acid sequence without obvious biochemical consequences.

The small size of the mtDNA molecule and the fact that it is completely sequenced in several species has focused attention on this genome even though it encodes less than 16% of the subunits of the respiratory chain. However, evidence for nuclear encoded defects has recently been sought in the mitochondrial disorders. Several families with autosomal dominant mitochondrial myopathy have now been described and found to be associated with the presence of multiple deletions of mtDNA (pleioplasmy) in skeletal muscle (Zeviani *et al.*, 1989). This clearly shows that nuclear genomic defects can produce qualitative defects of mtDNA. MtDNA depletion presents as early onset hypotonia and muscle weakness, occasionally in association with liver or kidney failure (Mazziotta *et al.*, 1992). Severe mtDNA depletion (<1% of control age matched values) results in death before the first year of life. The apparently normal fetal development and the subsequent rapid decline after birth might suggest a developmental defect in controlling mtDNA copy number. Family histories, where informative, suggest autosomal inheritance. Nuclear involvement in mtDNA depletion was recently confirmed using cell complementation studies and the demonstration that mtDNA levels can be restored to normal in a control nuclear environment (Bodnar *et al.*, 1993). Finally, a defect of mitochondrial import of a nuclear encoded subunit implies another type of nuclear genomic mutation in the mitochondrial myopathies (Schapira *et al.*, 1990a).

All of the considerations discussed with reference to inborn metabolic defects of the respiratory chain and mutations of mtDNA are important in the interpretation of the relevance of mitochondrial defects observed in neurodegenerative diseases.

7.4 Parkinson's disease

The main pathological correlate of the clinical and pharmacological characteristics of PD is the death of dopaminergic neurones in the substantia nigra. The cause and even the mode of death of these cells are as yet unknown or unresolved. Other chapters in this book will focus on the evidence that iron and free radicals play important roles in the death of nigral cells in PD. This chapter will concentrate on the involvement of mitochondrial dysfunction.

7.4.1 Mitochondrial activity in PD substantia nigra

Several studies have now reported a selective deficiency of complex I activity in PD substantia nigra (Schapira *et al.*, 1989, 1990b,c; Mann *et al.*, 1992a; Reichmann *et al.*, 1994). This equates to a 30–37% decrease in enzyme activity, and is highly statistically significant. The activities of complexes II–IV have been reported as no different from matched controls in all studies. The complex I deficiency is derived from nigral homogenates, and so incorporates both glial and neuronal components. As dopaminergic neurones constitute less than 3% of the nigral cell population in end-stage PD, it is likely that the complex I deficiency observed represents defective enzyme activity in both neurones and glia.

Immunoblotting of PD and control nigral homogenates with holoenzyme and subunit specific antisera failed to demonstrate any alteration in those complex I subunits that cross-reacted (Schapira *et al.*, 1990b). Another study used polyclonal holoenzyme antibodies in an immunohistochemical analysis of nigral sections (Hattori *et al.*, 1991). The authors found a decrease in complex I staining in a greater proportion of PD neurones than control. However, some of the neurones failed to stain at all for complex I and yet appeared normal morphologically. There was some decrease in complex II staining in a proportion of sections, but staining with complexes III and IV was no different from controls. The staining of glia in both PD and control sections was too weak to make any comparison.

MtDNA analysis with Southern blotting failed to identify the presence of any deletion greater than 200 base pairs present in any substantial concentrations (Lestienne *et al.*, 1990; Schapira *et al.*, 1990d). Polymerase chain reaction (PCR) amplification did not reveal the presence of the 5 kb 'common' deletion in PD nigra, but this was present at the same concentrations as age-matched controls and at a very low level (0.1–0.01%) (Lestienne *et al.*, 1991; Mann *et al.*, 1992b).

7.4.2 Other brain areas

Mitochondrial activity has been assessed in PD caudate, putamen, globus pallidum, tegmentum, cerebral cortex and cerebellum. In all reports, the activities of complexes I–IV have been found to be no different from controls (Mizuno *et al.*, 1990; Schapira *et al.*, 1990c; Mann *et al.*, 1992a; Cooper *et al.*, 1994). One report did describe a 36% decrease in the striatum of five PD patients (Mizuno *et al.*, 1990).

MtDNA analysis for the 5 kb deletion suggested an increase in PD striatum, but samples were not age matched (Ikebe *et al.*, 1990; Ozawa *et al.*, 1990) and, when this was taken into account, no excess of deletions was detected (Lestienne *et al.*, 1990).

7.4.3 Other tissues

Several studies have now reported on respiratory chain function in PD skeletal muscle. The majority of these studies have found normal mitochondrial activity (Mann *et al.*, 1992a; Anderson *et al.*, 1993; Di Donato *et al.*, 1993; Di Monte *et al.*, 1993; Taylor *et al.*, 1994), employing polarographic and spectrophotometric assays of freshly isolated mitochondria, magnetic resonance spectroscopy and enzyme assays in muscle homogenates. Other studies, however, have shown a variety of respiratory chain defects, usually involving a severe deficiency (80–90%) of complex I activity (Bindoff *et al.*, 1991; Shoffner *et al.*, 1991; Cardellach *et al.*, 1993). Thus, this subject remains controversial (DiMauro, 1993), but given the experience with complex I deficiency in mitochondrial myopathies:

(1) Defects of the order of 80–90% in complex I activity are invariably associated with severe muscle weakness. This is not a clinical feature normally associated with PD.
(2) Severe respiratory chain defects are usually associated with a lactic acidosis at rest or on exercise, but this has never been seen in PD (Di Monte *et al.*, 1991; Bravi *et al.*, 1992; Hattori *et al.*, 1992).
(3) Complex I defects are usually, but not always, found with typical ragged red fibres and other abnormalities on biopsy. No significant morphological abnormality has been observed in any of the PD patients' biopsies.

Platelet mitochondria have also been used as a means to determine the tissue distribution of the mitochondrial defect in PD. Again, the majority of studies have shown either no abnormality (Bravi *et al.*, 1992; Mann *et al.*, 1992a) or a mild (16%) deficiency of complex I (Krige *et al.*, 1992; Yoshino *et al.*, 1992). The overlap between controls and patients in these studies was such that a complex I defect was not discriminatory between the two groups and so could not be used as a diagnostic test. One study also used an enriched preparation of platelet mitochondria but showed a marked deficiency of complex I activity (Parker *et al.*, 1989).

Those results suggesting complex I deficiency in skeletal muscle and platelets in PD might seem at odds with the specificity of the effect within the brain. Why, for instance, should there be a mitochondrial abnormality in muscle but none in caudate or putamen? Whilst this may be difficult to rationalize for muscle, there are certain features of platelets which might render them susceptible to a complex I abnormality. In particular, platelets have MAO-B and uptake mechanisms for concentrating MPP^+ (Da Prada *et al.*, 1988). Thus, platelet complex I could be affected by toxins with similar characteristics to MPTP. However, because of the rapid turnover of platelets such a toxin would either have to be constantly circulating in the blood or be sequestered in the bone marrow with access to the megakaryocyte.

An alternative hypothesis that would accommodate these conflicting results of tissue distribution of the mitochondrial defect is that PD is a disorder with a variety of causes. In some patients a single cause might be operating whilst in others a combination of factors may be necessary to precipitate the disease. The interplay of genetic and environmental factors might differ from one platelet to another. Thus, a minority of patients may indeed have a systemic complex I defect as a result of either a genetic defect or toxin exposure. Such patients have not as yet had multiple tissue analysis ante- or post-mortem to confirm this hypothesis.

7.4.4 Specificity of the complex I deficiency for PD

An obvious explanation for the complex I defect in PD is that it simply represents a secondary phenomenon resulting from neuronal degeneration. Multiple system atrophy (MSA) is a term used to encompass a variety of 'Parkinson-plus' akinetic rigid syndromes characterized by, amongst other features, severe neuronal degeneration in the substantia nigra. Thus, if the complex I deficiency is merely a result of degeneration, it would be expected that the same abnormality would be present in MSA nigra. In fact, respiratory chain function in MSA substantia nigra is no different from controls (Schapira *et al.*, 1990c). In postencephalitic parkinsonism, there is a decline in the activities of complexes I and IV, again suggesting that the pure complex I deficiency in PD is not the result of degenerating cells (Schapira, 1994).

7.4.5 Effect of L-DOPA

Another simple explanation for the complex I deficiency of PD would be that it is an effect of L-DOPA treatment. L-DOPA and dopamine are toxic to dopaminergic neuronal cells lines in culture (Michel and Hefti, 1990) and decrease dopaminergic cell survival in the tegmentum of rats pretreated with 6-hydroxydopamine (Blunt *et al.*, 1993). However, L-DOPA does not lead to dopaminergic cell loss in control rats (Perry *et al.*, 1984) nor accelerates the progression of PD (Markham and Diamond, 1981). There was no enhancement of nigrostriatal dopaminergic neuronal loss in one patient who did not have PD but was chronically treated with L-DOPA (Quinn *et al.*, 1986). Against this background is the absence of a complex I defect in MSA patients who had been taking L-DOPA in quantities and for durations comparable to PD patients.

Recently, Przedborski and colleagues (Przedborski *et al.*, 1993) have demonstrated that L-DOPA may induce a 25% deficiency of complex I in control rat striatum and substantia nigra. This defect is reversible on withdrawal of L-DOPA. The normal respiratory chain function in PD striatum after chronic L-DOPA use calls into question whether a similar mechanism to this rat model exists in PD. The question as to whether L-DOPA might accelerate a pre-existing complex I defect in substantia nigra cannot be confirmed or refuted. It may be that the abnormalities of the PD substantia nigra (e.g. excess iron, oxidative stress) are necessary to provide the toxic environment for this effect of L-DOPA to be seen. The evidence presented to date would indicate that L-DOPA alone cannot cause the severity of complex I deficiency in PD.

7.4.6 Oxidative stress and mitochondrial dysfunction

Under aerobic conditions, the mitochondrial respiratory chain is the most important intracellular source of hydrogen peroxide and free radicals, mainly superoxide. Inhibition of the respiratory chain, especially of complexes I and III, results in an increase in free radical production (Takeshige and Minakami, 1979; Hasegawa *et al.*, 1990; Cleeter *et al.*, 1992). Free radicals may also damage the respiratory chain and inhibit the function of the constituent complexes. *In vivo* studies have suggested that complex IV, followed by complex I, is the most vulnerable to oxidative damage. Depletion of reduced glutathione (GSH) results first in a complex IV deficiency followed by a less severe decrease in complex I activity (Benzi *et al.*, 1991). Iron loading of PC12 dopaminergic cells results in increased oxidative stress and a 21% decline in complex IV activity and an 11% reduction in complex I (Hartley *et al.*, 1993). Finally, vitamin E deficiency in rats produces a similar pattern of respiratory chain deficiency in skeletal muscle (Thomas *et al.*, 1993). *In vitro* studies with isolated mitochondria or submitochondrial particles suggest that complex I followed by complexes III and IV are the most vulnerable to free radical damage (Narabayashu *et al.*, 1982; Hillered and Ernster, 1983; Zhang *et al.*, 1990). The *in vivo* studies, but not the *in vitro* ones, would suggest that the pure complex I defect seen in PD would not be caused by free radicals alone. The oxidative stress and damage observed in the PD substantia nigra could, in contrast, be caused by the mitochondrial defect. However, the necessary ingredients are present in the PD nigra to generate a self-replicating cycle of complex I deficiency, free radical production and oxidative damage which may be accelerated by the presence of iron and possibly by the administration of L-DOPA.

7.5 Complex I deficiency and the cause of PD

Although not conclusive, evidence is accumulating that the complex I deficiency in PD may not only be an important factor in the death of nigral dopaminergic neurones, but is also closely related to the cause of PD. Before more detailed analysis of this concept, it is important to accept that it is becoming increasingly likely that the cause of PD is likely to be multifactorial and, as discussed above, may involve a different combination in different patients. Thus the search for a universal biochemical or genetic 'marker' of PD may prove fruitless as well as misleading. The potential pathways for inducing the complex I deficiency in PD are briefly reviewed below as they may have an impact on developing neuroprotective strategies in the future.

Much has been written on a genetic component to the cause of PD. Gowers observed that there seemed to be more PD patients with similarly affected family members than appeared likely from chance (Gowers, 1903). Twin studies are often valuable in defining whether a disease may or may not have a genetic basis. Reports thus far have indicated that there is no increase in concordance for PD amongst identical twins over non-identical twins (Ward *et al.*, 1983; Marsden, 1987). However, there are problems inherent in these studies which limit interpretation (Johnson *et al.*, 1990). One major problem is in disease ascertainment – an apparently normal co-twin may in fact have presymptomatic PD, only developing clinical features several years after completion of the study. Also, a genetic susceptibility to develop PD which may be expressed by both twins may combine with the necessary environmental trigger in only one. An attempt to circumvent these problems has been to use positron emission tomography (PET) to estimate fluoro-DOPA uptake by the striatum. The results of one such study have shown that some apparently normal co-twins had significantly reduced uptake. Correction for such a result in the interpretation of twin data has increased the likelihood of a genetic component for PD (Burn *et al.*, 1992).

Naturally, in view of the complex I deficiency, a mutation of either the nuclear or the mitochondrial genomes may be responsible for conferring a genetic component on PD. Twin studies cannot discriminate the potential contribution of either. To date, there is no published evidence for maternal inheritance of PD (Maraganore *et al.*, 1991). However, Golbe and colleagues (Golbe *et al.*, 1990) have published details of a large American–Italian kindred with autosomal dominant parkinsonism. There are some atypical features in this family in terms of age of onset and absence of tremor. However, affected members of this family do respond to L-DOPA, and post-mortem examination of one affected individual has shown evidence of Lewy body formation.

The lack of a clear genetic influence in PD suggests that, in the majority of patients genes can at best determine only susceptibility rather than full disease penetration. The discovery that a specific allele (2D6) of the cytochrome P_{450} system is found in statistically greater frequency (approximately 2.5-fold) amongst PD patients (Armstrong *et al.*, 1992; Smith *et al.*, 1992) is important in support of this. Bearing in mind the possibility of abnormal xenobiotic metabolism in PD (Steventon *et al.*, 1989; Green *et al.*, 1991), it may be that a variety of such associations will be found, perhaps including genes for superoxide dismutase, glutathione peroxidase, other enzymes involved in free radical metabolism, and genes for complex I. In this situation, it may be that susceptibility to developing PD is a consequence of an individual expressing a given number and combination of 'risk' genes.

Sporadic groupings of PD patients have been associated with certain environmental agents such as proximity to wood mills, rural living, well water drinking or exposure to certain known chemicals, e.g. MPTP or manganese (Salach *et al.*, 1984; Rajput *et al.*, 1987; Tanner, 1989; Koller *et al.*, 1990; Jiménez-Jiménez *et al.*, 1992; Semchuk *et al.*, 1992). These incidences account for only a minority of cases, and no clear environmental source has been implicated in the majority of sporadic PD patients. However, the MPTP model has created a precedent on which other neurotoxins may be modelled. The cell selectivity of this agent is the result of specific uptake and conversion mechanisms which have already been outlined above. The discovery of complex I deficiency in PD substantia nigra confirmed a direct biochemical link between the idiopathic disease and the MPTP model, and raised the possibility that the mitochondrial defect in PD might be the result of toxin exposure. As has been outlined above, such a toxin may be exogenous or endogenous, generated by abnormal xenobiotic metabolism.

7.6 Mechanisms of cell death in PD

To date, it has been assumed that the dopaminergic cells in the substantia nigra in PD are lost through cell necrosis, an event possibly precipitated by oxidative damage or complex I deficiency. Some recent evidence suggests that these same biochemical abnormalities may also induce apoptosis. Apoptosis refers to programmed cell death and is distinct from cell necrosis with characteristic morphological and DNA changes. Apoptosis may be precipitated by a variety of agents including inhibitors of protein synthesis and withdrawal of growth factors (Batistatou and Green 1991; Edwards *et al.*, 1991). Morphological changes include chromatin condensation and the formation of chromatin bodies. Nuclear DNA undergoes extensive cleavage with digestion at nucleosomes situated 185 base pairs apart. This produces the characteristic laddering with DNA fragments sized in multiples of 185 when viewed by ultraviolet light after separation by agarose gel electrophoresis. Apoptosis is employed during early development to eradicate redundant cells, especially neurones which have not formed synaptic connections. Thus, approximately 50% of anterior horn cell neurones are lost through apoptosis during early development. BCI_2 is a nuclear encoded protein, M_r 25 kDa, which is an inhibitor of apoptosis (McDonnell *et al.*, 1989). This protein is found on the mitochondrial outer membrane, the nuclear membrane and endoplasmic reticulum (Hockenbery *et al.*, 1990; Monaghan *et al.*, 1992). The mechanism of action of BCI_2 is unclear, although there is some evidence that it may exert its effect through antioxidant properties (Hockenberg *et al.*, 1993). Although prominently located on mitochondria, BCL_2 is able to function independently of the respiratory chain (Jacobsen *et al.*, 1993). Interestingly, mtDNA does not appear to undergo nucleosomal cleavage during apoptosis (Murgia *et al.*, 1992).

The first report suggesting that apoptosis may be relevant to PD came from a study of the action on MPP^+ on cerebellar granule cells in culture (Dipasquale *et al.*, 1991). MPP^+ induced apoptosis in these cells although its mechanism of action was unclear. These cells are glutamate-sensitive, and one possibility was that the respiratory chain inhibition induced by the MPP^+ may have resulted in excitotoxin mediated apoptosis.

In another series of experiments, Hartley *et al.* (1994) have shown that MPP^+ can induce apoptosis in a concentration-dependent manner; low concentrations of MPP^+

cause apoptosis whilst higher concentrations inhibit the respiratory chain and produce necrosis. This shift from apoptosis to necrosis according to the concentration of toxin may have profound implications for our understanding of the sequence of events, and potential mechanisms of cell death, in the development of PD. For instance, it is possible that a low concentration of a neurotoxin may begin the process of dopaminergic cell death through apoptosis. This may evolve over a period of years, leading to a gradual decline in the neuronal population. The toxin may precipitate apoptosis either through complex I inhibition or oxidative stress. As explained above, these may generate a vicious cycle of cellular damage. Over time, the toxin may accumulate and its effect shift from apoptosis to necrosis. At this stage the necrosis will stimulate gliosis and biochemical expression will be dominated by the effect of the toxin on mitochondrial function and oxidative stress.

The demonstration of dopaminergic cells undergoing apoptosis in PD nigra will have important implications not only for our understanding of the mechanisms of cell death in PD, but also for developing treatment strategies in transplantation. However, such a demonstration has some obvious obstacles. Apoptosis is a rapid process and even when involving a substantial proportion of a cell population, the typical morphological changes can easily be missed because of their brief duration. The appearance of DNA laddering may also be undetectable because of the small proportion of cells affected at any one time. In addition, it may be that if apoptosis does play a role in PD, it does so at the very beginning of the disease and may no longer contribute to cell death once other mechanisms, more severe complex I deficiency and oxidative damage have been established. Thus, defining the involvement of apoptosis in PD may require exhaustive morphological analysis of several patients at early and late stages of the disease.

7.7 Neuroprotective strategies

Neuroprotective strategies for PD in the context of mitochondrial dysfunction must be directed at preventing the development of complex I deficiency, reversing or inhibiting any further deterioration in activity, ameliorating or circumventing the toxic effects of the defect, or enhancing remaining respiratory chain function.

7.7.1 Prevention of complex I deficiency

The success of preventing the complex I defect and protecting dopaminergic neurones will clearly depend upon its contribution to cell death at the various stages of the disease. Furthermore, the strategies employed in prevention will depend upon the cause. If the complex I deficiency is solely due to a genetic defect (nuclear or mitochondrial) then the prospect of its prevention would seem remote. However, if complex I were inhibited by a exogenous or endogenous toxin then several protective strategies could be envisaged.

7.7.1.1 Avoidance of toxin

If complex I is inhibited by an exogenous toxin, it may prove possible to isolate and identify the toxin from PD tissue. In addition, the general structure of complex I inhibitors may give clues to potential candidates in the environment. Furthermore, based on the model for MPTP, the structural requirements necessary to allow targeting of the nigral dopaminergic neurones will provide additional indications of structure. Following identification, the avoidance of such chemicals, or their eradication from the environment, would have to be addressed.

7.7.1.2 Inhibition of proxotin metabolism

Again using MPTP as a model for a dopaminergic cell selection toxin, the metabolic requirements necessary for its selectivity of action may require enzymatic conversion from an inactive to an active compound. In the case of MPTP, this is performed by MAO. The question of the neuroprotective effect of MAO inhibition in PD will be discussed in detail elsewhere in this book. Whether MAO is involved in metabolizing a complex I inhibitor is unknown.

A complex I inhibitor might target nigral neurones via the dopamine uptake system, as indeed does MPP^+. Abnormalities of the reuptake system, e.g. increased affinity for potential toxins, may determine susceptibility to specific compounds. This mechanism would not only provide another clue to specific structural requirements but also a further means to reduce toxin access to its active site. A dopamine reuptake inhibitor, however, would have intrinsic parkinsonian effects and so could prove impossible to use in this context.

7.7.2 Reversal of complex I inhibition

A putative PD-causing complex I-binding toxin may, in its initial stages, be reversible. As outlined earlier, MPP^+ binds only weakly and reversibly to complex I. Second-phase inhibition, however, is irreversible and probably free radical mediated (Cleeter *et al.*, 1992). If the reversible phase of the inhibition of a toxin to complex I were sufficiently long it might prove possible to use a compound that might bind competitively to complex I and displace the inhibitor without itself affecting enzyme function.

7.7.3 Circumventing the complex I defect

7.7.3.1 Respiratory chain bypass

Attempts to bypass complex I deficiency in patients with mitochondrial myopathy have proved disappointing. Anecdotal reports of the use of succinate, ubiquinone, ascorbate, nicotinamide and riboflavin in various combinations and permutations have occasionally suggested clinical efficacy, but have not always been supported by objective data. In some respects this may be related to the penetration of these chemicals into muscle, but such a qualification is also likely to apply to brain as well.

7.7.3.2 Increasing mitochondrial mass

The benefit of increasing mitochondrial mass to circumvent the complex I defect in PD would depend entirely on the biochemical basis of the deficiency. Inborn metabolic defects of complex I such as occur in the mitochondrial myopathies are often associated with a significant rise in skeletal muscle mitochondrial numbers. However, despite this increase, there is still a significant biochemical deficiency on a basis of wet weight of tissue. However, a stimulus to mitochondrial proliferation in the presence of an inhibitor might be a potential mechanism to dilute the effect of the inhibition. Recent insight into the nuclear controls of mtDNA replication has been obtained from the study of patients with the mtDNA depletion syndrome (Bodnar *et al.*, 1993), and it is anticipated that the gene controlling mtDNA copy number will be cloned and characterized.

7.7.4 Enhancing complex I activity

The enhancement of complex I activity would involve either improving its efficiency of electron transport or increasing the concentration of respiratory chain proteins per unit mitochondrion. Electron carriers are unlikely to be capable of penetrating the complex I structure to enhance electron flow. Alternative substrates to pyruvate, glutamate or β-hydroxybutyrate may conceivably have higher affinity for complex I, but this would not be of value if the complex I defect in PD is distal to substrate binding. Increasing the rate of synthesis of respiratory chain proteins would have to involve the concerted action of both nuclear and mitochondrial genomes. In this respect, the mitochondrial myopathies may provide some important clues to the regulation of respiratory chain expression.

7.8 Neuroprotection and the cytotoxic theory

The elements of the cytotoxic cycle theory are explained earlier in this chapter. Essentially, it is suggested that the cycle begins with exposure either to an environmental toxin or the expression of a genetically determined deficiency of complex I activity, or both. In the case of toxin exposure, inhibition of complex I may be partial and result in apoptosis, or severe (e.g. MPTP) and produce necrosis. Both the putative toxins and the inborn defect result in increased superoxide production by complex I. This in turn produces oxidative stress and then damage, the latter affecting mitochondrial proteins and lipids at first, and then more generalized cellular free radical damage. Complex I, because of its inherent genetically determined defects or its toxin binding, is rendered most susceptible to this oxidative damage. The cycle of complex

I deficiency and oxidative damage is enhanced by the presence of free iron. Neuronal cell death may first be caused by apoptosis but, as the cycle proceeds and amplifies, death is induced by necrosis.

This cycle provides several opportunities for neuroprotection, some of which have been discussed above. Protection against free radical damage may prevent the generation or amplification of the cycle although it is likely that to be effective, antioxidants would have to be made available at an early stage. Prevention of apoptotic cell death, perhaps through selective BCI_2 expression, would also help to retard the development of the effects of the cytotoxic cycle.

7.9 Conclusions

Defining the role of complex I deficiency in the sequence of events that terminates in dopaminergic cell death in PD offers insight into the cause of this disease. In addition, new protective strategies may be developed which may either correct or ameliorate the effects of this biochemical deficiency. The efficacy of these strategies is likely to be dependent upon their early use. A means to identify patients at risk of developing PD, or in the presymptomatic stages of the disease, will therefore be of tremendous value. At present there is no reliable biochemical 'marker' for presymptomatic PD. Defining a pool of genetic risk factors probably offers the best chance of determining susceptibility.

References

Alston, T.A., Mela, L. & Bright, H.J. (1977) *Proc. Natl Acad. Sci. USA* **74**, 3767–3771.

Anderson, J.J., Bravi, D., Ferrari, R., Davis, T.I., Baronti, F., Chase, T.N. & Dagani, F. (1993) *Neurol. Neurosurg. Psychiatry* **56**, 477–480.

Armstrong, M., Daly, A.K., Cholerton, S., Bateman, D.N. & Idle, J.R. (1992) *Lancet* **339**, 1017–1018.

Batistatou, A. & Greene, L.A. (1991) *J. Cell. Biol.* **115**, 461–471.

Benzi, G., Curti, D., Pastoris, O., Marzatico, F., Villa, R.F. & Dagani, F. (1991) *Neurochem. Res.* **16**, 1295–1302.

Bindoff, L.A., Birch-Martin, M.A., Cartlidge, N.E.F., Parker Jr, W.D.F. & Turnbull, D.M. (1991) *J. Neurol. Sci.* **104**, 203–208.

Blunt, S.B., Jenner, P. & Marsden, C.D. (1993) *Mov. Disord.* **8**, 129–133.

Bodnar, A.G., Cooper, J.M., Holt, I.J., Leonard, J.V. & Schapira, A.H.V. (1993) *Am. J. Hum. Genet.* **53**, 663–669.

Bravi, D., Anderson, J.J., Dagani, F., Davis, T.L., Ferrari, R., Gillespic, M. & Chasc, T.N. (1992) *Mov. Disord.* **7**, 228–231.

Burn, D.J., Mark, M.H., Playford, E.D., Maraganore, D.M., Zimmerman, T.R., Duvoison, R.C., Harding, A.E., Marsden, C.D. & Brooks, D.J. (1992) *Neurology* **42**, 1894–1900.

Burns, R.S., Chiueh, C.C., Markey, S.P., Ebert, M.H., Jacobwitz, D.M. & Kopin, I.J. (1983) *Proc. Natl Acad. Sci. USA* **80**, 4545–4550.

Bus, J.S. & Gibson, J.E. (1984) *Environ. Health Perspect.* **55**, 37–76.
Cardellach, C., Marti, M.J., Fernandez-Sola, J., Marin, C., Hoek, J.B., Tolosa, E. & Urbano-Marquez, A. (1993) *Neurology* **43**, 2258–2262.
Cleeter, M.J.W., Cooper, J.M. & Schapira, A.H.V. (1992) *J. Neurochem.* **58**, 786–789.
Cooper, J.M., Daniel, S.E., Marsden, C.D. & Schapira, A.H.V. (1995) *Mov. Disord.* **10**, 295–297.
Da Prada, M., Cesura, A.M., Launay, J.M. & Richards, J.G. (1988) *Experientia* **44**, 115–126.
Davis, G.C., Williams, A.C., Markey, S.P., Ebert, M.H., Caine, E.D., Reichert, C.M. & Kopin, I.J. (1979) *Psychiatry Res.* **1**, 649–654.
DiDonato, S., Zeviani, M., Giovannini, P., Savarese, N., Rimoldi, M., Mariotti, C. & Girotti, F.T. (1993) *Neurology* **43**, 2262–2268.
Di Mauro, S. (1993) *Neurology* **43**, 2170–2172.
Di Monte, D.A. & Smith, T.M. (1988) *Rev. Neurosci.* **2**, 67–81.
Di Monte, D.A., Tetrud, J.W. & Langston, J.W. (1991) *Ann. Neurol.* **29**, 342–343.
Di Monte, D.A., Sandy, M.S., Jewell, S.A., Adornato, B., Tanner, C.T. & Langston, J.W. (1993) *Neurodegeneration* **2**, 275–281.
Dipasquale, B., Marini, A.M. & Youie, R.J. (1991) *Biochem. Biophys. Res. Commun.* **181**, 1442–1448.
Drapier, J.-C. & Hibbs Jr., H.B. (1988) *J. Immunol.* **140**, 2829–2838.
Earley, F.G.P., Patel, S.D., Ragan, C.I. & Attardi, G. (1987) *FEBS Lett.* **219**, 108–112.
Edwards, S.N., Buckmaster, A.E. & Tolkovsky, A.M. (1991) *J. Neurochem.* **57**, 2140–2143.
Frank, D.M., Arora, P.K., Bloomer, J.L. & Sayre, L.M. (1987) *Biochem. Biophys. Res. Commun.* **147**, 1095–1104.
Gill, P., Ivanov, P.L., Kimpton, C., Piercy, R., Benson, N., Tully, G., Evett, I., Hagelberg, E. & Sullivan, K. (1994) *Nature Genet.* **6**, 1130–1135.
Gluck, M.R., Krueger, M.K., Ramsay, R.R., Sablin, S.O., Singer, T.P. & Nicklas, W.J. (1994) *J. Biol. Chem.* **269**, 3167–3174.
Golbe, L.I., Di Iorio, G., Bonavita, V., Miller, D.C. & Duvoisin, R.C. (1990) *Ann. Neurol.* **27**, 276–282.
Gowers, W.R. (1903) *A Manuel of Diseases of the Nervous System*, 2nd edn., p. 636. Blakiston, Philadelphia.
Green, S., Buttrum, S., Molloy, H., Steventon, G., Sturman, S., Waring, R., Pall, H. & Williams, A. (1991) *Lancet* **338**, 120–121.
Hartley, A., Cooper, J.M. & Schapira, A.H.V. (1993) *Brain Res.* **627**, 349–353.
Hartley, A., Heron, C., Cooper, J.M. & Schapira, A.H.V. (1994) *Clin. Pharamcol.* **8**, 155–156.
Hasegawa, E., Takeshige, K., Oishi, T., Murai, Y. & Minikami, S. (1990) *Biochem. Biophys. Res. Commun.* **170**, 1049–1055.
Hattori, N.B., Tanaka, M., Ozawa, T. & Mizuno, Y. (1991) *Ann. Neurol.* **30**, 563–571.
Hattori, Y.N., Yoshino, H., Kondo, T., Mizuno, Y. & Horai, S. (1992) *J. Neurol. Sci.* **10**, 29–33.
Heikkila, R.E., Hess, A. & Duvoisin, R.C. (1984) *Science* **224**, 1451–1453.
Heikkila, R.E., Nicklas, W.J., Yvas, I. & Duvoison, R.C. (1985) *Neurosci. Lett.* **62**, 389–394.
Hillered, L. & Ernster, L. (1983) *J. Cer. Blood Flow Metab.* **3**, 207–214.
Hockenbery, D., Nuñez, G., Milliman, C., Schreiber, R.D. & Korsmeyer, S.J. (1990) *Nature* **348**, 334–336.
Hockenbery, D.M., Oltvai, Z.N., Yin, X.-M., Milliman, C.L. & Korsmeyer, S.J. (1993) *Cell* **75**, 241–251.
Ikebe, S., Tanaka, M., Ohno, K., Sato, W., Hattori, K., Kondo, T., Mizuno, Y. & Ozawa, T. (1990) *Biochem. Biophys. Res. Commun.* **170**, 1044–1048.
Jacobsen, M.D., Burne, J.F., King, M.P., Miyashita, T., Reed, J.C. & Raff, M.C. (1993) *Nature* **361**, 365–369.
Jiménez-Jiménez, F.J., Mateo, D. & Giménez-Roldán, S. (1992) *Mov. Disord.* **7**, 149–152.
Johannessen, J.N., Adams, J.D., Schuller, H.M., Bacon, J.P. & Markey, S.P. (1986) *Life Sci.* **38**, 743–749.
Johnson, W.G., Hodge, S.E. & Duvoisin, R. (1990) *Mov. Disord.* **5**, 187–194.

Koller, W., Vetere-Overfield, B., Gray, C., Alexander, C., Chin, T., Dolezal, J., Hassanein, R. & Tanner, C. (1990) *Neurology* **40**, 1218–1221.
Krige, D., Carroll, M.T., Cooper, J.M., Marsden, C.D. & Schapira, A.H.V. (1992) *Ann. Neurol.* **32**, 782–788.
Langston, J.W., Ballard, P., Tetrud, J.W. & Irwin, I. (1983) *Science* **219**, 979–980.
Lestienne, P., Nelson, J., Riederer, P., Jellinger, K. & Reichmann, H. (1990) *J. Neurochem.* **55**, 1810–1812.
Lestienne, P., Riederer, P. & Jellinger, K. (1991) *J. Neurochem.* **56**, 1819.
McDonnell, T.J., Deane, N., Platt, F.M., Nuñez, G., Jaeger, U., McKearn, J.P. & Korsmeyer, S.J. (1989) *Cell* **57**, 79–88.
Mann, V.M., Cooper, J.M., Krige, D., Daniel, S.E., Schapira, A.H.V. & Marsden, C.D. (1992a) *Brain* **115**, 333–342.
Mann, V.M., Cooper, J.M. & Schapira, A.H.V. (1992b) *FEBS Lett.* **299**, 218–222.
Maraganore, D.M., Harding, A.E. & Marsden, C.D. (1991) *Mov. Disord.* **6**, 205–211.
Markham, C.H. & Diamond, S.G. (1981) *Neurology* **31**, 125–131.
Marsden, C.D. (1987) *J. Neurol. Neurosurg. Psychiatry* **50**, 105–106.
Mazziotta, R.M.R., Ricci, E., Bertini, E., Vici, C.D., Servidei, S., Burling, A.B., Sabetta, G., Bartuli, A., Manfredi, G., Silvestri, G., Moraes, C.T. & DiMauro, S. (1992) *J. Pediatr.* **121**, 896–901.
Michel, P.P. & Hefti, F. (1990) *J. Neurosci. Res.* **26**, 428–435.
Mizuno, Y., Suzuki, K. & Ohta, S. (1990) *J. Neurol. Sci.* **96**, 49–57.
Monaghan, P., Robertson, D., Amos, T.A.S., Dyer, M.J.S., Mason, D.Y. & Greaves, M.F. (1992) *J. Histochem. Cytochem.* **40**, 1819–1825.
Murgia, M., Pizzo, P., Sandona, D., Zanovello, P., Rizzuto, R. & Virgilio, F. (1992) *J. Biol. Chem.* **267**, 10939–10941.
Nagatsu, T. & Yoshida, M. (1988) *Neurosci. Lett.* **87**, 178–182.
Narabayashu, M., Takeshige, K. & Minikami, S. (1982) *Biochem. J.* **202**, 97–105.
Nicklas, W.J., Vyas, I. & Heikkila, R.E. (1985) *Life Sci.* **36**, 2503–2508.
Niwa, T., Takeda, N., Kandea, N., Hashizume, Y. & Nagatsu, T. (1987) *Biochem. Biophys. Res. Commun.* **144**, 1084–1089.
Ozawa, T., Tanaka, M., Ikebe, S., Ohno, K., Kondo, T. & Mizuno, Y. (1990) *Biochem. Biophys. Res. Commun.* **172**, 483–489.
Pardo, B., Mena, M.A., Fahn, S. & Yebenes, J.G. (1993) *Mov. Disord.* **8**, 278–284.
Parker, W.D., Boyson, S.J. & Parks, J.K. (1989) *Ann. Neurol.* **26**, 719–723.
Perry, T.L., Yong, V.W., Ito, M., Foulks, I.M., Wall, J.G., Godin, D.V. & Clavier, R.M. (1984) *J. Neurochem.* **43**, 990–993.
Przedborski, S., Kostic, V., Jackson-Lewis, V., Naini, A.B., Simonetti, S., Fahn, S., Carlson, E., Epstein, C.J. & Cadet, J.L. (1992) *J. Neurosci.* **12**, 1658–1667.
Przedborski, S., Jackson-Lewis, V., Muthane, U., Jiang, H., Ferreira, M., Naini, A.B. & Fahn, S. (1993) *Ann. Neurol.* **34**, 715–723.
Quinn, N., Parkes, D., Janota, I. & Marsden, C.D. (1986) *Mov. Disord.* **1**, 65–68.
Rajput, A.H., Vitti, R.J., Stern, W., Laverty, W., O'Donnell, K., O'Donnell, D., Yuen, W.K. & Dus, A. (1987) *Can. J. Neurol. Sci.* **14**, 414–418.
Ramsay, R.R., Dadger, J., Trevor, A. & Singer, T.P. (1986) *Life Sci.* **39**, 581–588.
Ramsay, R.R., Mehlhorn, R.J. & Singer, T.P. (1989) *Biochem. Biophys. Res. Commun.* **159**, 983–990.
Ramsay, R.R., Krueger, M.J., Youngster, S.K., Gluck, M.R., Casida, J.E. & Singer, T.P. (1991) *J. Neurochem.* **51**, 1184–1190.
Reichmann, H., Janetzky, B., Hauck, S., Youdim, M.B.H. & Riederer, P. (1994) *New Trends Clin. Pharmacol.* **8**, 49.
Salach, J.I., Singer, T.P., Castagnoli, N. & Trevor, A. (1984) *Biochem. Biophys. Res. Commun.* **125**, 831–835.

Schapira, A.H.V. (1994) In *Mitochondria: DNA, Proteins and Disease* (eds Darley-Usmar, V. & Schapira, A.H.V.) Portland Press, London, pp. 241–278.

Schapira, A.H.V., Cooper, J.M., Dexter, D., Jenner, P., Clark, J.B. & Marsden, C.D. (1989) *Lancet* **i**, 1269.

Schapira, A.H.V., Cooper, J.M., Morgan-Hughes, J.A., Landon, D.N. & Clark, J.B. (1990a) *N. Engl. J. Med.* **323**, 37–42.

Schapira, A.H.V., Cooper, J.M., Dexter, D., Clark, J.B., Jenner, P. & Marsden, C.D. (1990b) *J. Neurochem.* **54**, 823–827.

Schapira, A.H.V., Mann, V.M., Cooper, J.M., Dexter, D., Daniel, S.E., Jenner, P., Clark, J.B. & Marsden, C.D. (1990c) *J. Neurochem.* **55**, 2142–2145.

Schapira, A.H.V., Holt, I.J., Sweeney, M., Harding, A.E., Jenner, P. & Marsden, C.D. (1990d) *Mov. Disord.* **5**, 294–297.

Semchuk, K.M., Love, E.J. & Lee, R.G. (1992) *Neurology* **42**, 1328–1335.

Shoffner, J.M., Watts, R.L., Juncos, J.L., Torrini, A. & Wallace, D.C. (1991) *Ann. Neurol.* **30**, 332–339.

Smith, C.A.D., Gough, A.C., Leigh, P.N., Summers, B.A., Harding, A.E., Maranganore, D.M., Sturman, S.G., Schapira, A.H.V., Williams, A.C., Spurr, N.K. & Wolf, C.R. (1992) *Lancet* **339**, 1375–1377.

Steventon, G.B., Heafield, M.T.E., Waring, R.H. & Williams, A.C. (1989) *Neurology* **39**, 883–887.

Stuehr, D.J. & Nathan, C.P. (1989) *J. Exp. Med.* **169**, 1543–1555.

Suzuki, K., Mizuno, Y. & Yoshida, M. (1989) *Biochem. Biophys. Res. Commun.* **162**, 1541–1545.

Takeshige, K. & Minakami, S. (1979) *Biochem. J.* **180**, 129–135.

Tanner, C.M. (1989) *TINS* **12**, 49–54.

Taylor, D.J., Krige, D., Barnes, P.R.J., Kemp, G.J., Carroll, M.T., Mann, V.M., Cooper, J.M., Marsden, C.D. & Schapira, A.H.V. (1994) *J. Neurol. Sci.* **125**, 77–81.

Thomas, P.K., Cooper, J.M., King, R.H.M., Workman, J.M., Schapira, A.H.V., Goss-Sampson, M.A. & Muller, D.P.R. (1993) *J. Anat.* **183**, 451–461.

Voos, W., Moczko, M. & Pfanner, N. (1994) In *Mitochodria: DNA, Proteins and Disease* (eds Darley-Usmar, V. & Schapira, A.H.V.) Portland Press, London, pp. 55–80.

Ward, C.D., Duvoisin, R.C., Ince, S.E., Nutt, J.D., Eldridge, R. & Calne, D.B. (1983) *Neurology* **33**, 815–824.

Yoshino, H., Nakagawa-Hattori, Y., Kondo, T. & Mizuno, Y. (1992) *J. Neural. Transmission* **4**, 27–34.

Zeviani, M., Servidei, S., Gellera, C., Bertini, E., DiMauro, S. & DiDonato, S. (1989) *Nature* **339**, 309–311.

Zhang, Y., Marcillat, O., Giulivi, C., Ernster, I. & Davies, K.J. (1990) *J. Biol. Chem.* **265**, 16330–16336.

CHAPTER 8

BIOENERGETICS AND EXCITOTOXICITY: THE WEAK EXCITOTOXIC HYPOTHESIS

James G. Greene and J. Timothy Greenamyre*

Departments of Neurology, Neurobiology and Anatomy, and Pharmacology, University of Rochester Medical Center, Rochester, NY 14642, USA
**Present address: Department of Neurology, Emory University, Atlanta, GA 30322, USA*

Table of Contents

8.1 Introduction

Neurodegenerative diseases are characterized by an insidious onset and a progressive, irreversible and relentless decline in motor and cognitive capabilities. Currently, in the USA, approximately 4 million people are afflicted with Alzheimer's disease (AD), 600 000 with Parkinson's disease (PD) and 10 000 with Huntington's disease (HD) (Bennett and Evans, 1992; McDowell and Cedarbaum, 1992). These numbers are sure to rise as the age of the population increases. Despite the prevalence of neurodegenerative disorders, causes and mechanisms of disease remain obscure. Recently,

Neurodegeneration and Neuroprotection
in Parkinson's Disease
ISBN 0-12-525445-8

it has been hypothesized that excitatory amino acids, such as glutamate, may play a significant role in the aetiology of neurodegenerative diseases (Greenamyre and Young, 1989; Albin and Greenamyre, 1992; Beal, 1992; Beal *et al.*, 1993c).

In some situations, glutamate acts not only as a neurotransmitter, but also as a toxin, or 'excitotoxin'. The concept of excitotoxicity derives from the finding that there is a rough correlation between the ability of glutamate and related analogues to cause neuronal excitation and their propensity to cause neurotoxic lesions (Olney *et al.*, 1971). Both the excitatory and toxic properties of glutamate are mediated by neuronal receptors that have been named for the agonist compounds that most specifically stimulate them: *N*-methyl-D-aspartate (NMDA), α-amino-3-hydroxy-5-methyl-4-isoxazolepropionate (AMPA) and kainate (Watkins and Evans, 1981; Dingledine *et al.*, 1988). Each of these glutamate receptors is a ligand-gated ion channel ('ionotropic receptor'); when glutamate binds to the receptor, the ion channel opens and permits influx of Na^+ or Ca^{2+} ions. In addition, glutamate acts on a receptor type linked to G proteins. Excessive activation of these 'metabotropic' receptors can also produce excitotoxicity (Sacaan and Schoepp, 1992). Both the excitatory and toxic properties of glutamate can be blocked by receptor antagonists.

Although the post-receptor mechanisms of NMDA receptor-mediated excitotoxicity have not yet been fully elucidated, the increased Ca^{2+} flux resulting from excessive receptor stimulation apparently plays a pivotal role (Rothman, 1992). This calcium influx can lead to generation of nitric oxide and other free radicals which may cause lipid peroxidation (Choi, 1988). Large increases in intracellular calcium may also cause mitochondrial dysfunction, resulting in local free radical formation in mitochondria, an inability to handle free calcium, and decreased energy production. Calcium also stimulates the production of arachidonic acid metabolites. Finally, the elevation of intracellular calcium may result in the non-specific activation of many enzymes, including those involved in cellular signalling and proteolysis; this may cause aberrant control of signalling pathways and decreased integrity of cytoskeletal elements (Nicotera and Orrenius, 1992). In general, it is hypothesized that excitotoxic insults disrupt the normally tightly controlled system for the regulation of intracellular calcium and that this disruption causes the resultant neuronal damage.

In acute neurological injuries, such as hypoglycaemia or ischaemia, there is a large, rapid increase in extracellular glutamate as a result of increased release and decreased reuptake (Beneviste *et al.*, 1984; Silverstein *et al.*, 1986). This causes prolonged stimulation of glutamate receptors and neuronal death. In experimental models of hypoglycaemia and ischaemia, NMDA receptor antagonists exert profound neuroprotective effects and prevent cell death (Rothman and Olney, 1986). Thus, there is compelling evidence that glutamate plays a central role in the pathogenesis of acute brain injuries.

Evidence implicating excitotoxicity in neurodegenerative disease is more speculative. In AD, PD and HD, there is a relatively selective loss of NMDA receptors as compared to receptors for other neurotransmitters, suggesting that neurones expressing the NMDA receptor may be particularly vulnerable in these diseases (Greenamyre *et al.*, 1986, 1992; Young *et al.*, 1988; Landwehrmeyer *et al.*, 1993). However, excitotoxicity cannot explain why there are different anatomical patterns of pathology in each of the chronic diseases, and the distribution of NMDA receptors is not a reliable predictor of where pathology will occur. In addition, contrary to the situation of acute neurological insults, there is no convincing evidence for elevated extracellular levels in neurodegenerative diseases.

Clearly, in this form, the excitotoxic hypothesis is not adequate to explain the pathogenesis of chronic neurodegenerative disease. As detailed in this review, however, relatively mild defects in neuronal energy metabolism predispose neurones to NMDA receptor-mediated excitotoxicity, even in the absence of elevated extracellular glutamate. Since there is evidence for metabolic defects in neurodegenerative diseases, this concept of 'secondary' excitotoxicity – 'the weak excitotoxic hypothesis' – may provide a unifying pathogenic mechanism for a variety of neurodegenerative disorders.

8.2 Neuronal energy metabolism

Neuronal energy metabolism is almost completely dependent on oxidative metabolism of glucose. Glucose crosses the blood–brain barrier into neurones, where it is phosphorylated by hexokinase and enters glycolysis. Pyruvate, the end-product of glycolysis, diffuses into the mitochondrial matrix, where it is transformed into acetyl-CoA by the pyruvate dehydrogenase complex; acetyl-CoA enters the tricarboxylic acid (TCA) cycle by combination with oxaloacetate to form citrate (Figure 1). Metabolism of 1 mol of glucose through glycolysis and the TCA cycle results in a net gain of 2 mol each of ATP and NADH in the cytoplasm from glycolysis and 8 mol of NADH, 2 mol of $FADH_2$, and 2 mol of GTP in the mitochondrial matrix from the TCA cycle. The reducing equivalents of NADH in the cytosol are carried into the mitochondrial matrix by either the glycerol 3-phosphate shuttle, which is irreversible and results in 1 mol of mitochondrial $FADH_2$ from 1 mol of cytoplasmic NADH, or the malate–aspartate shuttle, which is reversible and results in 1 mol of mitochondrial NADH from 1 mol of cytoplasmic NADH. Because the malate–aspartate shuttle is reversible and requires no net energy expenditure, this shuttle only functions if the $NADH/NAD^+$ ratio in the cytoplasm is higher than that in the mitochondrial matrix.

The potential energy contained in NADH and $FADH_2$ is transformed into ATP via oxidative phosphorylation. Electrons from NADH and $FADH_2$ are carried through a series of transport enzymes in the inner mitochondrial membrane and donated to oxygen, the final electron acceptor. The electron transport chain consists of four major complexes. Complex I (NADH dehydrogenase; NADH:ubiquinone oxidoreductase) is the entry point for electrons from NADH. Electrons are passed to complex III (ubiquinol – cytochrome-*c* oxidoreductase) via ubiquinone. Ubiquinone is the entry point for electrons derived from $FADH_2$. Most $FADH_2$ is generated by the TCA cycle via complex II (succinate dehydrogenase; succinate:ubiquinone oxidoreductase), but some comes indirectly from glycolysis via the glycerol 3-phosphate

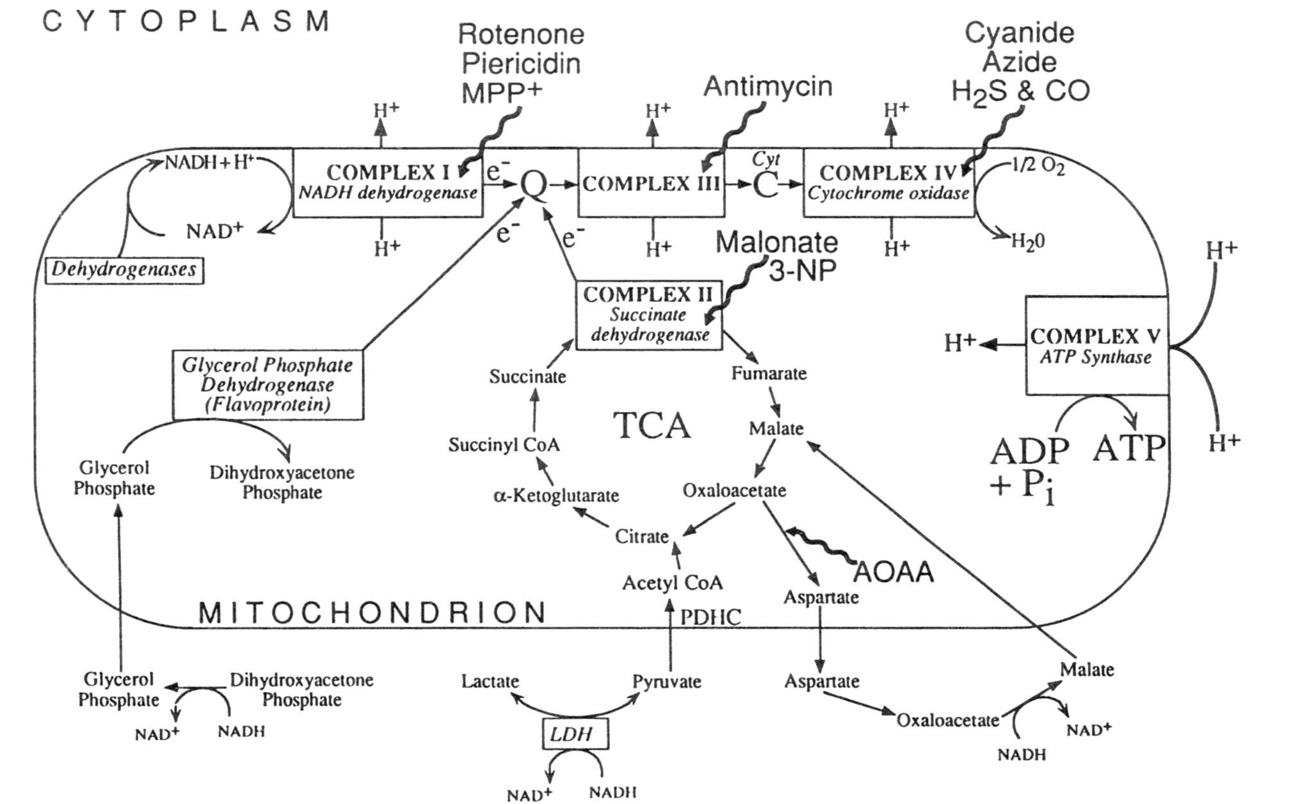

Figure 1 Schematic diagram of neuronal energy metabolism (*see text*).

shuttle. Complex III donates electrons via cytochrome *c* to complex IV (cytochrome oxidase), which donates electrons to molecular oxygen. Complexes I, III and IV pump protons from the inner mitochondrial matrix to the outer mitochondrial matrix; these protons flow back down their electrochemical gradient through ATP synthase (complex V), providing the energy to produce ATP. The flow of 2 mol of electrons through complexes I, III and IV provides enough energy for each enzyme complex to pump enough protons to make 1 mol of ATP. Therefore, electrons that enter at complex I (NADH) generate 3 mol of ATP, while electrons that enter at ubiquinone (from $FADH_2$) only generate 2 mol of ATP. Metabolism of 1 mol of glucose through oxidative phosphorylation generates 36 mol of ATP if the glycerol 3-phosphate shuttle is used, or 38 mol of ATP if the malate–aspartate shuttle is used. Under normal circumstances, glycolysis only accounts for about 5% of neuronal energy production, but it is necessary to provide precursors for the TCA cycle and oxidative phosphorylation. Under hypoxic conditions, NAD^+ may be regenerated via anaerobic mechanisms, such as lactate production, thereby increasing the energy production of anaerobic glycolysis. Regulation of neuronal energy metabolism is based on cellular energy charge, the need for anabolic precursors, and the supply of glucose and oxygen (Stryer, 1988; Pulsinelli and Cooper, 1989).

8.3 Metabolic impairment and neurodegenerative disease

8.3.1 Alzheimer's disease

Evidence of impaired energy metabolism has been consistently reported in patients with AD. Examination of biopsy and autopsy specimens from AD patients has revealed increased $^{14}CO_2$ production from [U-^{14}C]glucose, but cellular energy charge was normal. This suggests a partial uncoupling of energy metabolism, since increased carbon flux was required to maintain normal energy charge (Sims *et al.*, 1983). Decreases in the activities of the pyruvate dehydrogenase complex (Sorbi *et al.*, 1983; Sheu *et al.*, 1985) and the 2-ketoglutarate dehydrogenase complex (Gibson *et al.*, 1988; Mastrogiacomo *et al.*, 1993) have also been noted. These two enzymes play key roles in the regulation of the TCA cycle, suggesting that these defects may have severe consequences. Glycolytic enzymes may be affected, as well. Deficits have been reported in the activities of hexokinase (Iwangoff *et al.*, 1980; Sorbi *et al.*, 1990; Liguri *et al.*, 1990), transketolase (Gibson *et al.*, 1988) and aldolase (Iwangoff *et al.*, 1980). The prominent regulatory enzyme of glycolysis, phosphofructokinase (PFK), may also be impaired (Iwangoff *et al.*, 1980).

Deficits in the mitochondrial electron transport chain have also been demonstrated in AD brains. Point mutations of ND2 subunit of complex I of the electron transport chain are common in AD brains, but not in controls (Lin *et al.*, 1992). There may also be defective allosteric regulation of complex I in AD hippocampus (Higgins and Greenamyre, 1993). Deficient activity of complex IV has been reported in AD in frontal, temporal, and parietal cortex, but not in occipital cortex, a region relatively spared from AD pathology (Kish *et al.*, 1992). Parker *et al.* (1990b) demonstrated decreased complex IV activity in platelet mitochondria from patients with AD, but the existence of this systemic defect remains unproved (Van Zuylen *et al.*, 1992).

Positron emission tomography (PET) scans of AD patients demonstrate similar changes. PET scans using [^{18}F]2-deoxy-D-glucose demonstrate no differences correlating with increasing age, but there is a marked decrease in glucose metabolism in AD patients when compared to age-matched controls (Benson *et al.*, 1983; Duara *et al.*, 1984). In addition, lower glucose utilization is seen in AD patients before the onset of overt AD (Haxby *et al.*, 1986).

The metabolic defects reported in AD are varied and encompass the entire cellular complement of metabolic machinery. This might suggest that these changes are a result of AD, as opposed to a cause, but many of these defects have been noted in areas relatively free of pathology, and several markers of mitochondria (glutamate dehydrogenase, citrate synthase) are normal in AD (Sorbi *et al.*, 1983; Mastrogiacomo *et al.*, 1993). In addition, the cellular mechanisms of energy production are highly coordinated; a defect at a key regulatory point, or the deficiency of an important cofactor, may account for several enzyme activity changes. Finally, an interesting study performed by Blass *et al.* (1990) indicates that incubation of normal fibroblasts with an uncoupler of mitochondrial respiration, carbonyl cyanide *m*-chlorophenylhydrazone (CCCP), dramatically increases the proportion of cells expressing AD-related antigens. PHF-positive cells increased 10-fold, and ALZ-50-positive cells increased 157-fold. In addition, a report by Cheng and Mattson (1992) indicates that expression of neurofibrillary tangle associated antigens (ALZ-50, 5E2 and ubiquitin) is induced in hippocampal neurones by glucose deprivation. These reports suggest that mitochondrial impairment induces these antigens and that mitochondrial defects may play a role in the AD process.

8.3.2 Parkinson's disease

Mitochondrial dysfunction has also been strongly implicated in the pathogenesis of PD. In the early 1980s, several young adults presented with a parkinsonian syndrome caused by unintentional self-administration of 1-methyl-4-phenyl-1,2,5,6-tetrahydropyridine (MPTP), a toxic contaminant produced during the illicit synthesis of meperidine analogues (Langston *et al.*, 1983). The toxic effects of this compound have since been well described, and it has been shown that MPTP administration produces a fairly accurate clinical and neuropathological picture of Parkinson's disease in humans and non-human primates, including destruction of midbrain dopaminergic neurones (Kopin and Markey, 1988). MPTP freely crosses the blood–brain barrier; it is then metabolized by monoamine oxidase type B (presumably in glial cells) to its toxic metabolite, 1-methyl-4-phenylpyridinium (MPP^+) (Chiba *et al.*, 1984; Heikkila *et al.*, 1984; Langston *et al.*, 1984; Salach *et al.*, 1984). MPP^+ is actively taken up by the dopamine transporter and thereby concentrated in dopamine neurones, which accounts for its selective dopaminergic neurotoxicity (Javitch *et al.*, 1985). Within these neurones, MPP^+ is further concentrated in mitochondria by an energy-driven process, resulting in mitochondrial MPP^+ concentrations in the millimolar range (Ramsay *et al.*, 1986; Trevor *et al.* 1987).

The mechanism by which MPP^+ exerts its toxic effects appears to be inhibition of complex I of the mitochondrial electron transport chain (Figure 1). MPP^+ has been shown to inhibit mitochondrial respiration and complex I activity *in vitro*, possibly by interacting with the rotenone-sensitive binding site on that enzyme complex (Nicklas *et al.*, 1985; Ramsay *et al.*, 1991). Recent experiments from our laboratory indicate that MPP^+ displaces [^{3}H]dihydrorotenone binding to complex I in a competitive fashion, with an IC_{50} in the millimolar range (Greenamyre and Higgins, 1993). MPTP also inhibits mitochondrial respiration *in vivo* (Mizuno *et al.*, 1988). Furthermore, intoxication with MPTP in mice, or stereotaxic injection of MPP^+ in rats, has been shown to substantially reduce ATP levels (Chan *et al.*, 1991; Storey *et al.*, 1992).

The elucidation of MPTP as a mitochondrial toxin raised the possibility that a mitochondrial defect, particularly one involving complex I, may play a role in the pathogenesis of idiopathic PD. In fact, deficits in complex I activity have been demonstrated in the substantia nigra of PD patients (Schapira *et al.*, 1989, 1990a,b; Mann *et al.*, 1992). Abnormal immunostaining of complex I has also been seen in the nigra, but not other brain regions (Hattori *et al.*, 1991). Complex I defects have been reported in skeletal muscle (Bindoff *et al.*, 1991; Schoffner *et al.*, 1991) and platelets (Parker *et al.*, 1989; Krige *et al.*, 1992; Yoshino *et al.*, 1992) of PD patients, although the existence of these systemic deficits remains controversial (Mann *et al.*, 1992; Anderson *et al.*, 1993). The other complexes of the electron transport chain appear to be less affected by PD (Schapira *et al.*, 1989, 1990a,b; Hattori *et al.*, 1991; Mann *et al.*, 1992).

8.3.3 Huntington's disease

In HD, several defects in energy metabolism have been reported. PFK activity is decreased in HD caudate (Bird *et al.*, 1977). The activity of the pyruvate dehydrogenase complex is reduced in HD caudate and hippocampus, but not in other brain regions (Sorbi *et al.*, 1983; Butterworth *et al.*, 1985). Several deficits of the electron transport chain have been observed in HD. Parker and colleagues (1990a) have reported a deficiency of complex I activity in platelets from HD patients. In the caudate nucleus of HD brains, there is decreased complex II/III activity (Brennan *et al.*, 1985; Mann *et al.*, 1990). It has also been shown that mitochondrial respiration in HD caudate is reduced due to decreased complex IV activity and reduced cytochrome aa_3 levels; cytochromes b and cc_1 were normal in this study (Brennan *et al.*, 1985).

PET scans indicate decreased glucose metabolism in the basal ganglia and cerebral cortex; some studies have also noted differences between individuals at risk and controls (Kuhl *et al.*, 1982; Mazziota *et al.*, 1987; Kuwert *et al.*, 1990). Recently, *in vivo* nuclear magnetic resonance spectroscopy has demonstrated that lactate levels are elevated in the cerebral cortex in HD, further suggesting impaired oxidative metabolism in HD (Koroshetz *et al.*, 1992).

In addition to these post-mortem and clinical data, it has recently been demonstrated that several metabolic inhibitors, whether administered systemically or intrastriatally, mimic the behavioural and neuropathological results of HD (see Section 8.5).

A final interesting point about the involvement of metabolic defects in neurodegenerative diseases relates to the age-related onset of these diseases. The activities of several metabolic enzymes have been shown to decrease with increasing age (Trounce *et al.*, 1989; Bowling *et al.*, 1993). Furthermore, deletions of mitochondrial DNA, where several subunits of electron transport chain complexes are encoded, become more common with increasing age (Linnane *et al.*, 1990). It is possible that a small, but pathogenic, defect in energy production, acquired or inherited, causes no noticeable effect until it is superimposed upon the decline in metabolic capacity associated with normal ageing. This defect may then be of sufficient severity to cause significant metabolic impairment and expression of neurodegenerative disease.

8.4 Excitotoxicity and neurodegenerative disease

8.4.1 Alzheimer's disease

Although it is clear that excitotoxicity contributes to hypoxic/ischaemic brain damage, the exact role that the excitatory amino acids play in the pathogenesis of neurodegenerative diseases remains unclear. Many of the pathological findings of AD can be mimicked by the application of glutamate to neuronal cultures. First, incubation of human neuronal cultures with aspartate or glutamate can induce formation of paired helical filaments (PHFs) similar to those that comprise the neurofibrillary tangles characteristically found in AD brains (De Boni and Crapper-McLachlan, 1985). Second, application of glutamate in culture can also induce the expression of certain AD-related antigens, such as PHF, ALZ-50 and ubiquitin (Mattson, 1990; Cheng and Mattson, 1992). Third, disorders of RNA synthesis are well characterized in AD, and cognitive test scores correlate well with cytoplasmic RNA content in certain brain regions of AD patients (Neary *et al.*, 1986). In neuronal cultures, glutamate can reduce cytoplasmic RNA content, although the mechanism of this reduction is unknown (Dhindsa, 1983). Fourth, in AD, there is a progressive loss of neuronal dendrites and simplification of the dendritic tree (Mehraein *et al.*, 1975; Mann *et al.*, 1986); sublethal doses of glutamate cause an identical reaction in cultured neurones (Mattson, 1988; Mattson *et al.*, 1988). Finally, amyloid deposition in senile plaques is a pathological hallmark of AD, and this deposition has been hypothesized to play a pathogenic role in AD. It has been reported that amyloid greatly enhances the toxicity of glutamate in neuronal culture (Koh, *et al.* 1990; Mattson *et al.*, 1992). Recent experiments from our laboratory indicate that the toxicity of intracerebral NMDA injections is greatly enhanced in mice transgenic for the β-amyloid precursor protein compared to age-matched control animals (R.H.P. Porter *et al.*, unpublished observations). This finding suggests that amyloid may make neurones more susceptible to NMDA receptor-mediated excitotoxic injury in AD. In addition, the distribution of neurofibrillary tangles and senile plaques in AD is related to the distribution of glutamatergic corticocortical association pathways, in that the areas connected by these pathways are much more severely affected than those that are not (Pearson *et al.*, 1985).

8.4.2 Parkinson's disease

As discussed above, administration of MPTP inhibits complex I, selectively destroys nigrostriatal dopaminergic neurones, and provides an animal model of PD. Turski and colleagues (1991) reported that administration of NMDA antagonists prevents the neurotoxicity associated with stereotaxic intranigral injection of MPP^+ (Turski *et al.*, 1991); this finding has been confirmed in rat striatum (Storey *et al.*, 1992). It has also been demonstrated in a primate model that co-administration of NMDA antagonists with MPTP exerts a profound neuroprotective effect and prevents the onset of parkinsonism (Zuddas *et al.*, 1992). Although this finding remains controversial (Sonsalla *et al.*, 1992), it suggests that NMDA receptor-mediated excitotoxicity plays a significant role in neuronal death in this model and possibly in idiopathic PD.

8.4.3 Huntington's disease

HD neuropathology is characterized by selective loss of neurones in the striatum. γ-Aminobutyric acid (GABA)/substance P neurones are almost completely destroyed, but large, aspiny, acetylcholine neurones are spared. Also spared are somatostatin/neuropeptide Y neurones; this subset also stains positively for NADPH-diaphorase (NO synthase). Nerve terminals in the striatum are spared, as are the axons of passage through this region (for a review, see Ferrante *et al.*, 1993).

Excitotoxicity was first implicated in the pathogenesis of HD when it was discovered that injection of kainate into rat striatum produced a lesion that had several neurochemical and behavioural characteristics in common with HD (Coyle and Schwarcz, 1976; McGeer and McGeer, 1976). It has since been reported that injection of the NMDA agonist quinolinate provides a more accurate picture of HD pathology, indicating some role for glutamate and the NMDA receptor in HD (Beal *et al.*, 1986; Ferrante *et al.*, 1993).

8.5 The weak excitotoxic hypothesis

There is substantial evidence that both metabolic impairment and excitotoxicity are involved in the pathogenesis of neurodegenerative diseases. These two distinct mechanisms may be related by the unique properties of the NMDA receptor. At the normal resting potential of a neurone, the NMDA receptor calcium channel is blocked by

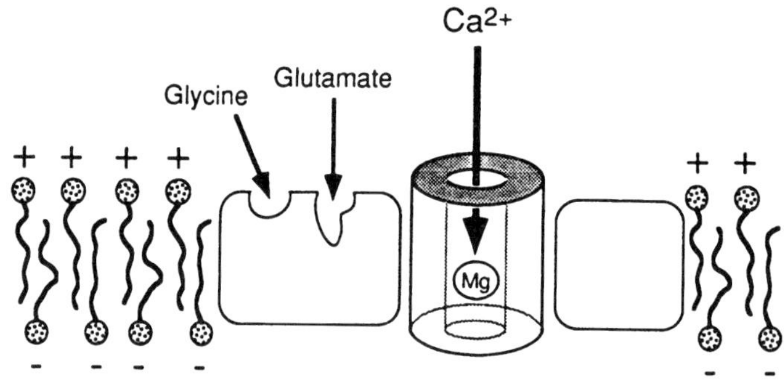

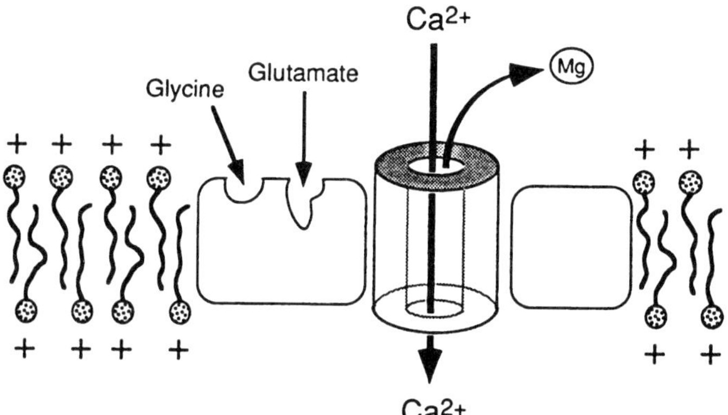

Figure 2 Schematic depicting the voltage-dependent nature of the NMDA receptor. At the resting membrane potential, Mg^{2+} prevents ion flux through the channel, even when glutamate and glycine are bound to the receptor complex. When the membrane is depolarized, the Mg^{2+} is extruded and Ca^{2+} can flow into the neurone.

extracellular magnesium. Under this condition it cannot transduce a signal, even if stimulated by agonist compounds. However, this 'gating' is voltage-dependent, and if the neurone becomes depolarized for any reason, the magnesium blockade is lifted, the receptor is activated, and calcium ions can flow through the channel (Figure 2; Nowak *et al.*, 1984).

The ability to maintain the membrane potential is dependent on adequate functioning of membrane ion pumps, particularly Na^+/K^+ ATPase. These pumps are, in turn, highly dependent on neuronal production of ATP to maintain their consistent high level of activity. Therefore, inadequate neuronal energy production impairs neuronal ability to maintain normal membrane potential. The inability to maintain membrane potential eases the magnesium blockade of the NMDA receptor and facilitates activation by glutamate; the resulting increase in intracellular calcium may lead eventually to neuronal death, as discussed previously. The first experimental evidence to support this scheme of neuronal death was the demonstration by Novelli and colleagues that inhibition of oxidative phosphorylation by cyanide or of the Na^+/K^+ ATPase by ouabain caused non-toxic concentrations of glutamate to become lethal *in vitro*; neuronal death was prevented by NMDA antagonists (Novelli *et al.*, 1988).

The proposed link between bioenergetics and NMDA receptor-mediated excitotoxicity permits several modifications of the excitotoxic hypothesis of neurodegenerative diseases. First, it indicates that glutamate levels need not be elevated to cause excitotoxic neuronal death. In the presence of normal, endogenous glutamate concentrations, the postsynaptic depolarization caused by a bioenergetic defect is sufficient to activate NMDA receptors. Second, the distribution of excitotoxic damage need not correlate with NMDA receptor density; the pattern of pathology may reflect some other defect, possibly a bioenergetic one. Finally, it can explain how excitotoxicity can be implicated in many diseases with different patterns of neuropathology. In this scheme, excitotoxicity is simply a common final pathway to neuronal death. This modification has been termed the 'weak excitotoxic hypothesis' by Albin and Greenamyre (1992).

Substantial evidence from experiments *in vitro* and *in vivo* supports the 'weak' excitotoxic hypothesis. In chick retina, hypoglycaemia and sodium cyanide (complex IV inhibitor) produced 'excitotoxic' lesions with no detectable increase in extracellular glutamate; these lesions were blocked by NMDA antagonists (Zeevalk and Nicklas, 1990). Zeevalk and Nicklas (1991) have also demonstrated that graded titration of membrane potential with K^+ mimics the effects of graded metabolic inhibition. Studies have also shown that relief of the Mg^{2+} blockade of the NMDA receptor ion channel resulted in excitotoxic damage similar to that caused by inhibition of metabolism (Zeevalk and Nicklas, 1992). In cultured hippocampal neurones, K^+ channel activators can prevent excitotoxic damage, presumably by inducing hyperpolarization of the neuronal membrane and maintaining the voltage-dependent blockade of the NMDA receptor (Abele and Miller, 1990). Also in cultured hippocamal neurones, chemical hypoxia (achieved by using sodium cyanide, sodium azide, oligomycin or dinitrophenol) greatly enhanced a glutamate-stimulated increase in intracellular calcium levels, presumably at least partly through the NMDA receptor ion channel. NMDA receptor antagonists diminished, but did not abolish, the increase in the calcium level (Dubinsky and Rothman, 1991).

The first *in vivo* evidence supporting the 'weak' excitotoxic hypothesis was generated from experiments involving amino-oxyacetic acid (AOAA), a non-specific inhibitor of transaminases. Two factors led to investigation of its neurotoxic properties. First, AOAA blocks *in vitro* kynurenate transaminase activity, leading to decreased concentrations of the endogenous glutamate antagonist kynurenic acid (Schwarcz *et al.*, 1990). Second, AOAA blocks the mitochondrial malate–aspartate shuttle, presumably by interfering with the transamination of oxaloacetate to aspartate (see Figure 1).

Intrastriatal injection of AOAA produces axon-sparing lesions that can be blocked by NMDA antagonists or prior decortication, which removes corticostriatal glutamatergic fibres (Beal *et al.*, 1991; Urbanska *et al.*, 1991). The neurotoxicity of AOAA is not due to direct NMDA receptor stimulation (Beal *et al.*, 1991; McMaster *et al.*, 1991). In addition, AOAA concentrations that caused marked striatal damage did not affect kynurenate transaminase activity, suggesting that the neurotoxicity of AOAA was not mediated by a decrease in kynurenic acid (Beal *et al.*, 1991). However, AOAA injection did result in decreased striatal ATP levels and increased striatal lactate formation, indicating that these concentrations of AOAA significantly impaired neuronal energy metabolism (Beal *et al.*, 1991). Finally, the lesion produced by intrastriatal injection of AOAA is nearly identical to that produced by quinolinic acid, an NMDA agonist (Beal *et al.*, 1991). As discussed above, this lesion is neurochemically similar to the pathological lesions of HD (Beal *et al.*, 1986; Ferrante *et al.*, 1993). These experiments suggest that AOAA neurotoxicity is mediated indirectly by the NMDA receptor by impairing neuronal energy metabolism and relieving the voltage-dependent Mg^{2+} blockade of the NMDA receptor.

This conclusion is further supported by the finding that the neurotoxicity of MPP^+ is mediated by the NMDA receptor. The dopaminergic neurotoxicity produced by intranigral injection of MPP^+ is attenuated by administration of NMDA antagonists (Turski *et al.*, 1991). NMDA antagonists are also protective following MPTP intoxication in primates (Zuddas *et al.*, 1992). The mechanism of MPP^+ excitotoxicity is apparently indirect NMDA receptor activation via inhibition of complex I. Intrastriatal injection of MPP^+ results in decreased striatal ATP levels and increased striatal lactate. These striatal lesions are prevented by prior decortication or treatment with NMDA antagonists (Storey *et al.*, 1992). NMDA antagonists also prevent the subsequent degeneration of dopaminergic neurones in the substantia nigra following intrastriatal injection of MPP^+ (Srivastava *et al.*, 1993). These findings implicate 'weak' excitotoxicity not only in the MPTP model of parkinsonism, but also in the pathogenesis of idiopathic PD.

Manganese ingestion by humans induces a parkinsonian syndrome and accompanying lesions in the basal ganglia (Bernheimer *et al.*, 1973; Yamada *et al.*, 1986; Barbeau, 1984). It has recently been demonstrated by Beal and colleagues that intrastriatal injection of manganese results in a marked depletion of striatal ATP and striatal lesions that are blocked by prior decortication or administration of MK-801, a non-competitive NMDA antagonist (Brouillet *et al.*, 1993b).

3-Nitropropionic acid (3-NP) is an irreversible, suicide inhibitor of succinate dehydrogenase (SDH; see Figure 1) (Alston, 1977; Coles *et al.*, 1979). When administered systemically, it produces selective striatal damage (Hamilton and Gould, 1987). Chronic administration of low doses of 3-NP produces a neurochemical and behavioural syndrome very similar to that seen in HD. It results in large decreases in striatal GABA and substance P, but spares large acetylcholine neurones and neurones positive for NADPH-diaphorase. It also spares dopaminergic input and produces morphological neuritic changes similar to those seen in HD. These lesions were prevented by prior decortication to remove glutamatergic input, and no increase in striatal glutamate levels was detected following 3-NP administration (Beal *et al.*, 1993a).

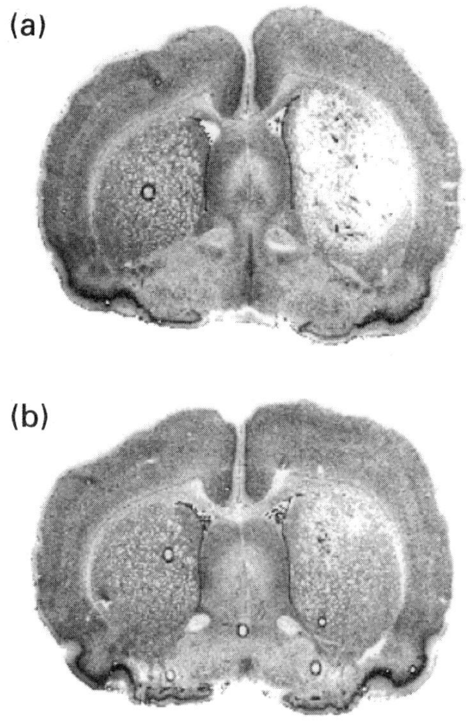

Figure 3 Digitized representative images of fresh-frozen, Nissl-stained sections from (a) one control and (b) one MK-801-treated rat injected intrastriatally with 2 μmol of malonate. Note the marked protective effect of MK-801 administration.

Both systemic and intrastriatal administration of 3-NP result in metabolic impairment, as evidenced by decreased striatal ATP levels and increased striatal lactate concentrations. It is also of interest that the lesions produced by 3-NP are age-dependent, in that young animals are relatively resistant to its neurotoxic effects (Brouillet *et al.*, 1993a).

Malonic acid is a reversible, competitive inhibitor of SDH (see Figure 1). Its intrastriatal injection is accompanied by marked neurotoxicity that is dependent on NMDA receptor activation (Figure 3; Beal *et al.*, 1993b; Greene *et al.*, 1993; Greene and Greenamyre, 1995a). The appearance of this lesion is preceded by decreases in striatal ATP levels and increases in striatal lactate (Beal *et al.*, 1993b). Co-injection of excess succinate with malonic acid abolishes its neurotoxicity, indicating that this toxicity is caused specifically by the inhibition of SDH. Like the toxicity produced by the irreversible SDH inhibitor 3-NP, malonic acid toxicity is age-dependent. This is intriguing, in light of the adult onset of neurodegenerative diseases in general, and HD in particular (Beal *et al.*, 1993b).

The above evidence suggests that inhibition of metabolism leads to neuronal death

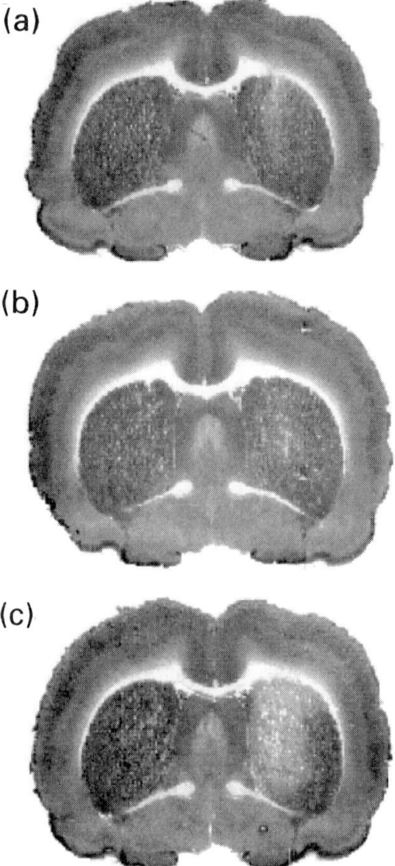

Figure 4 Digitized representative images of fresh-frozen, cytochrome oxidase-stained sections from rats injected intrastriatally with (a) 0.6 μmol of glutamate, (b) 0.6 μmol of malonate, and (c) 0.6 μmol of glutamate + 0.6 μmol of malonate. The combination of glutamate and malonate produces synergistic toxicity.

via a indirect excitotoxic mechanism. Metabolic impairment may lead to NMDA receptor activation without any increase in extracellular glutamate by relieving the voltage-dependent Mg^{2+} blockade of the associated ion channel. Recent experiments also indicate that mild metabolic inhibition combined with a modest increase in extracellular glutamate producc a synergistic toxicity. When relatively non-toxic doses of malonate and glutamate are co-administered, there is a massive potentiation of neuronal damage (Figure 4; Green and Greenamyre, 1995b).

8.6 Conclusions

The causes and mechanisms of pathogenesis of the neurodegenerative diseases remain obscure. It is possible that bioenergetics and excitotoxicity are intimately related and that this relationship plays a key role in the selective neurodegeneration that characterizes these disorders. Although the primary defect in any of these disorders may not be glutamatergic in nature, the glutamate system may be involved in a secondary manner. Metabolic defects may lead to slow depolarization, easing the voltage-dependent blockade of the NMDA receptor and facilitating its activation by endogenous glutamate. In this way, the neuropathology caused by a variety of defects in neuronal energy metabolism (or in other systems important for maintenance of membrane potential, e.g. ion pumps) may ultimately be mediated by excessive activation of the NMDA receptor, and excitotoxicity may be a final common pathway to neuronal death. If 'weak' excitotoxicity is an important pathogenic mechanism of neurodegenerative disease, NMDA antagonists and other anti-excitotoxic strategies potentially have broad therapeutic applications for these diseases.

Acknowledgements

Preparation of this manuscript was supported by PHS grants NS01487 (JTG) & GM08427 (JGG), the Hope A. Geoghegan Fund, the Gustavus and Louise Pfeiffer Foundation, and an unrestricted grant-in-aid from Fisons Pharmaceuticals, Rochester, NY.

References

Abele, A.E. & Miller, R.J. (1990) *Neurosci. Lett.* **115**, 195–200.
Albin, R.L. & Greenamyre, J.T. (1992) *Neurology* **42**, 733–738.
Alston, T.A. (1977) *Proc. Natl Acad. Sci. USA* **74**, 3767–3771.
Anderson, J.J., Bravi, D., Ferrari, R., Davis, T.L., Baronti, F., Chase, T.N. & Dagani, F. (1993) *J. Neurol. Neurosurg. Psychiatry* **56**, 477–480.
Barbeau, A. (1984) *Neurotoxicology* **5**, 13–36.
Beal, M.F. (1992) *Ann. Neurol.* **31**, 119–130.
Beal, M.F., Kowall, N.W., Ellison, D.W., Mazurek, M.F., Swartz, K.J. & Martin, J.B. (1986) *Nature* **321**, 168–172.
Beal, M.F., Swartz, K.J., Hyman, B.T., Storey, E., Finn, S.F. & Koroshetz, W. (1991) *J. Neurochem.* **57**, 1068–1073.
Beal, M.F., Brouillet, E., Jenkins, B.G., Ferrante, R.J., Kowall, N.W., Miller, J.M., Storey, E., Srivastava, R., Rosen, B.R. & Hyman, B.T. (1993a) *J. Neurosci.* **13**, 4181–4192.
Beal, M.F., Brouillet, E., Jenkins, B., Henshaw, R., Rosen, B. & Hyman, B.T. (1993b) *J. Neurochem.* **61**, 1147–1150.
Beal, M.F., Hyman, B.T. & Koroshetz, W. (1993c) *TINS* **16**, 125–131.
Benveniste, H., Drejer, J., Schousboe, A. & Diemer, N.H. (1984) *J. Neurochem.* **43**, 1369–1374.

Bennett, D.A. & Evans, D.A. (1992) *Dis. Mon.* **38**, 1–64.
Benson, D.F., Kuhl, D.E., Hawkins, R.A., Phelps, M.E., Cummings, J.L. & Tsai, S.Y. (1983) *Arch. Neurol.* **40**, 711–714.
Bernheimer, H., Birkmayer, K., Hornykiewicz, K., Jellinger, K. & Seitelberger, F. (1973) *J. Neurol. Sci.* **20**, 415–455.
Bindoff, L.A., Birch-Machin, M.A., Cartlidge, N.E., Parker, W.D. Jr. & Turnbull, D.M. (1991) *J. Neurol. Sci.* **104**, 203–208.
Bird, E.D., Gale, J.S. & Spokes, E.G. (1977) *J. Neurochem.* **29**, 539–545.
Blass, J.P., Baker, A.C., Ko, L. & Black, R.S. (1990) *Arch. Neurol.* **47**, 864–869.
Bowling, A.C., Mutisya, E.M., Walker, L.C., Price, D.L., Cork, L.C. & Beal, M.F. (1993) *J. Neurochem.* **60**, 1964–1967.
Brennan, W.A., Bird, E.D. & Aprille, J.R. (1985) *Neurochem.* **44**, 1948–1950.
Brouillet, E., Jenkins, B.G., Hyman, B.T., Ferrante, R., Kowal, N.W., Srivastava, R., Roy, D.S., Rosen, B.R. & Beal, M.F. (1993a) *J. Neurochem.* **60**, 356–359.
Brouillet, E.P., Shinobu, L., McGarvey, U., Hochberg, F. & Beal, M.F. (1993b) *Exp. Neurol.* **120**, 89–94.
Butterworth, J., Yates, C.M. & Reynolds, G.P. (1985) *J. Neurol. Sci.* **67**, 161–171.
Chan, P., DeLanney, L.E., Irwin, I., Langston, J.W. & Di Monte, D. (1991) *J. Neurochem.* **57**, 348–351.
Cheng, B. & Mattson, M.P. (1992) *Exp. Neurol.* **117**, 114–123.
Chiba, K., Trevor, A. & Castagnoli, N. Jr (1984) *Biochem. Biophys. Res. Commun.* **120**, 574–578.
Choi, D.W. (1988) *Neuron* **1**, 623–634.
Coles, C.J., Edmondson, D.E. & Singer, T.P. (1979) *J. Biol. Chem.* **254**, 5161–5167.
Coyle, J.T. & Schwarcz, R. (1976) *Nature* **263**, 244–246.
De Boni, U. & Crapper-McLachlan, D.R. (1985) *J. Neurol. Sci.* **68**, 105–118.
Dhindsa, K.S. (1983) *Acta Anat.* **116**, 201–205.
Dingledine, R., Boland, R.M., Chamberlin, N.L., Kawasaki, K., Kleckner, N.W., Traynelis, S.F. & Verdoon, T.A. (1988) *Crit. Rev. Neurobiol.* **4**, 1–96.
Duara, R., Grady, C., Haxby, J., Ingvar, D., Sokoloff, L., Margolin, R.A., Manning, R.G., Cutler, N.R. & Rapoport, S.I. (1984) *Ann. Neurol.* **16**, 702–713.
Dubinsky, J.M. & Rothman, S.M. (1991) *J. Neurosci.* **11**, 2545–2551.
Ferrante, R.J., Kowall, N.W., Cipolloni, P.B., Storey, E. & Beal, M.F. (1993) *Exp. Neurol.* **119**, 46–71.
Gibson, G.E., Sheu, K.-F.R., Blass, J.P., Baker, A., Carlson, K.C., Harding, B. & Perrino, P. (1988) *Arch. Neurol.* **45**, 836–840.
Greenamyre, J.T. & Higgins, D.S. (1993) *Soc. Neurosci. Abstr.*, 407.
Greenamyre, J.T. & Young, A.B. (1989) *Neurobiol. Aging* **10**, 593–602.
Greenamyre, J.T., Penney, J.B., Young, A.B., D'Amato, C.J., Hicks, S.P. & Shoulson, I. (1986) *Science* **227**, 1496–1499.
Greenamyre, J.T., Young, A.B. & Penney, J.B. Jr (1992) In *New Leads and Targets in Drug Research, Alfred Benzon Symposium 33* (eds Krogsgaard-Larsen, P., Christensen, S.B. & Kofod, H.), pp. 304–314. Munksgaard, Copenhagen.
Greene, J.G., Porter, R.H.P., Eller, R.V. & Greenamyre, J.T. (1993) *J. Neurochem.* **61**, 1151–1154.
Greene, J.G. & Greenamyre, J.T. (1995a) *J. Neurochem.* **64**, 430–436.
Greene, J.G. & Greenamyre, J.T. (1995b) *J. Neurochem.* **64**, 2332–2338.
Hamilton, B.F. & Gould. D.H. (1987) *Acta Neuropathol. (Berlin)* **72**, 286–297.
Hattori, N., Tanaka, M., Ozawa, T. & Mizuno, Y. (1991) *Ann. Neurol.* **30**, 563–571.
Haxby, J.V., Grady, C.L., Duara, R., Schlageter, N., Berg, G. & Rapoport, S.I. (1986) *Arch. Neurol.* **43**, 882–885.
Heikkila, R.E., Manzino, L., Cabbat, F.S. & Duvoisin, R.C. (1984) *Nature* **311**, 467–469.
Higgins, D.S. & Greenamyre, J.T. (1993) *Soc. Neurosci. Abstr.* 1044.
Iwangoff, P., Armbruster, R., Enz, A. & Meier-Ruge, W. (1980) *Mech. Ageing Dev.* **14**, 203–209.
Javitch, J.A., D'Amato, R.J., Strittmatter, S.M. & Snyder, S.H. (1985) *Proc. Natl Acad. Sci. USA* **82**, 2173–2177.

Kish, S.J., Bergeron, C., Rajput, A., Dozic, S., Mastrogiacomo, F., Chang, L.J., Wilson, J.M., DiStefano, L.M. & Nobrega, J.N. (1992) *J. Neurochem.* **59**, 776–779.
Koh, J.-Y., Yang, L.L. & Cotman, C.W. (1990) *Brain Res.* **533**, 315–320.
Kopin, I.J. & Markey, S.P. (1988) *Ann. Rev. Neurosci.* **11**, 81–96.
Koroshetz, W.J., Jenkins, B.G., Beal, M.F. & Rosen, B.R. (1992) *Neurology* **42**, 319.
Krige, D., Carroll, M.T., Cooper, J.M., Marsden, C.D. & Schapira, A.H.V. (1992) *Ann. Neurol.* **32**, 782–788.
Kuhl, D.E., Phelps, M.E., Markham, C.H., Metter, J., Riege, W.H. & Winter, J. (1982) *Ann. Neurol.* **12**, 425–434.
Kuwert, T., Lange, H.W., Langer, K.-J., Herzog, H., Aulich, A. & Feinendegen, L.E. (1990) *Brain* **113**, 1405–1423.
Landwehrmeyer, G.B., Standaert, D.G., Testa, C.B., Wüllner, U., Catania, M.V., Young, A.B. & Penney, J.B. (1993) *Soc. Neurosci. Abstr.*, 129.
Langston, J.W., Ballard, P.A., Tetrud, J.W. & Irwin, I. (1983) *Science* **219**, 979–980.
Langston, J.W., Irwin, I., Langston, E.B. & Forno, L.S. (1984) *Science* **225**, 1480–1482.
Liguri, G., Taddei, N., Nassi, P., Latorraca, S., Nediani, C. & Sorbi, S. (1990) *Neurosci. Lett.* **112**, 338–342.
Lin, F.H., Lin, R., Wisniewski, H.M., Hwang, Y.W., Grundke-Iqbal, I., Healy-Louie, G. & Iqbal, K. (1992) *Biochem. Biophys. Res. Commun.* **182**, 238–246.
Linnane, A.W., Baumer, A., Maxwell, R.J., Preston, H., Zhang, C. & Marzuki, S. (1990) *Biochem. Int.* **22**, 1067–1076.
McDowell, F.H. & Cedarbaum, J.M. (1992) In *Clinical Neurology* (ed. Joynt, R.J.), Vol. 3, Chap. 38. Lippincott, Philadelphia.
McGeer, E.G. & McGeer, P.L. (1976) *Nature* **263**, 517–519.
McMaster, O.G., Du, F., French, E.D. & Schwarcz, R. (1991) *Exp. Neurol.* **113**, 378–385.
Mann, D.M.A., Yates, P.O. & Marcyniuk, B. (1986) *J. Neurol. Neurosurg. Psychiatry* **49**, 310–312.
Mann, V.M., Cooper, J.M., Javoy-Agid, F., Agid, Y., Jenner, P. & Schapira, A.H.R. (1990) *Lancet* **336**, 749.
Mann, V.M., Cooper, J.M., Krige, D., Daniel, S.E., Schapira, A.H. & Marsden, C.D. (1992) *Brain* **115**, 333–342.
Mattson, M.P. (1988) *Brain Res. Rev.* **13**, 179–212.
Mattson, M.P. (1990) *Neuron* **4**, 105–117.
Mattson, M.P., Dou, P. & Kater, S.B. (1988) *J. Neurosci.* **8**, 2087–2100.
Mattson, M.P., Cheng, B., Davis, D., Bryant, K., Lieberburg, I. & Rydel, R.E. (1992) *J. Neurosci.* **12**, 376–389.
Mastrogiacomo, F., Bergeron, C.& Kish, J.J. (1993) *J. Neurochem.* **61**, 2007–2014.
Mazziotta, J.C., Phelps, M.E., Pahl, J.I., Huang, S.-C., Baxter, L.R., Riege, W.H., Hoffman, J.M., Kuhl, D.E., Lento, A.B., Wapenski, J.A. & Markham, C.H. (1987) *N. Engl. J. Med.* **316**, 356–362.
Mehraein, P., Yamada, M. & Tarnowska-Dziduszko, E. (1975) *Adv. Neurol.* **12**, 453–458.
Mizuno, Y., Suzuki, K., Sone, N. & Saitoh, T. (1988) *Neurosci. Lett.* **91**, 349–353.
Neary, D., Snowden, J.S., Mann, D.M.A., Bowen, D.M., Sims, N.R., Northen, B., Yates, P.O. & Davison, A.N. (1986) *J. Neurol. Neurosurg. Psychiatry* **49**, 229–237.
Nicklas, W.J., Vyas, I. & Heikkila, R.E. (1985) *Life Sci.* **36**, 2503–2508.
Nicotera, P. & Orrenius, S. (1992) *Ann. NY Acad. Sci.* **648**, 17–27.
Novelli, A., Reilly, J.A., Lysko, P.G. & Henneberry, R.C. (1988) *Brain Res.* **451**, 205–212.
Nowak, L., Bregestovski, P., Ascher, P., Herbet, A. & Prochiantz, A. (1984) *Nature* **307**, 462–465.
Olney, J.W., Ho, O.L. & Rhee, V. (1971) *Exp. Brain Res.* **14**, 61–76.
Parker, W.D. Jr, Boyson, S.J. & Parks, J.K. (1989) *Ann. Neurol.* **26**, 719–723.
Parker, W.D. Jr, Boyson, S.J., Luder, A.S. & Parks, J.K. (1990a) *Neurology* **40**, 1231–1234.
Parker, W.D. Jr, Filley, C.M. & Parks, J.K. (1990b) *Neurology* **40**, 1302–1303.
Pearson, R.C., Esiri, M.M., Hiorns, R.W., Wilcock, G.K. & Powell, T.P.S. (1985) *Proc. Natl Acad. Sci. USA* **82**, 4531–4534.

Pulsinelli, W.A. & Cooper, A.J.L. (1989) In *Basic Neurochemistry* (eds Siegel, G., Agranoff, B., Albers, R.W. & Molinoff, P.), pp. 765–781. Raven Press, New York.
Ramsay, R.R., Dadgar, J., Trevor, A. & Singer, T.P. (1986) *Life Sci.* **39**, 581–588.
Ramsay, R.R., Krueger, M.J., Youngster, S.K., Gluck, M.R., Casida, J.E. & Singer, T.P. (1991) *J. Neurochem.* **56**, 1184–1190.
Rothman, S.M. (1992) *Ann. NY Acad. Sci.* **648**, 132–139.
Rothman, S.M. & Olney, J.W. (1986) *Ann. Neurol.* **19**, 105–111.
Sacaan, A.L. & Schoepp, D.D. (1992) *Neurosci. Lett.* **139**, 77–82.
Salach, J.I., Singer, T.P., Castagnoli, N. Jr, & Trevor, A. (1984) *Biochem. Biophys. Res. Commun.* **125**, 831–835.
Schapira, A.H.V., Cooper, J.M., Dexter, D., Jenner, P., Clark, J.B. & Marsden, C.D. (1989) *Lancet* **i**, 1269.
Schapira, A.H.V., Copper, J.M., Dexter, D., Clark, J.B., Jenner, P. & Marsden, C.D. (1990a) *J. Neurochem.* **54**, 823–827.
Schapira, A.H.V., Mann, V.M., Cooper, J.M., Dexter, D., Daniel, S.E., Jenner, P., Clark, J.B. & Marsden, C.D. (1990b) *J. Neurochem.* **55**, 2142–2145.
Schwarcz, R., Speciale, C. & Turski, W.A. (1990) In *Parkinsonism and Aging* (eds D.B. Calne, G. Comi, D. Crippa, R. Horowski & M. Trabucchi), pp. 97–105. Raven Press, New York.
Sheu, K.F., Kim, Y.T., Blass, J.P. & Weksler, M.E. (1985) *Ann. Neurol.* **17**, 444–449.
Shoffner, J.M., Watts, R.L., Juncos, J.L., Torroni, A. & Wallace, D.C. (1991) *Ann. Neurol.* **30**, 332–339.
Silverstein, F.S., Buchanan, M. & Johnston, M.V. (1986) *J. Neurochem.* **47**, 1614–1619.
Sims, N.R., Bowen, D.M., Neary, D. & Davison, A.N. (1983) *J. Neurochem.* **41**, 1329–1334.
Sonsalla, P.K., Zeevalk, G.D., Manzino, L., Giovanni, A. & Nicklas, W.J. (1992) *J. Neurochem.* **58**, 1979–1982.
Sorbi, S., Bird, E.D. & Blass, J.P. (1983) *Ann. Neurol.* **13**, 72–78.
Sorbi, S., Mortilla, M., Piacentini, S., Tonini, S. & Amaducci, L. (1990) *Neurosci. Lett.* **117**, 165–168.
Srivastava, R., Brouillet, E., Beal, M.F., Storey, E. & Hyman, B.T. (1993) *Neurobiol. Aging* **14**, 295–301.
Storey, E., Hyman, B.T., Jenkins, B., Brouillet, E., Miller, J.M., Rosen, B.R. & Beal, M.F. (1992) *J. Neurochem.* **58**, 1975–1978.
Stryer, L. (1988). In *Biochemistry*, pp. 349–426. W.H. Freeman, New York.
Trevor, A.J., Castagnoli, N. Jr., Caldera, P., Ramsay, R.R. & Singer, T.P. (1987) *Life Sci.* **40**, 713–719.
Trounce, I., Byrne, E. & Marzuki, S. (1989) *Lancet* **i**, 637–639.
Turski, L., Bressler, K., Rettig, K.J., Loschmann, P.A. & Wachtel, H. (1991) *Nature* **349**, 414–418.
Urbanska, E., Ikonomidou, C., Sieklucka, M. & Turski, W.A. (1991) *Synapse* **9**, 129–135.
Van Zuylen, A.J., Bosman, G.J., Ruitenbeek, W., Van Kalmthout, P.J. & De Grip, W.J. (1992) *Neurology* **42**, 1246–1247.
Watkins, J.C. & Evans, R.H. (1981) *Annu. Rev. Pharmacol. Toxicol.* **21**, 165–204.
Yamada, M., Ohno, S., Okayasu, I., Okeda, R., Hatakeyama, S., Watanabe, H., Ushio, K. & Tsukagoshi, H. (1986) *Acta Neuropathol.* **70**, 273–278.
Yoshino, H., Nakagawa-Hattori, Y., Kondo, T. & Uno, Y. (1992) *J. Neural Transm.* **4**, 27–34.
Young, A.B., Greenamyre, J.T., Hollingsworth, Z., Albin, R., D'Amato, C., Shoulson, I. & Penney, J.B. (1988) *Science* **241**, 981–983.
Zeevalk, G.D. & Nicklas, W.J. (1990) *J. Pharmacol. Exp. Ther.* **253**, 1285–1292.
Zeevalk, G.D. & Nicklas, W.J. (1991) *J. Pharmacol. Exp. Ther.* **257**, 870–878.
Zeevalk, G.D. & Nicklas, W.J. (1992) *J. Neurochem.* **59**, 1211–1220.
Zuddas, A., Oberto, G., Vaglini, F., Fascetti, F., Fornai, F. & Corsini, G.U. (1992) *J. Neurochem.* **59**, 733–739.

CHAPTER 9

CALCIUM IONS IN NECROTIC AND APOPTOTIC CELL DEATH

Pierluigi Nicotera and Sten Orrenius

Institute of Environmental Medicine, Division of Toxicology, Karolinska Institutet, Box 210, S-171 77 Stockholm, Sweden

Table of Contents

9.1 Introduction

The role of the calcium ion as intracellular regulator of many physiological processes is now well established. Thus, transient and often oscillatory increases in the level of cytosolic Ca^{2+} mediate the effects of a variety of hormones and growth factors (for reviews, see Carafoli, 1989; Berridge, 1991). Ca^{2+} release from intracellular stores followed by extracellular Ca^{2+} entry result in such rapid intracellular Ca^{2+} transients. While Ca^{2+} can directly promote modifications of structural proteins and enzymes, most of the regulatory effects of this ion are mediated by Ca^{2+}-binding proteins (e.g. calmodulin) and are achieved by alterations of the phosphorylation state of target proteins. It is therefore apparent that alterations in Ca^{2+} signalling can also affect cell functions and play a determinant role in a variety of pathological and toxicological processes.

Neurodegeneration and Neuroprotection
in Parkinson's Disease
ISBN 0-12-525445-8

The early observations that Ca^{2+} accumulates in necrotic tissue and that a disruption of intracellular Ca^{2+} homeostasis was associated with the early development of cell injury (Schanne *et al.*, 1979; Jewell *et al.*, 1982; Fleckenstein *et al.*, 1983) led to the formulation of the calcium hypothesis of cell injury, proposing that perturbation of intracellular Ca^{2+} homeostasis may be a common step in the development of cytotoxicity. Support for this hypothesis has come from a large number of studies demonstrating that the calcium ion plays a critical role in cytotoxicity and cell killing in many tissues, including the central nervous system and the immune system (for a review, see Nicotera *et al.*, 1992).

9.2 Intracellular Ca^{2+} homeostasis and signalling

Studies using selective indicators have shown that the Ca^{2+} concentration in the cytosol of unstimulated cells is maintained between 0.05 and 0.2 μM (Carafoli, 1989). Extracellular Ca^{2+} levels are approximately four orders of magnitude higher (1.3 mM). This produces a large, inwardly directed, electrochemical driving force that is primarily balanced by active Ca^{2+} extrusion through the plasma membrane and by the coordinated activity of Ca^{2+}-sequestering systems located in the mitochondrial, endoplasmic reticular and nuclear membranes. In the excitable tissues, different types of voltage- and receptor-operated Ca^{2+} channels are involved in Ca^{2+} entry during neurotransmitter or hormone stimulation. Although isolated mitochondria can accumulate large amounts of Ca^{2+}, the affinity of the uniport carrier for Ca^{2+} uptake is low, and the mitochondria appear to play a minor role in buffering cytosolic Ca^{2+} under normal conditions. Electron probe X-ray microanalysis of rapidly frozen liver sections has shown that mitochondria contain little Ca^{2+} *in situ* (about 1 nmol Ca^{2+} per mg protein), whereas the endoplasmic reticulum represents the major intracellular Ca^{2+} store (Somlyo *et al.*, 1985). However, it should be noted that several physiological constituents (e.g. polyamines) can potentially increase the affinity of the uniport carrier for Ca^{2+}, at least to the level that is reached in the cytosol during agonist stimulation (about 400 nM) (Rottenberg and Marbach, 1990). Thus, the role of mitochondria in the modulation of the agonist-stimulated Ca^{2+} transients deserves further consideration.

Studies in our laboratory have shown that liver nuclei possess an ATP-stimulated Ca^{2+} uptake system responsible for intranuclear Ca^{2+} accumulation (Nicotera *et al.*, 1989), and that Ca^{2+} can be released from a nuclear compartment in response to intracellular messengers (Nicotera *et al.*, 1990a,b), suggesting the possibility that the nucleus may have self-regulating mechanisms to control its Ca^{2+} level and to modulate intranuclear Ca^{2+} responses to hormones and growth factors. Others have obtained similar results, and several studies in individual cells have shown that nuclear Ca^{2+} levels can be selectively regulated during cell activation (for a review, see Nicotera and Rossi, 1994).

The mechanisms whereby Ca^{2+}-mobilizing hormones elicit Ca^{2+} transients have been extensively studied in recent years (Berridge, 1987). The signal transduction pathway leading to an elevation of cytosolic Ca^{2+} can be summarized as follows: upon binding of the agonist to its plasma membrane receptor, a specific phospholipase C is activated via stimulation of a G protein, resulting in the hydrolysis of phosphatidylinositol 4,5-bisphosphate and the generation of two second messengers, inositol 1,4,5-trisphosphate (Ins(1,4,5)P_3) and diacylglycerol. Diacyglycerol is a potent activator of protein kinase C, whereas Ins(1,4,5)P_3 binds to specific receptors and stimulates Ca^{2+} release from non-mitochondrial intracellular stores.

In addition to mobilizing Ca^{2+} from intracellular stores, hormones can stimulate Ca^{2+} influx from the extracellular compartment through specific receptor-operated Ca^{2+} channels (Barritt *et al.*, 1981). With the development of techniques to study Ca^{2+} changes in single cells, it has become possible to study the spatial and temporal distribution of transients within the cell. This led to the observation that at low, close-to-threshold concentrations of Ca^{2+}-mobilizing hormones, many cells respond to these agents with relatively rapid, oscillating spikes (Woods *et al.*, 1986). It has been suggested that such oscillatory patterns may carry a frequency-encoded message (Berridge, 1991). Furthermore, slow propagating Ca^{2+} waves have been detected after agonist stimulation in some cell types, suggesting that Ca^{2+} fluctuations are not uniform in time and spatial distribution. However, the possible implications of these phenomena to cell regulation have not been identified.

9.3 Interference with cell signalling

Xenobiotics can interfere with signal transduction at different levels with a resulting loss of normal Ca^{2+} responses to hormones and growth factors. Chemical, bacterial and viral toxins can interact with receptors, G proteins and other enzymes involved in cell signalling, or can directly affect intracellular Ca^{2+} homeostasis by interfering with Ca^{2+} pumps or Ca^{2+} channels. It has long been recognized that bacterial toxins can inhibit the generation of intracellular messengers linked to the phosphoinositide pathway and consequently cause a loss of cellular responses to hormone and growth factor stimulation. Toxic chemicals can also inhibit the generation of inositol polyphosphates (Bellomo *et al.*, 1987), whereas oxidative stress may result in either stimulation or inhibition of protein kinase C (Kass *et al.*, 1989). Although the implications of these effects for cell survival have yet to be established, it is becoming increasingly apparent that the inhibition of hormonal responses may result not only in the loss of a trophic stimulus, but also in the activation of a programme for cell self-deletion in some instances (Kyprianou *et al.*, 1990).

Other environmental toxins that interfere with cell signalling include various metal ions (Nicotera *et al.*, 1992). Several metals have been shown to interfere with intracellular Ca^{2+} transport systems and with Ca^{2+} channels and to compete with Ca^{2+} for Ca^{2+}-binding proteins, including calmodulin. The influx of Ca^{2+} through these channels is inhibited by Cd^{2+}, Co^{2+}, Ni^{2+}, Mn^{2+} and Mg^{2+}. In addition, mercury as well as organotin compounds can affect the transit of sodium or calcium ions through the respective membrane channels affecting cell depolarization and neurotransmitter release (Rossi *et al.*, 1993; Viviani *et al.*, 1994). Experiments using synaptosomal preparations have indicated that the effects of certain metals within a given class on intracellular Ca^{2+} levels can be directly correlated with their neurotoxic effects *in vivo*. For example, the neurotoxic organotin derivative trimethyltin is much more potent than the closely related non-neurotoxic mono- and dimethyltin in causing increases in intracellular Ca^{2+} (Komulainen and Bondy, 1987). Finally, at extremely low concentrations, certain metals can directly affect protein kinase C activity and inositol polyphosphate generation (Markovac and Goldstein, 1988).

9.4 Effects of cellular Ca^{2+} overload

While short-term intracellular Ca^{2+} changes modulate several physiological functions, disturbances in the Ca^{2+} homoeostatic control can lead to intracellular Ca^{2+} accumulation and trigger lethal processes. Since Ca^{2+} is an activator of several enzymes involved in the catabolism of proteins, phospholipids and nucleic acids, a sustained increase in cytosolic free Ca^{2+} concentration above the physiological level can result in uncontrolled breakdown of macromolecules of vital importance for the maintenance of cell structure and function. Sudden and massive intracellular Ca^{2+} overload can rapidly result in cell necrosis. Proteolysis and disruption of the cytoskeletal organization, mitochondrial dysfunction and DNA damage can all be caused by Ca^{2+} accumulation within the cytoplasm or other intracellular compartments. Furthermore, there is now compelling evidence that supports a role for Ca^{2+} in the activation of apoptosis.

9.4.1 Phospholipases

Phospholipases catalyse the hydrolysis of membrane phospholipids. They are widely distributed in biological membranes and generally require Ca^{2+} for activation. A specific subset of phospholipases, collectively known as phospholipase A_2, has been proposed to participate in the detoxication of phospholipid hydroperoxides by releasing fatty acids from peroxidized membranes (van Kuijk *et al.*, 1987). However, phospholipase activation can also mediate pathophysiological reactions by stimulating membrane breakdown or by generating toxic metabolites. One of the best known examples of phospholipase-mediated toxicity is probably that caused by phospholipase-based snake and bee venoms. Therefore, phospholipase activation has been proposed as an important mechanism of cell killing. Phospholipase A_2 is Ca^{2+}- and calmodulin-dependent, and thus it is susceptible to activation following an increase in cytosolic Ca^{2+} concentration. Hence, it has been suggested that a sustained increase in cytosolic Ca^{2+} can result in enhanced breakdown of membrane phospholipids and, in turn, in mitochondrial and cell damage. Although a number of studies have indicated that accelerated phospholipid turnover occurs during anoxic or toxic cell injury (Chien *et al.*, 1979), the importance of phospholipase activation in the development of cell damage remains to be established.

9.4.2 Proteases

During the past 10 years the involvement of non-lysosomal proteolysis in several cell processes has become progressively clear. Proteases which have a neutral pH optimum include the ATP- and ubiquitin-dependent proteases and the calcium-dependent proteases, or calpains. Calpains are present in virtually all mammalian cells, and they appear to be largely associated with membranes in conjunction with a specific inhibitory protein (calpastatin). The extralysosomal localization of this proteolytic system allows the protease to participate in several specialized cell functions, including cytoskeletal and cell membrane remodelling, receptor cleavage and turnover, enzyme activation, and modulation of cell mitosis.

Cellular targets for these enzymes include cytoskeletal elements and membrane integral proteins (Mirabelli *et al.*, 1989). Thus, the activation of Ca^{2+} proteases has been shown to cause modification of microfilaments in platelets and to be involved in cell degeneration during muscle dystrophy and in the development of ischaemic injury in nervous tissue. Studies from our laboratory have suggested the involvement of Ca^{2+}-activated proteases in oxidant injury in liver (Nicotera *et al.*, 1986). Although the substrates for protease activity during cell injury remain largely unidentified, it appears that cytoskeletal proteins may be a major target for Ca^{2+}-activated proteases during chemical toxicity.

9.4.3 Endonucleases

More recently, it has become apparent that Ca^{2+} overload may also result in DNA degradation. This includes single-strand breakage as well as endonuclease-mediated cleavage of DNA in cells undergoing apoptosis (Arends *et al.*, 1990). Finally, Ca^{2+} overload may also stimulate other enzymatic processes that result in DNA damage. Elevated Ca^{2+} levels can lock topoisomerase II in a form that cleaves, but does not religate, DNA; topoisomerase II-mediated DNA fragmentation has been implicated in the cytotoxic action of some anticancer drugs (Udvardy *et al.*, 1986).

9.4.4 Mitochondrial damage

Work from several laboratories has indicated that mitochondrial damage may represent a common event in cell injury caused by toxic agents and other insults, e.g. ischaemia. Mitochondrial damage is initially manifested by a decrease in the mitochondrial membrane potential followed by ATP depletion. Protons are constantly pumped from the matrix into the intermembrane space in mitochondria of living cells. Since the inner mitochondrial membrane is relatively impermeable to anions, a considerable proportion of the energy resulting from the proton concentration gradient is stored as membrane electric potential. The proton gradient and the transmembrane potential represent the electrochemical forces that are employed for ATP synthesis as well for other metabolic activities, including the maintenance of Ca^{2+} homeostasis within mitochondria. Calcium ions can be actively transported into mitochondria via an electrophoretic uniporter. The driving force for the continuous Ca^{2+} pumping is provided by the transmembrane potential. However, studies performed in isolated mitochondria have demonstrated that during Ca^{2+} uptake the membrane potential decreases and the extent of the decrease is proportional to the amount of Ca^{2+} taken up by the mitochondria (Gunther and Pfeiffer, 1990). Thus, it appears that, under conditions which cause massive amounts of Ca^{2+} to accumulate in the mitochondria, their membrane potential would collapse.

The existence of different Ca^{2+} uptake and release pathways in mitochondria provides a basis for Ca^{2+} cycling. This process continuously utilizes energy which is supplied by the membrane potential. Oxidation of intramitochondrial NAD(P)H can activate the release route and accelerate Ca^{2+} cycling across the mitochondrial membrane (Lehninger *et al.*, 1978). This condition is associated with a decrease in the mitochondrial membrane potential that parallels the rate of Ca^{2+} cycling. Moreover, chelation of extramitochondrial Ca^{2+} with EGTA, or inclusion of ruthenium red in the incubation medium to abolish the reuptake of the released Ca^{2+}, completely prevents the collapse of the membrane potential (Moore *et al.*, 1983). Evidence that this mechanism is also operational in intact cells has been obtained using video imaging analysis in cultured hepatocytes loaded with rhodamine 123 (Nicotera *et al.*, 1990a).

9.4.5 Cytoskeletal alterations

One of the early signs of cell injury is the appearance of multiple surface protrusions (blebs) (Jewell *et al.*, 1982). The events leading to bleb formation have not yet been fully elucidated, and several mechanisms may independently contribute to their formation. However, it is generally accepted that a perturbation of cytoskeletal organization and of the interaction between the cytoskeleton and the plasma membrane plays an important role. Evidence for this assumption is provided by the observation that agents which modify the cytoskeleton, such as cytochalasins and phalloidin, stimulate bleb formation and by the demonstration that the bundles of actin filaments present at the base of the bleb appear to be totally dissociated from the bleb-forming portion of the plasma membrane (Phelps *et al.*, 1989). The finding that treatment of cells with a Ca^{2+} ionophore was able to induce similar blebbing, and that this was prevented by the omission of Ca^{2+} from the incubation medium, led to the proposal that Ca^{2+} is involved in the cytoskeletal alterations associated with the formation of surface blebs during cell injury (Jewell *et al.*, 1982). A number of studies have since clearly demonstrated the importance of various Ca^{2+}-dependent events in the control of cytoskeletal organization and function. Ca^{2+}-dependent cytoskeletal alterations include modification of the association between actin and actin-binding proteins, disturbances of microtubular organization and Ca^{2+}-dependent proteolysis of actin-binding proteins (Mirabelli *et al.*, 1989; Phelps *et al.*, 1989).

However, other mechanisms of cytoskeletal damage and bleb formation have also been implied in cell injury. ATP depletion is probably sufficient to cause depolymerization of actin, breakdown of the actomyosin network and a change in the lipid order of the plasma membrane. Oxidative stress in hepatocytes can cause the oxidation of thiol groups in actin and other cytoskeletal proteins which results in the formation of protein aggregates and in the relocalization of microfilaments, generating weak sites in the periphery of the cell for potential bleb formation (Bellomo *et al.*, 1990).

Finally, it is becoming clear that disturbances in protein phosphorylation may also affect the cytoskeletal structure. Several cytoskeletal proteins are substrates for protein kinases and phosphatases, and it has been reported that inhibition of cytoskeletal protein dephosphorylation can lead to rapid microfilament reorganization and blebbing in hepatocytes (Eriksson *et al.*, 1989). These studies also suggested that an altered balance of protein kinase/protein phosphatase activity can result in abnormal phosphorylation of cytoskeletal and other proteins, ultimately leading to cell death. Calyculin-A, another inhibitor of protein phosphatases, has been found to cause enhanced phosphorylation of vimentin and 20 kDa myosin light chain in 3T3 fibroblasts and to promote their detachment from the substratum (Chartier *et al.*, 1991), providing further support for this proposal.

9.5 Role of Ca^{2+} in apoptotic cell killing

Programmed cell death during development, cell turnover and possibly during ageing is in most cases characterized by nuclear membrane blebbing, organelle relocation and compaction and chromatin condensation. This form of cell death is known as apoptosis and occurs also in certain diseases or after exposure to chemical or biological agents. Typical early morphological changes occur in apoptotic cells, including chromatin redistribution, cell shrinkage, compacting of organelles, polarization and, finally, condensation and fragmentation of the chromatin (Arends *et al.*, 1990). Genome fragmentation, by a multistep process that eventually ends with the cleavage of cell chromatin into oligonucleosomes, is considered the most typical biochemical feature of apoptosis (Wyllie, 1980).

The findings that removal of extracellular Ca^{2+}, or pretreatment with intracellular Ca^{2+} chelators, can prevent both the nuclear changes typical of apoptosis (i.e. DNA degradation and apoptotic body formation) and cell lysis, have provided support for the idea that Ca^{2+} signals can initiate apoptosis in some systems (Nicotera *et al.*, 1994). Evidence that Ca^{2+} influx can be sufficient to trigger apoptosis has come from studies with specific Ca^{2+} channel blockers, which abrogate apoptosis in the regressing prostate following testosterone withdrawal (Martikainen and Isaacs, 1990) and in pancreatic β cells treated with serum from patients with type I diabetes (Juntii-Berggren *et al.*, 1993). Further evidence for a critical role of Ca^{2+} in triggering apoptosis comes from a recent study by Dowd *et al.* (1992), demonstrating that the stable overexpression of calbindin-D28K in a thymoma cell line protected the cells from apoptosis normally produced by exposure to dexamethasone, forskolin and calcium ionophore, A23187.

At least three different Ca^{2+}-dependent mechanisms may be involved in triggering the nuclear alterations seen in apoptosis. These include the activation of a Ca^{2+}-

dependent endonuclease activity responsible for DNA fragmentation, the modification of chromatin conformation and its susceptibility to cleavage by the former or other nucleases, and alterations in gene expression.

In many *in vitro* models of apoptosis the loss of intracellular Ca^{2+} homeostasis is accompanied by the activation of a Ca^{2+}-dependent endonuclease activity (Arends *et al.*, 1990). The Ca^{2+} ionophore A23187 stimulates Ca^{2+}-dependent DNA fragmentation and apoptosis in thymocytes (McConkey *et al.*, 1989), and characteristic endonuclease activity in isolated nuclei is dependent on Ca^{2+} (Jones *et al.*, 1989). Ca^{2+} increases resulting in the triggering of endonuclease activity can also be compartmentalized. For example, exposure of human adenocarcinoma cells to tumour necrosis factor α causes an initial intranuclear Ca^{2+} accumulation, which is followed by endonuclease activation and apoptosis (Bellomo *et al.*, 1992). A more direct role for Ca^{2+} in the formation of condensed chromatin and apoptotic bodies in thymocyte apoptosis is shown by recent work using thapsigargin (Jiang *et al.*, 1994). In thymocytes, in the presence of extracellular Ca^{2+}, thapsigargin causes a sustained intracellular Ca^{2+} elevation, associated with the formation of apoptotic bodies. The duration of the Ca^{2+} increase can be modulated in this system by withdrawal of extracellular Ca^{2+} at various intervals. By this approach it has been possible to establish that it takes at least 90 minutes for the $[Ca^{2+}]_i$ elevation to induce chromatin condensation and DNA fragmentation in thymocytes. Ca^{2+} chelation before this time was sufficient to prevent the formation of condensed chromatin patches and apoptotic bodies.

A role for Ca^{2+} and Ca^{2+}-dependent proteins (i.e. calmodulin) in the regulation of gene expression has long been postulated. Early work had shown that the synthesis of prolactin and growth hormone in rat pituitary GH3 cells requires Ca^{2+} and subsequent studies showed that Ca^{2+} exerts its effect at the transcriptional level via calmodulin (White and Bancroft, 1987). In addition, Ca^{2+} can directly regulate the expression of certain genes, including c-*fos* (Morgan and Curran, 1986) and control the expression of others (i.e. the interleukin-2 gene) during cell activation (Crabtree, 1989). Moreover, nuclear ion changes may also favour some chromatin conformational changes required for gene activation. For example, chromatin unfolding initiated by ion changes in the nuclear interior may facilitate H1 redistribution (Osheroff and Zechiedrich, 1988) or topoisomerase II activation (Villaponteau *et al.*, 1986).

Several genes – whose involvement in the cell death program has been suggested – may be directly or indirectly regulated by Ca^{2+}. Examples include calmodulin (Dowd *et al.*, 1991), *c-fos* (Smeyne *et al.*, 1993) and, possibly, growth-arrest genes such as *gadd153* (Bartlett *et al.*, 1992). In addition, the role of calmodulin and Ca^{2+} in the regulation of the cell cycle is well established (Chafouleas *et al.*, 1984). Expression of a constitutive form of Ca^{2+}/calmodulin-stimulated protein kinase II can cause cell cycle arrest in the G_2 phase independently from phosphorylation of cdc2 kinase (Planas-Silva and Means, 1992); in turn, cell cycle alterations may result in apoptosis.

Despite being now considered a mid-to-late event in apoptosis, internucleosomal DNA cleavage by Ca^{2+}- and Mg^{2+}-dependent endonucleases remains a characteristic feature of apoptotic cell death. Ca^{2+} can activate internucleosomal DNA cleavage by endonucleases (Arends *et al.*, 1990) and is also required for the preceding high molecular weight DNA fragmentation (Zhivotovsky *et al.*, 1994). Thus, while Ca^{2+} removal prevents high molecular weight and internucleosomal DNA fragmentation in thymocytes exposed to glucocorticoids or topoisomerase II inhibitors, thapsigargin promotes both types of DNA cleavage in the same cells (Zhivotovsky *et al.*, 1994). The mechanism by which Ca^{2+} modulates DNA digestion during apoptosis is not yet clear. On one hand it is obvious that the Ca^{2+}/Mg^{2+}-dependent endonuclease responsible for DNA cleavage would require Ca^{2+} for activation. An alternative possibility is that Ca^{2+} could play a role in modifying chromatin conformation through topoisomerase II activation, making chromatin regions accessible to enzymes such as DNase I or other endonucleases.

Thus, in several models of apoptosis, the calcium ion plays a critical role in triggering this process. However, there are several experimental systems which fail to show any Ca^{2+} dependency (Iseki *et al.*, 1993), and others where elevations of cytosolic Ca^{2+} have been found to even prevent cell death (Koike *et al.*, 1989). The latter observation has indicated that Ca^{2+} elevations may have a different role in immature and mature cell systems. In particular, in developing sympathetic neurones a Ca^{2+} increase may function to prevent death by growth factor deprivation, whereas in post-mitotic neurones when intracellular Ca^{2+} has reached a 'set point' and cells are less dependent on growth factor stimulation, further Ca^{2+} increases will result in cell death.

9.6 Ca^{2+} overload in neuronal cell death

The evidence for a role of Ca^{2+} in cell killing is particularly strong in the central nervous system. Thus, Ca^{2+} appears to mediate the neurotoxicity of cyanide, chlordecone and various heavy metals, including lead, mercury and organotin compounds (Komulainen and Bondy, 1988). Further, intracellular Ca^{2+} overload due to excessive stimulation of excitatory amino acid receptors and enhanced Ca^{2+} influx through membrane channels appears to play an important role in ischaemic brain damage (Manev *et al.*, 1990). Ca^{2+} overload has also been reported to contribute to anoxic injury of mammalian central white matter (Stys *et al.*, 1990) and to mediate the cytotoxic effect of human immunodeficiency virus (HIV) type 1 coat protein gp 120 in retinal cultures (Dreyer *et al.*, 1990). An ample number of studies have recently focused on glutamate-induced excitotoxicity and its contribution to brain damage in various disease conditions (Choi, 1992). The ability of glutamate or related compounds to induce neuronal death by receptor overstimulation has long been recognized. The calcium ion plays a critical role in this process, and intracellular Ca^{2+} overload appears to mediate the lethal effects of NMDA receptor overactivation. This mechanism is responsible not only for the brain damage induced by certain neurotoxins, but excitotoxicity is also strongly implicated in neuronal death following insults such as ischaemia and trauma. Thus, *N*-methyl-D-aspartate (NMDA) receptor antagonists do not only block Ca^{2+} influx and neuronal death elicited by glutamate or NMDA *in vitro* but can also reduce the volume of infarction produced by focal ischemia *in vivo*. Similarly, antisense oligodeoxynucleotides to NMDA-R1 receptor channel have been reported to protect cortical neurons from NMDA excitotoxicity *in vitro* and to reduce focal ischemic infarctions in the rat *in vivo* (Wahlestedt *et al.*, 1993). Further, cell permeant Ca^{2+} chelators have recently been found to reduce early excitotoxic and ischaemic neuronal injury both *in vitro* and *in vivo* (Tymianski *et al.*, 1993). Finally, antagonists of α-amino-3-hydroxy-5-methyl-4-isoxazolepropionate (AMPA)–kainate receptors have also demonstrated neuroprotective effects, particularly in experimental models of global ischaemia in which NMDA antagonists are ineffective (Choi, 1992).

Excitotoxicity may also play an important role in other diseases associated with impairment of cellular energy metabolism and in neurodegenerative disorders. Glutamate has been implicated in the neurodegeneration that occurs in several diseases, including Huntington's chorea (Coyle and Schwarcz, 1976) and Alzheimer's disease (Maragos *et al.*, 1987). Further, there is experimental evidence that 1-methyl-4-phenyl-1,2,5,6-tetrahydropyridine (MPTP)-induced destruction of nigrostriatal dopaminergic neurones can be attentuated by NMDA antagonists and that the β-amyloid protein that accumulates in Alzheimer's disease can potentiate excitotoxic degeneration; the β-amyloid protein can form calcium channels in bilayer membranes which may contribute to its neurotoxic effects (Aripse *et al.*, 1993). Finally, there is evidence from experiments with gp120 in neuronal cultures that activation of both voltage-dependent Ca^{2+} channels and NMDA receptor-operated channels contribute to HIV-related neuronal injury. In fact, it appears that both gp120 and glutamate-related molecules are necessary and act synergistically to produce neuronal damage (Lipton, 1992).

Thus, there is accumulating evidence that Ca^{2+} overload may play a critical role in ischaemia as well as in various neurodegenerative disorders. Less is known, however, about the biochemical mechanisms by which the Ca^{2+} overload causes neuronal cell death. As discussed above, sustained increases in intracellular Ca^{2+} can lead to the activation of degradative enzymes, such as phospholipases, proteases and endonucleases, and to mitochondrial dysfunction and perturbation of cytoskeletal organization. Whereas little is known about the possible contribution of phospholipase activation to neurodegeneration, Ca^{2+}-activated degradation of cytoskeletal proteins appears to be an early and important component of the postischaemic response in hippocampal neurones which can contribute to neuronal death (Lee *et al.*, 1991). Widespread activation of Ca^{2+}-dependent neutral proteases in the brain has also been reported in Alzheimer's disease (Saito *et al.*, 1993).

Another potential mechanism for Ca^{2+}-triggered neurotoxicity is represented by the nitric oxide (NO) synthase of NADPH-diaphorase-containing neurones. Recent studies have suggested that the generation of NO triggered by a Ca^{2+} rise caused by glutamate receptor overstimulation in these neurones may be lethal to neighbouring cells, whereas the NO synthase-containing neurones themselves seem to be less sensitive to NO toxicity (Dawson *et al.*, 1991). The mechanism of NO-induced cell death has been suggested to be related to DNA damage and subsequent activation of poly(ADP-ribose) synthetase leading to energy depletion (Zhang *et al.*, 1994). Further, neuroprotective effects of FK506 and cyclosporin A have been linked to the inhibition of the calcineurin-mediated dephosphorylation of NO synthase required for its activation (Dawson *et al.*, 1993). Although a subsequent study has suggested a neuroprotective effect of a NO synthase inhibitor in a rat model of focal cerebral ischaemia (Boisson *et al.*, 1992), the possible clinical implications of these observations are unclear and require further studies.

There are several recent studies suggesting that cell killing in neurodegenerative disorders may occur by apoptosis. Thus, Kure *et al.* (1991) found that glutamate could trigger internucleosomal DNA cleavage in neuronal cultures *in vitro* and in hippocampal neurones *in vivo*. Both DNA fragmentation and neuronal death were prevented by a nuclease inhibitor, aurintricarboxylic acid. Further, 1-methyl-4-phenylpyridinium (MPP^+), the neurotoxic metabolite of the parkinsonian agent MPTP, can trigger internucleosomal DNA cleavage and apoptosis in cerebellar granule neurons (Dipasquale *et al.*, 1991). The relationship, if any, between the mitochondrial damage due to inhibition of NADH dehydrogenase by MPP^+ (Singer and Ramsay, 1990) and nuclear DNA fragmentation is presently not clear, although MPTP has been reported to cause intracellular Ca^{2+} accumulation in hepatocytes (Kass *et al.*, 1988).

The possible involvement of Ca^{2+} in the pathogenesis of Parkinson's disease is suggested by several observations. An accumulation of iron has been observed in the substantia nigra and the zona compacta during the progression of the disease. The increased iron level in these areas is believed to induce oxidative stress, associated with alterations in mitochondrial Ca^{2+} metabolism, glutathione depletion and cell death (Youdim *et al.*, 1993b). Support for this hypothesis has come from the observations that 6-hydroxydopamine toxicity in the rat (a model for Parkinson's disease) is linked to the liberation of iron from ferritin, resulting in mitochondrial dysfunction, mitochondrial Ca^{2+} release and oxidative stress, and by the protective effect of iron-chelating agents (Youdim *et al.*, 1993a). Further, it has been shown that $^{45}Ca^{2+}$ uptake is increased in nigral catecholaminergic neurones in animals treated with ferric lactate (Hirsh, 1992) and that the drug memantine – used in the treatment of Parkinson's disease – functions as a NMDA channel blocker (Chen *et al.*, 1992). A role for Ca^{2+} in Parkinson's disease has also been suggested by studies indicating that neurones which contain high levels of Ca^{2+}-binding proteins (i.e. calbindin-D28K) are spared from MPTP toxicity (German *et al.*, 1992). However, the possible involvement of Ca^{2+} in Parkinson's disease is far from being established and is still matter of intense debate.

A possible role of Ca^{2+} and apoptosis in Alzheimer's dementia is also the subject of intense investigation. Forloni *et al.* (1993) have recently reported that the neurotoxicity induced by chronic application of a synthetic peptide homologous to residues 25–35 of the β-amyloid protein to rat hippocampal neurones occurs by apoptosis, suggesting that amyloid fibrils may induce neuronal death by apoptosis in Alzheimer's disease. Similar findings were reported by Loo *et al.* (1993). Moreover, low levels of K^+ have been found to induce apoptosis in cerebellar granule neurones (D'Mello *et al.*, 1993), whereas hippocampal neurones cultured in a high-oxygen atmosphere have been reported to die by apoptosis (Enokido and Hatanaka, 1993). Conversely, evidence supporting a role for apoptosis in focal cerebral ischaemia in rats has recently been published (Linnik *et al.*, 1993). Finally, the proto-oncogene *bcl-2* has been found to prevent neuronal cell death induced by glutathione depletion by decreasing the net cellular generation of active oxygen species, although it is unclear whether *bcl-2* protection in neuronal cells relates to apoptotic or necrotic cell death (Kane *et al.*, 1993).

Therefore, it appears that intracellular Ca^{2+} overload can result in the activation of a variety of degradative functions which can contribute to neuronal death. In several experimental models, cell killing can be prevented by Ca^{2+} channel blockers or glutamate receptor antagonists, although it has not been conclusively shown that the protective effects of glutamate receptor antagonists depend solely on prevention of Ca^{2+} overload. Nimodipine, a member of the 1,4-dihydropyridine class of Ca^{2+} channel blockers, has been found to influence neuronal Ca^{2+} metabolism and to protect from damage in a number of experimental systems, demonstrating the importance of L-type channel blockade in neuroprotection (Bär *et al.*, 1990). In several experimental models of neurodegeneration, Ca^{2+} overload appears to be due to the combined effects of the activation of both receptor-operated and L-type Ca^{2+} channels.

It is well established that the sensitivitity to excitotoxic amino acids differs between different neuronal populations, and that the sensitive populations in the hippocampus and striatum are vulnerable also in Alzheimer's disease and Huntington's chorea (Mattson *et al.*, 1991). In view of the calcium hypothesis of cell killing, one would expect that resistant and vulnerable neurones would exhibit differences in their Ca^{2+} metabolism or, alternatively, that resistant neurones would be less susceptible to injurious effects of Ca^{2+} overload. In fact, there is emerging experimental evidence in support of this assumption. Thus, selective neuronal vulnerability to excitotoxic amino acids seems to be related not only to the density and types of glutamate receptors but also to the capacity of neuronal Ca^{2+} buffering and extrusion systems, e.g. the efficacy of the plasma membrane Na^{+}/ Ca^{2+} exchanger (Mattson *et al.*, 1989). There are also several reports linking neuronal vulnerability in neurodegenerative disorders to the expression of various Ca^{2+}-binding proteins. Since some of these proteins have considerable Ca^{2+} buffering capacity, it is reasonable to assume that their presence may influence the ability of the cell to withstand Ca^{2+} overload. Much attention has been devoted to calbindin-D28K and calretinin, two Ca^{2+}-binding proteins expressed in largely different populations of neurones, which have been demonstrated to exert cytoprotective effects in *in vitro* models of excitotoxicity (Heizmann and Braun, 1992). A positive correlation has also been found between the level of parvalbumin or calbindin-D28K in populations of hippocampal neurones and their resistance to seizure-induced neuronal damage. On the other hand, a reduction of calbindin-D28K mRNA and protein levels has been observed in brain areas particularly affected in Alzheimer's, Parkinson's and Huntington's diseases (Mattson *et al.*, 1991). Since calbindin-D28K-positive neurones have been shown to be better able to reduce intracellular Ca^{2+} levels than calbindin-D28K-negative neurones, these findings suggest that differential expression of calbindin-D28K may be one determinant of neuronal vulnerability in neurodegenerative disorders.

9.7 Concluding remarks

Our understanding of the mechanisms responsible for cell death in various neurodegenerative disorders has improved during recent years but is still far from complete. It is clear that intracellular Ca^{2+} overload is an important factor in various *in vitro* models, and results from clinical studies appear to support this hypothesis. The Ca^{2+} overload seems to result from increased Ca^{2+} influx through both receptor-operated and L-type Ca^{2+} channels, as indicated by the neuroprotective effects of glutamate receptor antagonists and L-type channel blockers. Various Ca^{2+}-dependent degradative processes have been found to contribute to Ca^{2+}-mediated cell killing in *in vitro* studies, although their relative contribution is unclear; available evidence would suggest that perturbation of cytoskeletal organization and mitochondrial dysfunction may be of particular importance. Finally, although cell death in neurodegenerative disorders has been assumed to be of the necrotic type, recent studies have indicated that it may also occur by apoptosis. The role of the two modes of cell death in neurodegeneration remains to be established as well as the implications that a more selective single-cell deletion has for neuronal function as compared to undiscriminated necrotic cell death.

References

Arends, M.J., Morris, R.G. & Wyllie, A.H. (1990) *Am. J. Pathol.* **136**, 593–608.

Arispe, N., Rojas, E. & Pollard, H.B. (1993) *Proc. Natl Acad. Sci. USA* **90**, 567–571.

Aw, T.Y., Nicotera, P., Manzo, L. & Orrenius, S. (1990) *Arch. Biochem. Biophys.* **283**, 46–50.

Bading, H., Ginty, D.D. & Greenberg, M.E. (1993) *Science* **260**, 181–186.

Bär, P.R., Traber, J., Schuurman, T. & Gispen, W.H. (1990) *J. Neural Transm. (Suppl.)* **31**, 55–71.

Barnes, D.S. (1988) *Science* **242**, 1510–1511.

Barritt, G.J., Parker, J.C. & Wadsworth, J.C. (1981) *J. Physiol.* **312**, 29–55.

Bartlett, J.D., Luethy, J.D., Carlson, S.G., Sollett, S.J. & Holbrook, N.J. (1992) *J. Biol. Chem.* **267**, 20465–20470.

Bellomo, G., Thor, H. & Orrenius, S. (1987) *J. Biol. Chem.* **262**, 1530–1534.

Bellomo, G., Mirabelli, F., Richelmi, P., Malorni, W., Iosi, F. & Orrenius, S. (1990) *Free Radic. Res. Commun.* **8**, 391–399.

Bellomo, G., Perotti, M., Taddei, F., Mirabelli, F., Finardi, G., Nicotera, P. & Orrenius, S. (1992) *Cancer Res.* **52**, 1342–1346.

Berridge, M.J. (1987) *Annu. Rev. Biochem.* **56**, 159–193.

Berridge, M.J. (1991) *J. Biol. Chem.* **265**, 9583–9586.

Boisson, A., Plotkine, M. & Boulu, R.G. (1992) *Br. J. Pharmacol.* **106**, 766–767.

Carafoli, E. (1989) *Annu. Rev. Biochem.* **56**, 395–433.

Chafouleas, J.G., Lagace, L., Bolton, W.E., Boyd, A.E. & Means, A.R. (1984) *Cell* **36**, 73–81.

Chartier, L., Rankin, L.L., Allen, R.E., Kato, Y., Fusetani, N., Karaki, H., Watabe, S. & Harsthorne, D.J. (1991) *Cell Motil. Cytoskeleton* **18**, 26–40.

Chen, H.S., Pellegrini, J.W., Aggarwal, S.K., Lei, S.Z., Warach, S., Jensen, F.E. & Lipton, S.A. (1992) *J. Neurosci.* **12**, 4427–4436.

Chien, K.R., Pfau, R.G. & Farber, J.L. (1979) *Am. J. Pathol* **97**, 505–530.

Choi, D.W. (1992) *Science* **258**, 241–243.

Coyle, J.T. & Schwarcz, R. (1976) *Nature* **263**, 244–246.

Crabtree, G.R. (1989) *Science* **243**, 355–361.

Dawson, V.L., Dawson, T.M., London, G.D., Bredt, D.S. & Snyder, S.H. (1991) *Proc. Natl Acad. Sci. USA* **88**, 6368–6371.

Dawson, T.M., Stiner, J.P., Dawson, V.L., Dinerman, J.L., Uhl, G.R. & Snyder, S.H. (1993) *Proc. Natl Acad. Sci. USA* **90**, 9808–9812.

Dipasquale, B., Marini, A.M. & Youle, R.J. (1991) *Biochem. Biophys. Res. Commun.* **181**, 1442–1448.

D'Mello, S.R., Galli, C., Ciotti, T. & Calissano, P. (1993) *Proc. Natl Acad. Sci. USA* **90**, 10989–10993.

Dowd, D.R., MacDonald, P.N., Komm, B.S., Maussler, M.R. & Miesfeld, R. (1991) *J. Biol. Chem.* **266**, 18423–18426.

Dowd, D.R., MacDonald, P.N., Komm, B.S., Haussler, M.R. & Miesfeld, R.L. (1992) *Mol. Endocrinol.* **6**, 1843–1848.

Dreyer, E.B., Kaiser, P.K., Offerman, J.T. & Lipton, S.A. (1990) *Science* **248**, 364–367.

Enokido, Y. & Hatanaka, H. (1993) *Neuroscience* **57**, 965–972.

Eriksson, J.E., Paatero, G.I.L., Meriluoto, J.A.O., Codd, G.A., Kass, G.E.N., Nicotera, P. & Orrenius, S. (1989) *Exp. Cell Res.* **185**, 86–100.

Fasolato, C., Pizzo, P. & Pozzan, T. (1990) *J. Biol. Chem.* **265**, 20351–20355.

Fleckenstein, A., Frey, M. & Fleckenstein-Grün, G. (1983) In *Mechanisms of Hepatocyte Injury and Death* (eds Keppler, D., Popper, H., Bianchi, L. *et al.*), pp. 321–335. MTP Press, Lancaster.

Forloni, G., Chiesa, R., Smiroldo, S., Verga, L., Salmona, M., Tagliavini, F. & Angeretti, N. (1993) *NeuroReport* **4**, 523–526.

German, D.C., Manaye, K.F., Sonsalla, P.K. & Brooks, B.A. (1992) *Ann. NY Acad. Sci.* **648**, 42–62.

Gunther, T.E. & Pfeiffer, D.R. (1990) *Am. J. Physiol.* **258**, c755–c786.

Heizmann, C.W. & Braun, K. (1992) *Trends Neurosci.* **15**, 259–264.

Hirsch, E.C. (1992) *Ann. Neurol.* **32** (Suppl.), S88–S93.

Iseki, R., Kudo, Y. & Iwata, M. (1993) *J. Immunol.* **151**, 5198–5207.

Jewell, S.A., Bellomo, G., Thor, H., Orrenius, S. & Smith, M.T. (1982) *Science* **217**, 1257–1259.

Jiang, S., Chow, S.C., Nicotera, P. & Orrenius, S. (1994) *Exp. Cell. Res.* **212**, 84–92.

Jones, D.P., McConkey, D.J., Nicotera, P. & Orrenius, S. (1989) *J. Biol. Chem.* **264**, 6398–6403.

Juntti-Berggren, L., Larsson, O., Rorsman, P. Ämmälä, C., Bokvist, K., Wahlander, K., Nicotera, P., Dypbukt, J., Orrenius, S., Hallberg, A. & Berggren, P.-O. (1993) *Science* **261**, 86–90.

Kane, D.J., Sarafian, T.A., Anton, R., Hahn, H., Butler Gralla, E., Selverstone Valentine, J., Örd, T. & Bredesen, D.E. (1993) *Science* **262**, 1274–1277.

Kass, G.E.N., Wright, J.M., Nicotera, P. & Orrenius, S. (1988) *Arch. Biochem. Biophys.* **260**, 789–797.

Kass, G.E.N., Duddy, S.K. & Orrenius, S. (1989) *Biochem. J.* **260**, 499–507.

Koike, T., Martin, D.P. & Johnson, E.M. (1989) *Proc. Natl Acad. Sci. USA* **86**, 6421–6425.

Komulainen, H. & Bondy, S.C. (1987) *Toxicol. Appl. Pharmacol.* **88**, 77–86.

Komulainen, H. & Bondy, S.C. (1988) *Trends Pharmacol. Sci.* **9**, 154–156.

Kure, S., Tominaga, T., Yoshimoto, T., Tada, K. & Narisawa, K. (1991) *Biochem. Biophys. Res. Commun.* **179**, 39–45.

Kyrprianou, N., English, H.F. & Isaacs, J.T. (1990) *Cancer Res.* **50**, 3748–3753.

Lee, K.S., Frank, S., Vanderklish, P., Arai, A. & Lynch, G. (1991) *Proc. Natl Acad. Sci. USA* **88**, 7233–7237.

Lehninger, A.L., Vercesi, A. & Bababunmi, E.A. (1978) *Proc. Natl Acad. Sci. USA* **75**, 1690–1694.

Linnik, M.D., Zobrist, R.H. & Hatfield, M.D. (1993) *Stroke* **24**, 2002–2008.

Lipton, S.A. (1992) *Trends Neurosci.* **15**, 75–79.

Loo, D.T., Copani, A., Pike, C.J., Whittemore, E.R., Walencewicz, A.J. & Cotman, C.W. (1993) *Proc. Natl Acad. Sci. USA* **90**, 7951–7955.
McConkey, D.J., Hartzell, P., Duddy, S.K., Håkansson, H. & Orrenius, S. (1988) *Science* **242**, 256–258.
McConkey, D.J., Hartzell, P., Nicotera, P. & Orrenius, S. (1989) *FASEB J.* **3**, 1843–1849.
Manev, H., Costa, E., Wroblewski, J.T. & Guidotti, A. (1990) *FASEB J*, **4**, 2789–2797.
Maragos, W.F., Greenamyre, J.T., Penney, J.B. & Young, A.B. (1987) *Trends Neurosci.* **10**, 65–68.
Markovac, J. & Goldstein, G.W. (1988) *Nature* **334**, 71–73.
Martikainen, P. & Isaacs, J. (1990) *Prostate* **17**, 175–187.
Mattson, M.P., Guthrie, P.B. & Kater, S.B. (1989) *FASEB J.* **3**, 2519–2526.
Mattson, M.P., Rychlik, B., Chu, C. & Christakos, S. (1991) *Neuron* **6**, 41–51.
Mirabelli, F., Salis, A., Vairetti, M., Bellomo, G., Thor, H. & Orrenius, S. (1989) *Arch. Biochem. Biophys.* **270**, 478–488.
Moore, G.A., Jewell, S.A., Bellomo, G. & Orrenius, S. (1983) *FEBS Lett.* **153**, 289–292.
Morgan, J.I. & Curran, T. (1986) *Nature* **322**, 552–555.
Nicotera, P., Hartzell, P., Baldi, C., Svensson, S.A., Bellomo, G. & Orrenius, S. (1986) *J. Biol. Chem.* **261**, 14628–14635.
Nicotera, P., McConkey, D.J., Jones, D.P. & Orrenius, S. (1989) *Proc. Natl Acad. Sci. USA* **86**, 453–457.
Nicotera, P., Bellomo, G. & Orrenius, S. (1990a) *Chem. Res. Toxicol.* **3**, 484–494.
Nicotera, P., Orrenius, S., Nilsson, T. & Berggren, P.-O. (1990b) *Proc. Natl Acad. Sci. USA* **87**, 6858–6862.
Nicotera, P., Bellomo, G. & Orrenius, S. (1992) *Annu. Rev. Pharmacol. Toxicol.* **32**, 449–470.
Nicotera, P., Zhivotovsky, B., Bellomo, G. & Orrenius, S. (1994) In *Apoptosis* (eds Schimke, R. & Mihich, T.E.), Plenum Press, New York.
Osheroff, N. & Zechiedrich, L.E. (1988) *Biochemistry* **26**, 4304–4309.
Phelps, P.C., Smith, M.W. & Trump, B.F. (1989) *Lab. Invest.* **60**, 630–642.
Planas-Silva, M.D. & Means, A.R. (1992) *EMBO J.* **11**, 507–517.
Rossi, A.D., Manzo, L., Orrenius, S., Vahter, M., Berggren, P. & Nicotera, P. (1993) *FASEB J.* **7**, 1547.
Rottenberg, H. & Marbach, M. (1990) *Biochim. Biophys. Acta* **1016**, 77–86.
Saito, K.I., Elce, J.S., Hamos, J.E. & Nixon, R.A. (1993) *Proc. Natl Acad. Sci. USA* **90**, 2628–2632.
Schanne, F.A.X., Kane, A.B., Young, E.E. & Farber, J.L. (1979) *Science* **206**, 700–702.
Singer, T.P. & Ramsay, R.R. (1990) *FEBS Lett.* **274**, 1–8.
Smeyne, R.J., Vendrell, M., Hayward, M., Baker, S.J., Miao, G.G., Schilling, K., Robertson, L.M., Curran, T. & Morgan, J.I. (1993) *Nature* **363**, 166–169.
Somylo, A.P., Bond, M. & Somylo, A.V. (1985) *Nature* **314**, 622–625.
Stys, P.K., Ransom, B.R., Waxman, S.G. & Davis, P.K. (1990) *Proc. Natl Acad. Sci. USA* **87**, 4212–4216.
Tymianski, M., Wallace, M.C., Spigelman, I., Uno, M., Carlen, P.L., Tator, C.H. & Charlton, P. (1993) *Neuron* **11**, 221–235.
Udvardy, A., Schedl, P., Sander, M. & Hsieh, T. (1986) *J. Mol. Biol.* **191**, 231–246.
van Kuijk, F.J.G.M., Sevanian, A., Handelman, G.J. & Dratz, E.A. (1987) *Trends Biochem. Sci.* **12**, 31–34.
Villaponteau, B., Pribil, T.M., Grant, M.H. & Martinson, H.G. (1986) *J. Biol. Chem.* **261**, 10359.
Wahlestedt, C., Golanov, E., Yamamoto, S., Yee, F., Ericson, H., Yoo, H., Inturrisi, C.E. & Reis, D.J. (1993) *Nature* **363**, 260–263.
Waring, P., Eichner, R.D., Mullbacher, A. & Sjaarda, A. (1988) *Science* **263**, 18493–18499.
White, B.A. & Bancroft, C. (1987) *Calcium and Cell Functions* (ed. Cheung, W.Y.), pp. 109–132. Academic Press, New York.
Woods, N.M., Cuthbertson, K.S.R. & Cobbold, P.E. (1986) *Nature* **319**, 600–602.

Wyllie, A.H. (1980) *Nature* **284**, 555–556.
Youdim, M.B., Ben-Shachar, D., Eshel, G., Finberg, J.P. & Riederer, P. (1993a) *Adv. Neurol.* **60**, 259–266.
Youdim, M.B., Ben-Shachar, D. & Riederer, P. (1993b) *Mov. Disord.* **8**, 1–12.
Zhang, J., Dawson, V.L., Dawson, T.M. & Snyder, S.H. (1994) *Science* **263**, 687–689.
Zhivotovsky, B., Cedervall, B., Jiang, S., Nicotera, P. & Orrenius. S. (1994) *Biochem. Biophys. Res. Commun.* **202**, 120–127.

CHAPTER 10

NEUROTROPHIC FACTORS: TOWARDS A RESTORATIVE THERAPY OF PARKINSON'S DISEASE

C. Anthony Altar, Stanley J. Wiegand, Ronald M. Lindsay and Jesse M. Cedarbaum

Regeneron Pharmaceuticals, Inc., 777 Old Saw Mill River Road, Tarrytown, NY 10591-6707, USA

Table of Contents

Neurodegeneration and Neuroprotection in Parkinson's Disease
ISBN 0-12-525445-8

10.1 Introduction

More than any other neurodegenerative condition, Parkinson's disease (PD) has yielded to the advances of modern neuropharmacology. Yet the tenets upon which traditional neuropharmacological approaches are based – neurotransmitter replacement and agonism or antagonism of specific neurotransmitter receptors or transporters – are but 'finger in the dike' approaches for the PD patient. While often effective, these treatments of chemical imbalances in the brain side-step the fundamental issue of how to retard or arrest the progressive loss of neurones in the brains of PD victims. One challenge facing the present generation of neuroscientists is to better understand the biological mechanisms which underlie neuronal survival and neural system repair by discovering the signals and processes by which neurones normally maintain or lose their vitality. The purpose of this review is to present information that can help create protective or restorative treatments of neurodegenerative disease, in particular PD.

10.1.1 PD, a disorder affecting multiple neuronal systems

PD is primarily characterized pathologically by the progressive loss of dopaminergic neurones of the substantia nigra, and clinically by motor and cognitive disturbances consequent to such neuronal loss. It is estimated that the symptoms of PD emerge when anywhere from 50 to 90% of these neurones have become dysfunctional or have died (Bernheimer *et al.*, 1973). In the early stages of the disease, neuronal plasticity may result in partial compensation for the loss of functioning neurones. For example, two mechanisms which maintain complex sensorimotor functions despite a considerable loss of dopamine are increases in striatal dopamine receptor number (Lee *et al.*, 1978) and accelerated striatal dopamine turnover and release, which may preserve extracellular dopamine concentrations at near-normal levels (Altar *et al.*, 1987). The symptoms of PD begin to appear when the loss of dopamine or of other neurotransmitter systems is so great that the brain can no longer compensate (Agid *et al.*, 1992). Dopamine precursor or agonist therapy can prolong the period of functional compensation in the nigrostriatal system.

In addition to dopamine neurones, other types of neurones are depleted by the disease process; for example, cholinergic neurones in the basal forebrain, serotonergic and noradrenergic neurones of the brainstem, and the preganglionic neurones of the intermediolateral columns of the spinal cord. The degeneration of these non-dopaminergic systems in PD contribute to cognitive, affective and autonomic symptoms in more advanced stages of the disease (Bernheimer *et al.*, 1973; Perry *et al.*, 1985). While present-day treatments can enhance dopaminergic function, these other symptoms of PD can be exacerbated by dopaminergic agents, yielding increasingly troublesome complications of therapy as PD advances. At present, little can be done to treat the consequences of degeneration of non-dopaminergic systems in PD.

10.1.2 Current therapies for PD

10.1.2.1 Pharmacotherapy

The current era of pharmacological treatment of PD began when Cotzias and colleagues (Cotzias *et al.*, 1969) demonstrated that patients could be effectively treated with large oral doses of the dopamine precursor, L-DOPA. They found that L-DOPA produced a 'complete sustained disappearance or marked reduction' in the clinical signs and symptoms of PD. L-DOPA is effective because it is converted in the striatum to dopamine, which stimulates denervated dopamine receptors. To date, most effective antiparkinsonian drugs work by essentially the same mechanism, i.e. activation of the denervated postsynaptic dopamine receptors by either dopamine derived from the precursor L-DOPA or by a dopamine receptor agonist such as bromocriptine or pergolide (Lieberman and Goldstein, 1982).

Unfortunately, the net effectiveness of dopamine precursor or agonist treatments diminishes over time. The progressive loss of striatal dopamine nerve terminals reduces the storage capacity for L-DOPA and its product, dopamine, resulting in decreasing duration of responsiveness to individual doses of L-DOPA. In the majority of patients, a supersensitivity of postsynaptic dopamine receptors leads to the development of symptoms related to receptor overstimulation, such as drug-induced involuntary movements (dyskinesias) and psychiatric complications such as hallucinations and delusions (Lee *et al.*, 1978; Cedarbaum and McDowell, 1986; Cedarbaum and Olanow, 1992). The use of dopamine precursors or receptor agonists for PD also fails to address the fact that neuronal systems other than nigrostriatal dopamine neurones can also degenerate in PD. For example, commonly used anticholinergic adjuvants to L-DOPA can exacerbate cognitive deficits in PD patients, which are presumably due in part to cholinergic neurone degeneration. Thus, desirable characteristics of yet-to-be discovered neuroprotective or neurorestorative agents for PD might include (1) restoration of the balance of pre- and postsynaptic dopaminergic functions, to avoid secondary functional effects such as receptor supersensitivity and dyskinesias, (2) effective augmentation of the function of non-dopaminergic systems in the brain of the PD patient, and (3) a slowing or arresting of the degeneration of both dopaminergic and non-dopaminergic neurones in the basal ganglia.

10.1.2.2 Experimental surgical approaches to treating PD

In recent years, neurosurgical and new transplantation procedures have been employed with some initial success in the treatment of PD. The neurosurgical approaches mainly include pallidotomy (Bergman *et al.*, 1990) and ventral thalamic lesions or stimulation to interrupt non-dopaminergic components of the basal ganglia that are overactive in PD. Neural transplantation experiments have employed fetal nigral tissue and paraneuronal tissues or cells (e.g. adrenal medulla) to restore striatal dopamine. Both transplantation approaches have shown promise, in that some restoration of neurological function has been attained in dopamine-deficient experimental animals and PD patients (for reviews, see Brundin and Bjorklund, 1987; Kordower *et al.*, 1991). However, the recovery achieved has been incomplete, at least in part because of the generally poor survival of the engrafted cells and their limited anatomical integration into the host brain.

10.1.3 Directions for future therapy

Neurotrophic factors comprise several unique classes of endogenous proteins which can support the survival and differentiation of central and peripheral neurones during development (Thoenen, 1991; Lindsay, 1993; Lindsay *et al.*, 1993, 1994). During adult life, neurotrophic factors can also evoke metabolic responses in neurones which help them resist injury or degenerative processes. It is now clear that the survival or function of developing and mature dopamine neurones can be enhanced by several distinct proteinaceous growth factors. Thus, it is hoped that treatment with a specific neurotrophic factor, or combinations of factors, might slow or arrest the neuronal loss which underlies PD and other neurodegenerative disorders.

This review presents recent progress in the biology of neurotrophic factors and discusses the status of their preclinical development and potential clinical use in (1) retarding neuronal loss or restoring the functioning of residual dopaminergic and non-dopaminergic neurones in PD, and (2) promoting the survival and function of engrafted fetal nigral tissues or transplanted cells whose presence in the striatum would be expected to ameliorate the signs and symptoms of PD.

10.2 Neurotrophic factors

The development of the nervous system involves a precisely timed sculpting process during which there is an initial overproduction of neurones, followed by the formation of neuronal connections, and finally the elimination of excess neurones by a process of naturally occurring, neuronal death (for a review, see Oppenheim, 1991). The 'neurotrophic hypothesis' was based initially on observations that a soluble factor present in the embryonic chick limb bud appeared to augment the survival of spinal cord motor neurones. Thus, it was hypothesized that neuronal survival during development depended upon the transfer of soluble, 'trophic' or growth-supporting factors from their target (for a review, see Purves, 1988). The discovery of nerve growth factor (NGF) confirmed the target-derived neurotrophic factor hypothesis (for a review, see Levi-Montalcini and Angeletti, 1968). However, we now know that NGF is but one member of a family of at least four structurally related molecules, the neurotrophins (for reviews, see Thoenen, 1991; Lindsay *et al.*, 1994).

At least three kinds of approaches have been used to discover novel neurotrophic molecules. The most basic approach searches in tissue extracts or in conditioned media from cultured cells for novel molecules that will support the survival *in vitro* of specific neurones whose degeneration is characteristic of particular human neurodegenerative diseases, e.g. motor neurones in amyotrophic lateral sclerosis (Sendtner *et al.*, 1990). This strategy led directly to the discovery and subsequent molecular cloning of ciliary neurotrophic factor (CNTF) (Adler *et al.*, 1979; Lin *et al.*, 1989; Manthorpe *et al.*, 1989, 1992; Stöckli *et al.*, 1989), brain-derived neurotrophic factor (BDNF) (Barde *et al.*, 1982) and glial cell-line-derived neurotrophic factor (GDNF) (Lin *et al.*, 1993, 1994). A second strategy relies on molecular cloning techniques to identify additional members of gene families of known neurotrophic molecules or other proteins. This second approach resulted in the cloning of neurotrophin-3 (NT-3) (Maisonpierre *et al.*, 1990), first *Xenopus* and then human neurotrophin-4/5 (NT-4/5) (Hallbook *et al.*, 1991; Ip and Yancopoulos, 1992), and neurotrophin-6 (NT-6) in fish (Gotz *et al.*, 1994). A third strategy is to search for ligands that bind to neuronal receptors of the kind generally used by growth factors. In this approach, structural motifs that are shared by a known growth factor receptor family are identified. These novel, 'orphan receptor' proteins (e.g. receptor kinases) are identified by homology cloning, and then can be used to isolate, purify, identify and clone novel endogenous ligands that bind these orphan receptors with high affinity, as recently exemplified by the cloning of thrombopoietin.

Regardless of their method of discovery, neurotrophic factors can be loosely classified into two categories based upon the range of their biological actions within the organism. The first category consists of protein factors that act primarily on cells of the central and peripheral nervous systems and that are largely produced either by neurones or glial cells within the nervous system or by the primary targets of peripheral nerves (e.g. muscle, glands or sensory end-organs in the skin). These nervous system-specific trophic molecules include the NGF-related neurotrophins, namely BDNF, NT-3 and NT-4/5 as well as NGF itself, and possibly the transforming growth factor β (TFG-β) family member GDNF.

A second group of growth factors with neurotrophic activity consists of widely distributed proteins which, in addition to their trophic actions on neurones, also act as mitogens and growth factors on non-neural tissues. Examples of such families of pleiotrophic growth factors include the fibroblast growth factor (FGF) family, insulin and the insulin-like growth factors (IGFs), epidermal growth factor (EGF), platelet-derived growth factor (PDGF), members of the TGF family, and certain neuroactive cytokines (e.g. CNTF, interleukin-6 (IL-6) and leukaemia inhibitory factor (LIF)). (It should be noted that the name given a growth factor most often represents its first discovered biological action and may bear little relation to its predominant mode(s) of action in the body.)

Finally, in addition to the known neurotrophic factors, several neurotrophic *activities* have been identified in tissue extracts by *in vitro* methods or with *in vivo* lesion paradigms. For example, the survival and differentiation of high-affinity dopamine uptake and tyrosine hydroxylase capacities of cultured dopamine neurones is enhanced by soluble extracts from sciatic nerve (Collier *et al.*, 1990), neostriatum (Diporzio *et al.*, 1980; Prochiantz *et al.*, 1981; Tomozawa and Appel, 1986; Carvey *et al.*, 1989; Niijima *et al.*, 1990), muscle (Iacovetti and Lyandvert, 1989), type I astrocytes (O'Malley *et al.*, 1991) and glial cell lines (Engele and Bohn, 1991). The soluble factors responsible for these dopaminergic activities have not been definitively identified, with the exception of GDNF (Lin *et al.*, 1993), which corresponds to the dopaminergic neurotrophic activity identified in the B49 rat glial cell line by Engele and Bohn (1991).

10.2.1 Nervous system-specific growth factors

10.2.1.1 The neurotrophins

a. NGF

Because of its ability to support the survival of cultured peripheral sympathetic neurones and protect against chemical- or immuno-sympathectomy, it was once hoped that NGF would be a survival factor for central catecholaminergic neurones. Interestingly, this turned out not to be the case. In the brain, NGF acts as a survival factor for a restricted population of neurones that includes the cholinergic neurones of the basal forebrain (Hefti, 1986; Hefti and Will, 1987) and striatum (for a review, see Altar, 1991). For this reason, there has been considerable interest in NGF as a potential treatment for Alzheimer's disease (at least for that major component of the disease which is attributable to cholinergic dysfunction). On the other hand, NGFs lack of activity towards non-cholinergic central nervous system (CNS) neurones, including noradrenergic and dopaminergic neurones (e.g. Schwab *et al.*, 1979; Hyman *et al.*, 1991) has stimulated the search for other factors which might have trophic actions on other neuronal systems in the brain.

b. The NGF-related neurotrophins: BDNF, NT-3, NT-4/5 and NT-6

BDNF was initially identified as a neurotrophic activity in pig brain which supported the survival of cultured dorsal root ganglion neurones and whose activity could not be antagonized with antibodies to NGF (Barde *et al.*, 1982). In contrast to NGF, this activity also supported the *in vitro* survival of nodose ganglion neurones but was inactive on sympathetic neurones (Lindsay *et al.*, 1985). Upon cloning of the gene (Liebrock *et al.*, 1989), BDNF was found to be structurally related to NGF, with which it shares a 55% amino acid sequence homology. BDNF has since been shown to act on a broad spectrum of peripheral nervous system (PNS) and CNS neurones (for a review, see Lindsay, 1993; Lindsay *et al.*, 1994) including neural placode-derived sensory (Lindsay *et al.*, 1985; Davies *et al.*, 1986), hippocampal (Ip *et al.*, 1993b), cerebellar (Segal *et al.*, 1992; Lärkfors *et al.*, 1995), retinal ganglion (Johnson *et al.*, 1986) and motor neurones (Henderson *et al.*, 1993; Wong *et al.*, 1993).

Of particular relevance to this review, BDNF has potent survival and differentiating activity towards nigral dopaminergic neurones (Hyman *et al.*, 1991; Knüsel *et al.*, 1991), basal forebrain cholinergic neurones (Alderson *et al.*, 1990; Knüsel *et al.*, 1991), and medium spiny neurones of the striatum (Mizuno *et al.*, 1994; Ventimiglia *et al.*, 1995). The relatives of BDNF, NT-3 and NT-4/5, which share the ability to act through the TrkB receptor, also exert actions *in vitro* on these same populations of neurones (Alderson *et al.*, 1990; Hyman *et al.*, 1994; Ventimiglia *et al.*, 1993; Mizuno *et al.*, 1994), supporting cell survival and enhancing the expression of aspects of the neuronal phenotype such as the expression of neurotransmitter synthetic enzymes.

The receptors for the neurotrophins comprise a family of structurally related transmembrane receptor tyrosine kinases known as the Trks (for reviews, see Meakin and Shooter, 1992; Glass and Yancopoulos, 1993). TrkA is a high-affinity receptor for NGF, TrkB is a high-affinity receptor for BDNF and NT-4/5 (Glass *et al.*, 1991), and TrkC is a high-affinity receptor for NT-3, although, with somewhat lower affinity, NT-3 can bind to TrkB (Klein *et al.*, 1990; Glass *et al.*, 1991; Squinto *et al.*, 1991).

BDNF and NT-3 mRNA are present at extremely low or negligible levels in the adult rat striatum (Ernfors *et al.*, 1990; Hofer *et al.*, 1990; Maisonpierre *et al.*, 1990). Many dopamine neurones in the ventral tegmental area (VTA) and pars compacta contain BDNF or NT-3 mRNA (Gall *et al.*, 1992; Seroogy *et al.*, 1993b, 1994). Essentially all neurotrophin mRNA-positive VTA and pars compacta neurones are dopaminergic, and thus it is not surprising that most of the neurotrophin mRNA in the VTA and pars compacta is lost following extensive dopamine lesions (Numan and Seroogy, 1994).

The mRNAs for *trk*B and *trk*C, the high-affinity receptors for BDNF and NT-3, respectively, are also present in nigral neurones (Merlio *et al.*, 1992; Altar *et al.*, 1993, 1994b). High-affinity binding sites for [^{125}I]NT-3 are found in the pars compacta (Altar *et al.*, 1993), and the ability of BDNF as well as NT-3 to displace this binding is consistent with the presence of TrkB and TrkC receptors on these neurones. Intrastriatal infusions of BDNF and NT-3 or their iodinated counterparts result in a retrograde transport and accumulation of both ligands within dopaminergic somata (Wiegand *et al.*, 1992; Mufson *et al.*, 1995). Together, these observations raise the interesting speculation that BDNF and NT-3 (as well as other factors) might function in an autocrine or paracrine fashion, rather than as classical 'target-derived' neurotrophic substances in the adult nervous system (Lindsay *et al.*, 1993). In addition, BDNF or NT-3 made in nigral neurones could be released locally into the pars reticulata of the substantia nigra to promote the survival or maintenance of intrinsic GABAergic neurones (GABA, γ-aminobutyric acid) or striatonigral peptidergic neurones (Hyman *et al.*, 1991; Ventimiglia *et al.*, 1993, 1995; Arenas *et al.*, 1995).

10.2.1.2 GDNF

Recently, GDNF, a new member of the TGF-β superfamily of proteins was isolated, purified and cloned from a rat glial cell line. It was named on the basis of its ability to support the survival of dopaminergic neurones *in vitro* (Lin *et al.*, 1993, 1994). The mRNA for GDNF is present in rat striatum by the 17th day of gestation and at birth, but declines thereafter to reach low or undetectable levels by the fourth week of life (Lindqvist *et al.*, 1994; Schmidt-Kastner *et al.*, 1994). The absence of detectable mRNA levels in the adult striatum or nigra for GDNF (Strömberg *et al.*, 1993; Lindqvist *et al.*, 1994) raises the question as to whether GDNF ordinarily functions as an autocrine or target-derived trophic factor for adult dopamine neurones. Interestingly, GDNF mRNA in the olfactory tubercle persists into adulthood (Lindqvist *et al.*, 1994). Although initially discovered for its ability to promote the survival and differentiation of cultured dopamine neurones, GDNF has since been found in areas outside of the basal ganglia and shown to support the survival of non-dopaminergic neurones. For example, GDNF is also present in adult Schwann cells and skeletal muscle (Springer *et al.*, 1994; Henderson *et al.*, 1994) and promotes the survival of cultured motor neurones and prevents the atrophy of axotomized facial motor neurones (Henderson *et al.*, 1994). Unlike the neurotrophins, there are as yet no reports that describe the high-affinity binding or retrograde transport of GDNF, nor has the receptor for GDNF been identified.

10.2.2 Pleiotrophic growth factors and neurotrophic activity

In addition to the neurotrophins, GDNF and CNTF, a considerable number of growth factors which were initially identified by and named for their growth-promoting or mitogenic activities on non-neuronal cells are active in the CNS. These factors and their receptors are widely distributed in neural and non-neural tissue (for a review, see Lindsay *et al.*, 1993). Some of these factors are expressed in the basal ganglia and promote the survival or differentiation of dopamine neurones.

10.2.2.1 CNTF and neuroactive cytokines

CNTF was first described as an activity present in an extract of chick eye which promoted the survival of parasympathetic neurones of the ciliary ganglion (Adler *et al.*, 1979). High levels of CNTF activity were subsequently found in the rat sciatic nerve (Stöckli *et al.*, 1991). CNTF is not related to the neurotrophins, but rather displays a structural homology with the cytokines IL-6 and LIF (Bazan, 1991). CNTF is a potent neurotrophic factor for spinal motor neurones (Arawaka *et al.*, 1990; Wong *et al.*, 1993), certain sensory and sympathetic neurones (Manthorpe and Varon, 1985; Ernsberger *et al.*, 1989; Saadat *et al.*, 1989), hippocampal neurones (Ip *et al.*, 1991), and cerebellar Purkinje cells (Lärkfors *et al.*, 1994). The neurobiology of CNTF has been reviewed by Ip and Yancopoulos (1992).

CNTF produces neurotrophic effects on a wide variety of CNS and PNS neurones by binding to a tripartite receptor that shares two molecular components with the receptor complexes for the cytokines LIF and IL-6, a glycoprotein of 130 kDa (gp130) and the B subunit of the LIF receptor (LIFRβ; Davis *et al.*, 1991; Ip and Yancopoulos, 1992; Ip *et al.*, 1993a; Stahl *et al.*, 1993). The third component of the receptor, designated CNTFRα, confers specificity for CNTF. The CNTF receptor also shares structural motifs with the receptors for haemopoetic growth factors including erythropoietin, as well as the growth hormone receptor (Davis *et al.*, 1991).

CNTFRα is widely distributed in the CNS of the adult rat, where it is expressed abundantly on most neurones and in the retina (Ip and Yancopoulos, 1992). The mRNA for CNTFRα is prominently expressed by neurones of the pars reticulata of the substantia nigra, and is also seen at lower levels within the pars compacta (Lindsay *et al.*, 1993). In contrast to the widespread neuronal distribution of CNTFRα, CNTF protein and mRNA are present at very low levels in most brain regions, where expression is largely restricted to non-neuronal cells (Stöckli *et al.*, 1991; Dobrea *et al.*, 1992; Ip *et al.*, 1993a).

10.2.2.2 Other mitogenic growth factors

a. FGFs

Both acidic and basic FGF (aFGF and bFGF) mRNA are co-localized with tyrosine hydroxylase in dopamine neurones of the pars compacta (Bean *et al.*, 1991; Cintra *et al.*, 1991; Tooyama *et al.*, 1992) and become elevated within 1 week after an MPTP lesion of mouse nigrostriatal dopamine neurones (Leonard *et al.*, 1993). The ability of [^{125}I]bFGF to undergo retrograde transport from the neostriatum to pars compacta dopamine neurones (Ferguson and Johnson, 1991) demonstrates that increased striatal FGF levels also could augment FGF within nigral dopamine neurones. Interestingly, the number of bFGF-positive terminals is greatly diminished in the striatum in PD, as is bFGF staining within the surviving dopaminergic neurones (McGeer *et al.*, 1992).

b. IGFs

The IGF family members IGF-I and IGF-II are present in the adult rat substantia nigra (Haselbacher *et al.*, 1985; Rotwein *et al.*, 1988; Garcia-Segura *et al.*, 1991).

c. EGF and TGF-α

Immunoreactivity for EGF is present throughout the pallidal system, and is particularly dense in the substantia nigra (Fallon *et al.*, 1984). EGF receptor mRNA is expressed by the dopaminergic cells of the pars compacta and VTA, and is lost following extensive lesions of midbrain dopamine neurones (Seroogy *et al.*, 1995). During development, the mRNA for TGF-α, which binds to the EGF receptor, is abundantly expressed in striatal neurones and many other brain regions (Wilcox and Derynck, 1988; Seroogy *et al.*, 1993a,b). Interestingly, striatal TGF-α and EGF receptor mRNAs decline markedly from their early postnatal peaks to reach the lowest postnatal levels in the adult, whereas TGF-α mRNA levels increase during this time to appear in white matter (Seroogy *et al.*, 1993b).

10.3 Neurotrophic factors and dopaminergic neurons

10.3.1 *In vitro* actions of neurotrophic factors

10.3.1.1 Survival and differentiation

Many growth factors have been shown to promote the survival or differentiation of cultured dopamine neurones (Table 1). EGF promotes terminal arborization of mesencephalic precursor cells (Mytilineou *et al.*, 1992) or dopamine neurones (Casper *et al.*, 1991), as does TGF-α (Alexi and Hefti, 1992; Mayer *et al.*, 1993a). Survival and differentiation-promoting effects are also obtained with EGF and TGF-α and also

Table 1 Activities of Neurotrophic Factors on Dopaminergic Neurons

Ligand	Receptor	*In vitro* Survival	*In vivo* Effects	Protection or Rescue from Lesion
NGF	TrkA	none	none	none
BDNF	TrkB	+++; neurite outgrowth; Protects from MPP+ and 6-OHDA toxicity	Increases striatal dopamine turnover and nigral dopamine neuron firing rates; spontaneous and amphetamine-induced behaviors	Axotomy: preserves TOH phenotype, DA neuron survival. Partial DA lesion with 6-OHDA; behavior and DA turnover normalizes
NT-3	TrkC/TrkB	++	Increases spontaneous behaviors; Modest effects on amphetamine-induced behaviors or DA turnover	Axotomy: Preserves TOH, but weaker effects on survival and less functional recovery after partial 6-OHDA lesion than with BDNF
NT4/5	TrkB	++++; extensive neurite outgrowth	Increases striatal dopamine turnover and release; amphetamine-induced rotations	Axotomy: TOH phenotype, survival
GDNF	Unknown	+++; no protection from 6-OHDA cell loss but protects from MPP+ terminal loss	Increases nigral, striatal TOH; striatal dopamine content, turnover and release; motor behavior	Axotomy: Preserves TOH phenotype and DA neuron survival. MPTP: protection & recovery of TOH, dopamine metabolites, & behaviors. 6-OHDA: DA neuron and terminal rescue
CNTF	CNTFr/ LIFrB/gp130	none	Unknown	Axotomy: cell survival, no preservation of TOH
FGF	FGFr	+++; protects from MPP+; both effects require glia	Unknown	MPTP toxicity: DA cell survival, TOH phenotype phenotype and dopamine levels preserved
IGF-1/IGF-2	IGFr	++/+; no protection from MPP+	Unknown	Unknown
PDGF	PDGFr	++ or no effect	Unknown	Unknown
EGF	EGFr	++++; protects from MPP+, glia are required	Striatal DA, DOPAC, and TOH levels	Axotomy: DA cell survival, TOH phenotype and behavior preserved. MPTP: Striatal DA, DOPAC, and TOH levels
TGF-a	EGFr	+++	Unknown	Unknown

The Table summarizes the studies cited in the text and several references cited by R.M. Lindsay (Nature, **373**: 289–290, 1995).

with bFGF (Knüsel *et al.*, 1990; Casper *et al.*, 1991; Hartikka *et al.*, 1992; Mayer *et al.*, 1993a), PDGF (Nikkah *et al.*, 1993), IGF-I (Hartikka *et al.*, 1992), IGF-II (Knüsel *et al.*, 1991) and midkine (Kikuchi *et al.*, 1993) but not with PDGF or IL-1β (Engele and Bohn, 1991). In contrast to many of these factors, the ability of BDNF, NT-3, NT-4/5 (Knüsel *et al.*, 1990; Hyman *et al.*, 1991, 1994; Spina *et al.*, 1992; Strüder *et al.*, 1995) and GDNF (Lin *et al.*, 1993, 1994; Beck *et al.*, 1995) to enhance the survival and differentiation of cultured dopamine neurones can occur even at low cell densities and in glial-free cultures. The ability of BDNF to augment GABAergic or serotonergic neurone numbers or phenotype-specific markers in cultures of fetal rat or human mesencephalon (Hyman *et al.*, 1991; Spenger *et al.*, 1995) contrasts with GDNF, which does not appear to affect fetal rat serotonin or GABA neurones (Lin *et al.*, 1993).

10.3.1.2 Protection against neurotoxicity *in vitro*

In vitro systems can also be used to test the ability of neurotrophic factors to protect neurones from various types of injury. In addition to survival effects on dopaminergic neurones (Lin *et al.*, 1993; Hou and Mytilneou, 1994), GDNF also attenuates the loss of terminals of cultured dopamine neurones exposed to 1-methyl-4-phenylpyridinium (MPP^+), but not the cell loss induced by 6-hydroxydopamine (6-OHDA) (Hou and Mytilenou, 1994). BDNF can protect cultured dopamine neurones (or cells of the dopaminergic SH-SY5Y neuroblastoma line) from the neurotoxic effects of 6-OHDA or MPP^+ (Hyman *et al.*, 1991; Spina *et al.*, 1992; Fadda *et al.*, 1993). The protective effects of BDNF may involve antioxidative mechanisms, since BDNF increases the activity of the enzyme glutathione reductase and prevents the increase in oxidized glutathione which normally results from the exposure of these cells to 6-OHDA (Spina *et al.*, 1992). The mechanism of dopaminergic neuroprotection afforded by EGF (Park and Mytilineou, 1993) or bFGF (Hartikka *et al.*, 1992) against MPP^+ toxicity is unknown, although these protective actions appear to require the presence of glia. It is interesting to speculate that the glial cell requirement could involve the release of other supportive growth factors such as GDNF (Lin *et al.*, 1993, 1994) or CNTF (Rudge *et al.*, 1994).

10.3.2 *In vivo* actions of neurotrophic factors

10.3.2.1 Intact systems

The *in vivo* actions of some neurotrophic factors on dopamine neurones have been studied following chronic or acute infusion into the striatum or substantia nigra (Table 1). In otherwise intact animals, acute nigral injections of GDNF elevate general motor activity and increase nigral and striatal tyrosine hydroxylase immunoreactivity and striatal dopamine turnover (Hudson *et al.*, 1993). Similar effects have been observed in primates. An acute injection of GDNF into the macaque substantia nigra increases general activity, increases dopamine release in the caudate nucleus and putamen, and causes a proliferation of tyrosine hydroxylase positive neurites in the substantia nigra near the site of injection (Gerhardt *et al.*, 1994).

A single injection of BDNF into the substantia nigra augments the contraversive rotational responses to *d*-amphetamine for several months (Shults *et al.*, 1994). Continuous infusions of BDNF, NT-3 or NT-4/5 into the vicinity of the substantia nigra produce a more substantial increase in contraversive rotations and striatal dopamine turnover induced by the dopamine-releasing drug *d*-amphetamine (Martin-Iverson *et al.*, 1994). Supranigral infusions of BDNF or NT-4/5 also elevate striatal dopamine turnover (Altar *et al.*, 1992, 1994c). BDNF delivered in this manner also increases the number of pars compacta dopamine neurones that are spontaneously active, their average firing rate, and the number of action potentials during individual bursts of activity (Shen *et al.*, 1994). NT-3 appears to be less effective than BDNF or NT-4/5 both in its effects on striatal dopamine turnover and amphetamine-induced rotation. The greater effectiveness of BDNF and NT-4/5 suggests that the TrkB receptor may be a common mediator in these dopaminergic responses. This conclusion is supported by the greater potency of BDNF at the TrkB receptor compared with NT-3 (Squinto *et al.*, 1991) and, as described above, the preponderance of nigral TrkB receptor message and ligand binding compared with TrkC.

In contrast to the supranigral route of delivery, intrastriatal delivery of an equivalent daily amount of BDNF produces lesser effects on both rotational behaviour and striatal dopamine neurochemistry (Altar *et al.*, 1992). Supranigral infusions of BDNF may reach a greater proportion of nigrostriatal dopamine neurones than BDNF given into the striatum, in part because of its limited diffusion within the brain (Morse *et al.*, 1993; Anderson *et al.*, 1995).

10.3.2.2 Animal models of PD

The actions of a large number of growth factors have been studied in *in vivo* animal models of PD to determine their ability to promote the survival or function of dopamine neurones following injury (Table 1) or, as described in Section 10.5, following their transplantation to the denervated striatum.

a. NGF

NGF is not retrogradely transported from the striatum to the substantia nigra (Schwab *et al.*, 1979). This is consistent with the lack of TrkA, the high-affinity receptor for NGF, on nigrostriatal dopamine neurones (for a review, see Altar, 1991). Yet, unexpectedly, intraventricular (ICV) infusions of NGF have been reported to confer protection against 1-methyl-4-phenyl-1,2,5,6-tetrahydropyridine (MPTP) toxicity of dopamine neurones following intraventricular delivery of NGF (Garcia *et al.*, 1992). There are no other reports of the ability of NGF to protect injured dopamine neurones *in vivo* or *in vitro*.

b. BDNF, NT-3, NT-4/5 and CNTF

In a partial dopamine lesion model that mimics the intermediate stages of PD (Altar *et al.*, 1994a), nigral infusions of BDNF or NT-3 were each found to reverse amphetamine and apomorphine-induced rotational biases but not the loss of striatal dopamine nerve terminals induced by 6-OHDA. BDNF also augments basal and amphetamine-stimulated dopamine turnover in the surviving nigrostriatal nerve terminals.

A midbrain implantation of fibroblasts that contained a transgene for BDNF, but not those with the NGF transgene, attenuated the MPP^+-induced loss of pars compacta dopaminergic neurones (Frim *et al.*, 1994).

Hagg and colleagues (1994) found that following transection of the medial forebrain bundle, infusions of BDNF, NT-3 or NT-4/5 proximal to the knife cut attenuated the loss of pars compacta dopamine neurones, as determined by a preservation of cell number and the tyrosine hydroxylase phenotype of these cells. In the case of CNTF, dopamine neurone rescue in this model was manifest by the preservation of dopamine neuronal somata in the pars compacta but *not* their expression of tyrosine hydroxylase (Hagg and Varon, 1993). This finding corresponds to the inability of CNTF to prevent the loss of tyrosine hydroxylase or high-affinity dopamine uptake in cultured dopamine neurones, and suggests that CNTF may indeed prevent the loss of dopamine neurones *in vitro* independent of effects on the phenotype of these dopaminergic neurones.

c. GDNF

GDNF injections into the substantia nigra attenuate dopaminergic neurone losses induced by 6-OHDA injections into the striatum *or* substantia nigra (Kearns and Gash, 1994) or by axotomy (Beck *et al.*, 1995). Acute intranigral injections of GDNF attenuate both rotational responses to apomorphine and the loss in nigral dopamine content and tyrosine hydroxylase immunostaining produced by medial forebrain bundle infusions of 6-OHDA given weeks prior to GDNF treatment (Bowenkamp *et al.*, 1994). GDNF does not attenuate striatal dopamine losses in this model. In contrast, striatal injections of GDNF given to mice 1 day before or 7 days after MPTP administration was found by Tomac and colleagues (1995) to attenuate striatal as well as nigral dopamine depletions and to increase spontaneous motor activity.

d. FGF and EGF

The intraventricular infusion of bFGF and aFGF (Date *et al.*, 1990a) can each exert a modest attenuation of nigrostriatal neurone damage induced by MPTP. A similar protection has been obtained with ICV delivery of EGF (Pezzoli *et al.*, 1991), including a reduction in ipsiversive rotational behaviour after systemic *d*-amphetamine administration (Pezzoli *et al.*, 1991). The efficacy of EGF or bFGF in these models may be mediated via non-neuronal cells, since both growth factors depend upon the presence of glia for their *in vitro* trophic effects on dopamine neurones.

e. Interleukins

A slow release of IL-1 from polymer pellets implanted in the striatum several weeks after a selective loss of pars compacta neurones can reduce amphetamine-induced rotations and enhance DA terminal densities in the ventromedial and medial striatum (Wang *et al.*, 1994). Presumably, the behavioural compensations to IL-1 resulted from a growth of the mostly intact VTA dopamine neurones into the denervated striatum.

f. Gangliosides

Systemic injections of GM_1 ganglioside appear to protect against nigral cell losses or striatal dopamine losses, respectively, that are associated with hemitransection or MPTP-induced dopamine lesions in rats (Toffano *et al.*, 1983, 1984; Agnati *et al.*, 1984; Tilson *et al.*, 1988), mice (Schneider and Yuwiler, 1989; Weihmuller *et al.*, 1989), cats (Schneider, 1992) and primates (Schneider *et al.*, 1992; Herero *et al.*, 1993). In these studies, GM_1 prevented up to 30% of the loss of dopamine neurones, as measured by striatal dopamine levels or by tyrosine hydroxylase (TOH) immunocytochemistry, and decreased behavioural and neurological signs of nigrostriatal damage regardless of the species tested. The mechanism by which GM_1 gangliosides protect neurones in these or any other study remains unknown, and a related ganglioside preparation, bovine brain total gangliosides (tGS), produces essentially no improvement in neurochemical or behavioural deficits in rats with unilateral nigrostriatal lesions (Nishino *et al.*,1990). Interestingly, however, GM_1 ganglioside can render subthreshold concentrations of BDNF effective in protecting cultured dopamine neurones from 6-OHDA toxicity (Fadda *et al.*, 1993).

g. Other agents

Among other protective factors, iron chelation with desferrioxamine (Ben-Shachar *et al.*, 1991, 1992), ICV injections of neurotensin (Jolicoeur *et al.*, 1991), systemic injections of the adrenocorticotrophic hormone/melanocyte-stimulating hormone agonist analogue Org 2766 (Antonawich *et al.*, 1993), vitamin E (Cadet *et al.*, 1989), or the vigilance-enhancing drug modafinil (Fuxe *et al.*, 1992) can each protect against degeneration of dopamine cell bodies and their striatal terminals. The ability of vitamin E and iron to scavenge free radicals or prevent free radical formation by electron transfer, respectively, may contribute to the protection against 6-OHDA toxicity. The mechanisms of neurotensin, Org 2766 or modafinil remain unknown.

h. Unidentified factors active in dopamine lesion models

It is important to consider the role that up-regulation of *endogenous* factors or other responses to injury may play when invasive procedures are used to deliver a protective agent to the brain of lesioned animals or PD patients. In fact, a variety of *in vivo* studies demonstrate the ability of surgical trauma or peripheral tissue transplants to promote regenerative sprouting and improve neurological functions following partial dopamine neurone loss (for a review, see Kordower *et al.*, 1991). In rats or mice, these manipulations include sham transplantation as well as intrastriatal implants of liver, striatum (Przedborski *et al.*, 1991), sciatic nerve (Date *et al.*, 1990b; van Horne *et al.*, 1991), adrenal medulla, or non-chromaffin tissue (Bing *et al.*, 1979; Pezzoli *et al.*, 1988; Bohn *et al.*, 1987; Bohn and Kanuicki, 1990; Date *et al.*, 1990b), activated leukocytes (Ewing *et al.*, 1992), and even non-viable tissue (Bohn *et al.*, 1990). Similar effects have been observed in MPTP-lesioned primates that have received intrastriatal transplants of adrenal gland (Fiandica *et al.*, 1988; Hansen *et al.*, 1988) or a wound cavity alone (Plunkett *et al.*, 1990). The diffusable endogenous factor(s) or other mechanisms responsible for these protective effects remain unknown. The potential for trauma to confer protection in dopamine lesion paradigms warrants the use of caution in identifying the source of efficacy when trophic factors, tissue transplants or both are employed in animal models. This is even more relevant to clinical trials of PD, where small sample sizes or lack of certain controls may limit attributions for the effects obtained.

10.4 Neurotrophic factors and non-dopaminergic systems that degenerate in PD

In addition to degeneration of the nigrostriatal dopamine system, the neuropathology of PD also extends to cholinergic, noradrenergic and serotonergic systems (Bernheimer *et al.*, 1973; Perry *et al.*, 1985). Thus, it is interesting to note that the behavioural effectiveness of NT-3 in partial dopamine lesion models has been observed in the absence of changes in striatal dopamine metabolism (Altar *et al.*, 1994b), and supranigral infusions of BDNF have been shown to ameliorate amphetamine-induced rotation even in the virtual absence of nigrostriatal dopamine neurones (Wiegand *et al.*, 1993). Thus, non-dopaminergic neurones, such as peptidergic, GABAergic or serotonergic neurones, are likely to be involved in BDNF- and NT-3-induced behavioural compensations to nigrostriatal injury.

Neurotrophic factors tend not to be restricted in their actions to neurones of one particular phenotype. For example, infusions of bFGF attenuate the degeneration of axotomized basal forebrain cholinergic neurones (Anderson *et al.*, 1988; Schwaber *et al.*, 1991) as do BDNF and NT-4/5 (Morse *et al.*, 1993; Alderson *et al.*, 1995). In addition, BDNF increases the turnover of serotonin following midbrain, intrastriatal or intraventricular delivery (Martin-Iverson *et al.*, 1994; Siuciak *et al.*, 1994; Altar *et al.*, 1994a,c). BDNF and NT-3 also induce a local sprouting of serotonin neurones and attenuate serotonin axon losses caused by systemic injection of *p*-chloroamphetamine (Mamounas *et al.*, 1994).

Nigral or striatal infusions of BDNF or NT-4/5 also elevate the nigral content of substance P, substance K and several opioid peptides as well as their mRNAs in the striatum (Sauer *et al.*, 1994; Croll *et al.*, 1994; Arenas *et al.*, 1995). Increases in nigral GABAergic cell body size and glutamic acid decarboxylase (GAD) mRNA hybridization in the pars reticulata have also been observed during nigral infusions of BDNF to adult rats (Arenas *et al.*, 1995). One implication of these findings is that the activating effects of BDNF or NT-4/5 on dopamine neurone metabolism, release and firing rates may in part be mediated indirectly through striatonigral neuropeptide, nigral GABAergic systems, midbrain serotonergic neurones, or other neural systems. Changes in one or more of these could also contribute to the enhanced behavioural functions induced by nigral infusions of BDNF or NT-3 in rats with partial (Altar *et al.*, 1994a) or complete (Wiegand *et al.*, 1993) lesions of nigrostriatal dopamine neurones. A second implication of these findings is that neurotrophic factors such as BDNF, which affect survival and function of striatal neurones, might someday be of benefit in the treatment of movement disorders in addition to PD, such as

Huntington's disease and the striatonigral degenerations. Another corollary of these findings is that while dopamine neurones within a fetal ventral mesencephalon graft are likely to respond to trophic factors (see Section 10.5), neuropeptide, GABAergic and serotonergic neurones within such a graft may also be affected by neurotrophins.

10.5 The challenge of delivering protein growth factors to the brain

10.5.1 Limitations of delivery to the CNS

One of the obvious hurdles that may impede the use of neurotrophic factors as pharmacological agents is their inability to penetrate the blood–brain barrier. Even more obvious is the inability to utilize these molecules as oral agents even for peripheral indications due to their degradation within the gastrointestinal tract. A variety of strategies have been proposed to deal with these issues. Ideally, of course, one would like to have small-molecule drugs which could mimic the specific actions of a neurotrophic protein. However, despite advances in molecular modelling and a better understanding of structure–activity relationships of growth factors and their receptors, no small molecule has been discovered which can mimic the action of a protein growth factor, including the oldest or most clinically relevant proteins, such as insulin or growth hormone. In the absence of small-molecule mimetics, agents which up-regulate the endogenous production of neurotrophic factors is an appealing alternate strategy currently under study.

10.5.2 Methods of CNS delivery

10.5.2.1 ICV delivery

The most straightforward ways to deliver high molecular weight or small hydrophilic molecules into the CNS is by direct ICV or intrathecal injection into the cerebrospinal fluid (CSF). However, when the desired site of action lies deep within the substance of the brain, a steep concentration gradient must be established between the CSF and the affected neural tissues to drive the diffusion of drug to therapeutic levels (Pardridge, 1991). Once a steady state concentration of drug is reached in the CSF, the distance into the tissue through which a drug will diffuse is further limited by the receptors, reuptake sites, and degradative enzymes which can actively remove the drug from neural tissues (Pardridge, 1991; Anderson *et al.*, 1995).

Even though ICV administration is the usual method by which trophic factors are delivered to the CNS in animal experiments, surprisingly little is known of the eventual CNS distributions of such factors. The distribution of NGF and the related neurotrophins BDNF and NT-3 have been investigated most thoroughly. Following ICV administration, NGF diffuses relatively widely throughout the brain, and is retrogradely transported by cholinergic neurones within the septum, diagonal band and nucleus basalis (Ferguson *et al.*, 1991). In contrast, ICV-delivered BDNF shows only a limited entry into brain tissues, and negligible transport within forebrain cholinergic neurones (Morse *et al.*, 1993; Anderson *et al.*, 1995). The pattern of distribution seen with NT-3 is intermediate between that of NGF and BDNF. The different distribution profiles exhibited by these closely related factors is probably due to marked differences in the distribution and relative abundance of their respective high-affinity receptors (Yan *et al.*, 1994; Anderson *et al.*, 1995), including an especially dense concentration of TrkB within ventricular ependyma.

10.5.2.2 Targeted intraparenchymal delivery

The steep concentration gradient between the CSF and the targeted tissue that needs to be attained for ICV or intrathecal delivery becomes more of an issue as the size of the brain increases and when the target area lies deep within the brain. The direct infusion of the factor into the targeted brain area might be advantageous in these circumstances. Except for the neurotrophins, very little is known regarding the distribution of trophic factors following direct infusion into brain parenchyma. The neurotrophins exhibit the same relative differences in the extent of diffusion during intraparenchymal administration as they show with ICV delivery, with NGF diffusing a greater distance than NT-3 or BDNF (Anderson *et al.*, 1995). However, this disparity is smaller following intraparenchymal delivery than it is for ICV delivery. Furthermore, and in marked contrast to ICV delivery, BDNF is readily transported by TrkB-containing neurones following intraparenchymal administration (DiStefano *et al.*, 1992; Anderson *et al.*, 1995; Mufson *et al.*, 1995). It is particularly noteworthy that the higher local concentrations of BDNF achieved with targeted intraparenchymal infusion compared to ICV infusion is paralleled by increased efficacy in neurone rescue and behavioural models (Morse *et al.*, 1993; Siuciak *et al.*, 1994).

In addition to increased efficacy, local administration of trophic factors can also limit side-effects that might otherwise be obtained with ICV delivery. For example, ICV delivery of NGF produces a dose-dependent decrease in body weight (Williams, 1991). A similar anorectic effect can be seen following ICV delivery of the related neurotrophic factors BDNF, NT-3 and NT-4/5 (Lapchak and Hefti, 1992; Sauer *et al.*, 1993; Alderson *et al.*, 1995; Anderson *et al.*, 1995). In contrast, direct administration of equivalent amounts of these same factors into most regions of the CNS produces no effect, or only a small attenuation in weight gain (Altar *et al.*, 1992; Martin-Iverson *et al.*, 1994; Alderson *et al.*, 1995; Anderson *et al.*, 1995). Thus, targeted intraparenchymal administration may be an attractive alternative to ICV delivery of factors for the treatment of neurological injury or disease, especially if the affected structure lies deep within the brain. However, the practicality and long-term safety of directly delivering drugs into brain tissue, as opposed to the CSF, remains to be demonstrated.

10.5.2.3 Facilitated transport across the blood–brain barrier

Among the strategies which have been proposed for achieving delivery of systemically administered proteins to the brain is the use of carrier-mediated transendothelial transport (Pardridge, 1991). For example, NGF has been conjugated with antibodies to the transferrin receptor, which is relatively enriched on retinal and cerebral capillary endothelia (Pardridge, 1991). In the only convincing example of the effectiveness of this approach for a CNS indication, intravenous administration of NGF conjugated to the transferrin receptor antibody has been shown to enhance the survival of fetal basal forebrain cholinergic neurones transplanted into the anterior chamber of the eye (Friden *et al.*, 1993).

10.5.2.4 Cells as biological 'pumps'

An alternative to the use of a mechanical pump or cannula device for drug delivery to the CNS might include the implantation of cells into brain which secrete relatively large amounts of a particular neurotrophic factor, biosynthetic enzyme or transmitter. This approach includes the use of fibroblasts that have been genetically modified to secrete FGF (Gage *et al.*, 1988), BDNF (Frim *et al.*, 1994) or tyrosine hydroxylase (Bankiewicz *et al.*, 1994). Each approach has demonstrated behavioural efficacy in animal models of PD. However, the inability to precisely control or terminate the administration of the secreted protein may limit the appeal of such 'biological pumps'. The requirement of immunosuppression or the use of autologous cells as the source for such grafts further limits the appeal of this approach. As an alternative, transplantation of cells encapsulated within polymers that allow the diffusion of transmitters and trophic factors may largely overcome the above mentioned limitations. For example, transplantation of encapsulated bovine adrenal chromaffin cells to the rat striatum attenuates behavioural effects of nigrostriatal lesions, presumably through the release of dopamine produced by the encapsulated cells (Aebischer *et al.*, 1991).

10.6 Transplantation and trophic factors in PD

Transplantation of neural or paraneural tissues has shown considerable promise as an experimental therapy for PD (for reviews, see Lindvall, 1992; Kordower *et al.*, 1991; Brundin and Björklund, 1987). The incomplete nature of the recovery achieved with current methods of transplantation in experimental animals and human patients has been attributed principally to the suboptimal survival of the engrafted cells or their incomplete anatomical integration within the host striatum. For this reason, a great deal of effort has been directed towards the discovery of trophic factors which might support the survival of transplanted fetal dopamine neurones or paraneural tissues, or enhance their appropriate structural and functional integration into the recipient brain.

10.6.1 Adrenal chromaffin cell grafts and NGF

Among the paraneural tissues employed as biological sources of dopamine, the chromaffin cells of the adrenal medulla have been the most thoroughly studied. In early experiments, engraftment of autologous adrenal medullary tissue was reported to improve neurological function in both experimental animals and PD patients (for a review, see Kordower *et al.*, 1991). However, the neurological improvements proved to be transient. In these same studies, the long-term survival of chromaffin cells transplanted to the striatum was poor at best, and those cells that survived retained their endocrine phenotype.

Because cells of the sympathoadrenal lineage have long been known to respond to NGF, this trophic factor has been tested as an adjuvant to chromaffin cell transplantation to the dopamine-depleted rat striatum. An infusion of NGF into the vicinity of the transplant enhances chromaffin cell survival and induces a morphological transformation, such they exhibit a more neurone-like appearance and extend processes into contiguous regions of the host striatum (Strömberg *et al.*, 1985; Pezzoli *et al.*, 1988). The enhanced survival and phenotypic differentiation of the transplanted chromaffin cells is accompanied by a more complete and longer-lasting improvement in behaviour. This approach is now being employed in the clinic. To date, results have been published for only a single parkinsonian patient who received multiple chromaffin cell implants into the putamen and infusions of murine NGF (total dose 3.3 mg) administered for the first 23 days post-transplantation (Olson *et al.*, 1991). Unlike patients receiving adrenal medullary grafts alone, who typically show only an early, transient improvement in motor function, this individual exhibited an amelioration of symptoms that continued for at least 1 year. While it is difficult to draw conclusions from a single patient regarding the potential clinical efficacy of this approach, it is notable that the patient experienced no ill effects and did not develop antibodies against the infused murine NGF. This case also demonstrates that intracerebral delivery of trophic factors for at least several weeks is feasible using currently available technology.

10.6.2 Neurotrophic factors and transplantation of fetal dopaminergic neurones

Compared to paraneural tissues, transplantation of fetal dopamine neurones derived from the ventral mesencephalon (VM) has generally provided a more substantial and permanent amelioration of the neurological deficits exhibited by dopamine-depleted experimental animals (Björklund and Stenevi, 1979; Brundin and Björklund, 1987; Kordower *et al.*, 1991) and patients afflicted with PD (Lindvall *et al.*, 1990; Lindvall, 1992, Freed *et al.*, 1992; Olson *et al.*, 1991; Spencer *et al.*, 1992; Freeman *et al.*, 1994; Peschanski *et al.*, 1994). The greater efficacy obtained with grafts of fetal neural tissue is thought to occur because, unlike chromaffin cells, the engrafted fetal dopamine neurones survive indefinitely within the host brain and establish functionally active efferent connections with striatal neurones. Even so, functional recovery is incomplete, presumably due to the fact that relatively small numbers of fetal dopamine neurones survive the transplantation procedure and the extent of reinnervation of the host striatum is limited.

As for paraneural tissues, there has been considerable interest in the discovery of factors which might enhance the survival and phenotypic differentiation of engrafted fetal dopamine neurones. Unlike peripheral cathecholaminergic neurones, mesencephalic dopamine neurones do not express the high-affinity TrkA receptor for NGF, and consequently do not respond to NGF. Nevertheless, the existence of endogenous trophic molecules for dopamine neurones was indicated by the observation that co-culturing the embryonic VM with striatal extracts or tissue enhanced the survival and morphological differentiation of fetal dopamine neurones (Diporzio *et al.*, 1980; Hemmendinger *et al.*, 1981; Prochiantz *et al.*, 1981). Soluble molecules derived from adult sciatic nerve (Collier *et al.*, 1990), glial cell lines (Engele and Bohn, 1991) as well as a variety of other tissue sources (see Section 10.3.1.1) also were found to enhance the survival of fetal dopamine neurones in culture. The presence of dopaminotrophic molecules in these tissues was subsequently confirmed *in vivo*, by studies in which co-transplants of embryonic striatum (Brundin *et al.*, 1986; Yurek *et al.*, 1990) or sciatic nerve (van Horne *et al.*, 1991; Kordower *et al.*, 1991) were found to support the survival, differentiation and functional integration of fetal dopamine neurones placed into the dopamine-depleted striatum.

These important studies prompted direct assessments of various growth factors for their effects on VM transplants. For example, Giacobini and colleagues have examined the effects of PDGF and FGF on the development of fetal VM transplanted to the anterior chamber of the eye. Interestingly, PDGF dimers comprising distinct structural isoforms exerted distinct but complementary effects on the development of fetal dopamine neurones (Giacobini *et al.*, 1993a). Dimers of the AA isoform enhanced the outgrowth of dopaminergic fibres onto the host iris, but did not affect the dopamine cell number. In contrast, the survival but not neurite outgrowth of dopamine neurones was moderately enhanced by the PDGF BB isoform. A similar dichotomy of action was noted when different FGFs were compared; bFGF treatment increased the size of E16 VM grafts in this paradigm, and doubled the number of dopamine neurones but produced only a small effect on neurite outgrowth (Giacobini *et al.*, 1993b). Conversely, administration of aFGF produced a significant effect on neurite outgrowth, but did not increase the number of dopamine neurones or transplant size. Following transplantation of VM tissue into the dopamine-depleted striatum in hemiparkinsonian rats, administration of bFGF near the transplant site enhanced the amelioration of amphetamine-induced rotation produced by the transplants, and doubled the number of surviving dopaminergic neurones when compared to controls (Mayer *et al.*, 1993b).

GDNF appears to exert a much more profound effect on developing dopamine neurones than either PDGF or FGF (Strömberg *et al.*, 1993). GDNF administration doubled the size of the fetal VM grafts transplanted to the anterior chamber of the eye, and increased the number of dopamine neurones by at least 4–5-fold compared to control grafts. Interestingly, the density of TOH-immunopositive neurites within the transplants was only slightly increased, and there was no significant increase in fibre outgrowth onto the denervated host iris.

BDNF also has been reported to enhance the development of dopamine neurones within fetal substantia VM following transplantation to the striatum of dopamine-depleted rats (Yurek *et al.*, 1994). Compared to control grafted animals which received infusions of a phosphate-buffered saline vehicle (PBS), BDNF infusions of 4 weeks' duration greatly enhanced the effects of VM transplanted on amphetamine-induced rotation. Furthermore, BDNF treatment also increase the amount of dopamine released from the transplants into the host striatum following amphetamine challenge by up to 10-fold. In these animals, BDNF also increases by 4–5-fold the area of the host striatum which was reinnervated by TOH-immunoreactive fibres orginating from the transplants.

It is interesting to note that in a prior study, administration of a lower dose of BDNF for a shorter period of time (2 weeks) resulted in a much smaller effect on behaviour and no appreciable effect on the morphology of the transplanted dopamine neurones (Sauer *et al.*, 1993). This suggests that determination of the appropriate dose and duration of trophic factor treatment may be critical to achieving optimal transplant effects. Taken together, the above studies also suggest that even closely related trophic factors (e.g. the A versus B isoforms of PDGF) may differentially affect distinct aspects dopamine neurone development, e.g. cell survival versus neurite outgrowth. For this reason, optimal augmentation of neural transplantation procedures may require the application of multiple factors. On the other hand, it is also important to note that the effects of bFGF (Mayer *et al.*, 1993b) and BDNF (Yurek *et al.*, 1994) on dopamine cell number and neurite outgrowth, respectively, were retained for weeks following cessation of trophic factor treatment, as was the effect of such treatments on amphetamine-induced rotation. These observations suggest that once the graft is integrated with host tissue, the transplanted dopamine neurones may survive and function adequately in the absence of continued exogenous trophic support. Although the results of these experiments are encouraging, additional work in experimental animals is required before trophic factors can justifiably be applied as an adjuvant to the transplantation of human fetal dopamine neurones in PD patients.

It should also be noted that while transplant studies conducted to date have focused almost exclusively on the dopaminergic neurones contained within grafts of fetal VM, such grafts typically contain a variety of other cell types (e.g. GABAergic neurones of the pars reticulata) whose survival and integration with the host brain may also be beneficially affected by neurotrophic factors (see Section 10.4). The functional consequences of enhancing the survival of non-dopaminergic components of VM transplants are unknown.

10.7 Conclusion and future directions

The CNS actions of several families of nerve growth factors, in particular the neurotrophins and GDNF, suggest a role for these proteins in the potential treatment of PD. Preliminary preclinical studies demonstrate that GDNF, BDNF or NT-4/5 may be of potential clinical use in preventing the degeneration or augmenting the function of surviving dopaminergic and other afferent neuronal systems in PD. BDNF offers additional promise because of its ability to promote the outgrowth and functional effects of fetal rat VM transplants in rat models of PD (Yurek *et al.*, 1994). Because cultured fetal human ventral mesencephalic dopamine neurones also respond to BDNF (Spenger *et al.*, 1995), future clinical investigations may include BDNF or other factors as adjuvants to fetal nigral transplantation. The next several years offer tremendous promise that a growth factor may contribute significantly to the development of a restorative therapy for PD.

References

Adler, R., Landa, K.B., Manthorpe, M. & Varon, S. (1979) *Science* **204**, 1434–1436.

Aebischer, P., Tresco, P.A., Sagen, J. & Winn, S.R. (1991) *Brain Res.* **560**, 43–49.

Agid, Y., Javoy-Agid, F. & Ruberg, M. (1992) In *Movement Disorders 2* (eds Marsden, C.D. & Fahn, S.), pp. 166–230. Butterworth, Oxford.

Agnati, L.F., Fuxe, K., Calza, L., Goldstein, M., Toffano, G., Giardino, L. & Zoli, M. (1984) *Acta Physiol. Scand. Suppl.* **532**, 37–44.

Alderson, R.F., Alterman, A.L., Barde, Y.-A. & Lindsay, R.M. (1990) *Neuron* **5**, 297–306.

Alderson, R.F., Alterman, A., Morissey, D. & Lindsay, R.M. (1993) *Soc. Neurosci. Abstr.* **19**, 666.

Alderson, R.F., Wiegand, S.J., Cai, N., Anderson, K.A., Lindsay, R.M. & Altar, C.A. (1995) *Eur. J. Neurosci.* (in press).

Alexi, T. & Hefti, F. (1992) *Soc. Neurosci. Abstr.* **18**, 1292.

Altar, C.A. (1991) *Progr. Neuropsychopharmacol. Biol. Psychiatry* **15**, 157–169.

Altar, C.A., Marien, M.R. & Marshall, J.F. (1987) *J. Neurochem.* **48**, 390–399.

Altar, C.A., Boylan, C.B., Jackson, C., Hershenson, S., Miller, J., Wiegand, S.J., Lindsay, R.M. & Hyman, C. (1992) *Proc. Natl Acad. Sci. USA* **89**, 11347–11351.

Altar, C.A., Criden, M.R., Lindsay, R.M. & DiStefano, P.S. (1993) *J. Neurosci.* **13**, 733–743.

Altar, C.A., Boylan, C.B., Fritsche, M., Jackson, C. Lindsay, R.M. & Hyman, C. (1994a) *J. Neurochem.* **63**, 1021–1032.

Altar, C.A., Siuciak, J.A., Wright, P., Lindsay, R.M., Ip, N.Y. & Wiegand, S.J. (1994b) *Eur. J. Neurosci.* **6**, 1389–1405.

Altar, C.A., Boylan, C.B., Fritsche, M., Jackson, C., Lindsay, R.M. & Hyman, C. (1994c) *Exp. Neurol.* **130**, 31–40.

Anderson, K.J., Dam, D., Lee, S. & Cotman, C.W. (1988) *Science* **332**, 360–361.

Anderson, K.A., Alderson, R.F., Altar, C.A., DiStefano, P.S., Corcoran, T.L., Lindsay, R.M. & Wiegand, S.J. (1995) *J. Comp. Neurol.* **357**, 296–317.

Antonawich, F., Azmitia, E. & Strand, F. (1993) *Peptides* **14**, 1317–1424.

Arawaka, Y., Sendtner, M. & Thoenen, H. (1990) *J. Neurosci.* **10**, 3507–3515.

Arenas, E., Akerud, P., Wong, V., Boylan, C., Persson, H., Lindsay, R.M. & Altar, C.A. (1995) *Eur. J. Neurosci.* (in press).

Bankiewicz, K.S., Emborg, M., Gage, F., McLaughlin, W.W. & Sofroniew, M.V. (1994) *Abstr. Soc. Neurosci.* **20**, 1329.

Barde, Y.-A., Edgar, D. & Thoenen, H. (1982) *EMBO J.* **1**, 549–553.

Bazan, J.F. (1991) *Neuron* **7**, 197–208.

Bean, A.J., Elde, R., Cao, Y., Oellig, C., Tamminga, C., Goldstein, M., Pettersson, R.F. & Hökfelt, T. (1991) *Proc. Natl Acad. Sci. USA* **88**, 10237–10241.

Beck, K.D., Valverde, J., Alexi, T., Poulsen, K., Moffat, B., Vandlen, R.A., Rosenthal, A. & Hefti, F. (1995) *Nature* **373**, 339–341.

Ben-Shachar, D., Eshel, G., Finberg, J.P.M. & Youdim, M.B.H. (1991) *J. Neurochem.* **56**, 1441–1444.

Ben-Shachar, D., Eshel, D., Riederer, G. & Youdim, M.B.H. (1992) *Ann. Neurol.* **32**, S105–S110.
Bergman, H., Wichmann, T. & DeLong, M.R. (1990) *Science* **249**, 1436–1438.
Bernheimer, H., Birkmayer, W. & Hornykiewicz, O. (1973) *J. Neurol. Sci.* **20**, 415–455.
Bing, G.Y., Notter, M.F. & Hansen, J.T. (1979) *Brain Res. Bull.* **20**, 399–406.
Björklund, A. & Stenevi, U. (1979) *Brain Res.* **177**, 555–560.
Bohn, M.C. & Kanuicki, M. (1990) *J. Neurosci. Res.* **25**, 281–286.
Bohn, M.C., Cupit, L., Marciano, F. & Gash, D.M. (1987) *Science* **237**, 913–916.
Bowenkamp, K.E., Hoffman, A.F., Granholm, A.-C., Henry, M.A., Biddle, P., Huettl, P., Lin, L.-F., Martin, D., Hudson, J.L., Gerhardt, G.A. & Hoffer, B.J. (1994) *Soc. Neurosci. Abstr.* **20**, 1101.
Brundin, P. & Björklund, A. (1987) *Progr. Brain Res.* **71**, 293–308.
Brundin, P., Isacson, O., Gage, F.H. & Björklund, A. (1986) *Dev. Brain Res.* **24**, 77–84.
Cadet, J.L., Katz, M., Jackson-Lewis, V. & Fahn, S. (1989) *Brain Res.* **476**, 10–15.
Carvey, P.M., Ptak, L.R., Kao, L. & Klawans, H.L. (1989) *Clin. Neuropharmacol.* **12**, 425–434.
Casper, D., Mytilineou, C. & Blum, M. (1991) *J. Neurosci. Res.* **30**, 372–381.
Cedarbaum, J. & McDowell, F. (1986) *Adv. Neurol.* **45**, 469–472.
Cedarbaum, J. & Olanow, C. (1992) *The Scientific Basis for the Treatment of Parkinson's Disease*, pp. 119–137. Parthenon, London.
Cintra, A., Cao, Y., Oelling, C., Tinner, B., Bortolotti, F., Goldstein, M., Petterson, R.F. & Fuxe, K. (1991) *NeuroReport* **2**, 597–600.
Collier, T.J., Sladek, C.D., Gallagher, M.J., Gereau IV, R.W. & Springer, J.E. (1990) *J. Neurosci. Res.* **27**, 394–399.
Cotzias, C., Van Wort, M. & Schiffer, L. (1969) *N. Engl. J. Med.* **276**, 374–376.
Croll, S.D., Wiegand, S.J., Anderson, K.D., Lindsay, R.M. & Nawa, H. (1994) *Eur. J. Neurosci.* **6**, 1343–1353.
Date, I., Notter, M.F.D., Felten, S.Y. & Felten, D.L. (1990a) *Brain Res.* **526**, 156–160.
Date, I., Felten, S.Y., Olschowka, J.A. & Felten, D.L. (1990b) *Exp. Neurol.* **107**, 197–207.
Davies, A.M., Thoenen, H. & Barde, Y.-A. (1986) *J. Neurosci.* **6**, 1897–1904.
Davis, S., Aldrich, T.H., Valenzuela, D.M., Wong, V., Furth, M.E., Squinto, S.P. & Yancopoulas, G.D. (1991) *Science* **253**, 59–63.
Diporzio, U., Daguet, M.-C., Glowinski, J. & Prochiantz, A. (1980) *Nature* **288**, 370–373.
Di Stefano, P.S., Friedman, B., Radziejewski, C., Alexander, C., Boland, P., Schick, C.M., Lindsay, R.M. & Wiegand, S.J. (1992) *Neuron* **8**, 983–993.
Dobrea, G.M., Unnerstall, J.R. & Rao, M.S. (1992) *Dev. Brain Res.* **66**, 209–219.
Engele, J. & Bohn, M.C. (1991) *J. Neurosci.* **11**, 3070–3078.
Ernfors, P., Wetmore, C., Olson, L. & Persson, H. (1990) *Neuron* **5**, 511–526.
Ernsberger, V., Sendtner, M. & Rohrer, H. (1989) *Neuron* **2**, 1275–1284.
Ewing, S.E., Weber, R.J., Zauner, A. & Plunkett, R.J. (1992) *Brain Res.* **576**, 42–48.
Fadda, E., Negro, A., Facci, L. & Skaper, S.D. (1993) *Neurosci. Lett.* **159**, 147–150.
Fallon, J.H., Seroogy, K.B., Loughlin, S.E., Morrison, R.S., Bradshaw, R.A., Knauer, D.J. & Cunningham, D.D. (1984) *Science* **224**, 1107–1109.
Ferguson, I.A. & Johnson, E.M. Jr (1991) *J. Comp. Neurol.* **313**, 693–736.
Ferguson, I.A., Schweitzer, J.B., Bartlett, P.F. & Johnson, E.M. Jr (1991) *J. Comp. Neurol.* **313**, 680–692.
Fiandica, M.S., Kordower, J.H., Hansen, J.T., Jiao, S.-S. & Gash, D.M. (1988) *Exp. Neurol* **102**, 76–91.
Freed, C., Breeze, R.E., Rosenberg, N.L. *et al.* (1992) *N. Engl. J. Med.* **327**, 1549–1555.
Freeman, T.B., Hauser, R., Sandberg, P., Snow, B. & Olanow, C.W. (1994) *Neurology* **44**, A234.
Friden, P., Walus, L., Watson, P., Doctrow, S., Kozarich, J., Backman, C., Bergman, H., Hoffer, B., Bloom, F. & Granholm, A. (1993) *Science* **259**, 373–377.
Frim, D.M., Uhler, T.A., Galpern, W.R., Beal, M.F., Breakfield, X.O. & Isacson, O. (1994) *Proc. Natl Acad. Sci. USA* **91**, 5104–5108.
Fuxe, K., Janson, A.M., Rosén, L., Finnman, U.-B., Tanganelli, S., Morari, M., Goldstein, M. & Agnati, L.F. (1992) *Exp. Brain Res.* **88**, 117–130.
Gall, C.M., Gold, S.J., Isackson, P.J. & Seroogy, K.B. (1992) *Mol. Cell. Neurosci.* **3**, 56–63.

Garcia, E., Rios, C. & Sotelo, J. (1992) *Neurochem. Res.* **17**, 979–982.
Garcia-Segura, L.M., Perez, J., Pons, S., Rejas, M.T. & Torres-Aleman, I. (1991) *Brain Res.* **560**, 167–174.
Gerhardt, G.A., Cass, W.A., Zhang, Z., Ovadia, A., Cheesman, M., Hoffer, B.J., Huettl, P., Russell, D., Martin, D. & Gash, D.M. (1994) *Soc. Neurosci. Abstr.* **20**, 1101.
Giacobini, M.M.J., Almström, S., Funa, K. & Olson, L. (1993a) *Neuroscience* **57**, 923–929.
Giacobini, M.M.J., Strömberg, I., Almström, S., Cao, Y. & Olson, L. (1993b) *Dev. Brain Res.* **75**, 65–73.
Glass, D.J. & Yancopoulos, G.D. (1993) *Trends Cell Biol.* **3**, 262–268.
Glass, D.J., Nye, S.H., Hantzopoulos, P., Macchi, M.J., Squinto, S.P., Goldfarb, M. & Yancopoulos, G.D. (1991) *Cell* **66**, 405–413.
Götz, R., Koster, R., Winkler, C., Raulf, F., Lottspeich, F., Schartl, M. & Thoenen, H. (1994) *Nature* **372**, 266–269.
Hagg, T. (1994) *Abstr. Soc. Neurosci.* **20**, 441.
Hagg, T. & Varon, S. (1993) *Proc. Natl Acad. Sci. USA* **90**, 6315–6319.
Hallböök, F., Ibáñez, C.F. & Perssön, H. (1991) *Neuron* **6**, 247–252.
Hansen, J.T., Kordower, J.H., Fiandica, M.S., Jiao, S.-S., Notter, M.F.D. & Gash, D.M. (1988) *Exp. Neurol.* **102**, 65–75.
Hartikka, J., Staufenbiel & Lübbert, H. (1992) *J. Neurosci. Res.* **32**, 190–201.
Haselbacher, G.K., Schwab, M.E., Pasi, A. & Humbel, R.E. (1985) *Proc. Natl Acad. Sci. USA* **82**, 2153–2157.
Hefti, F. (1986) *J. Neurosci.* **8**, 2155–2162.
Hefti, F. & Will, B. (1987) *J. Neural Transm.* **24**, 309–315.
Hemmendinger, L.M., Garber, B.B., Hoffman, P.C. & Heller, A. (1981) *Proc. Natl Acad. Sci. USA* **78**, 1264–1268.
Henderson, C.E., Camu, W., Mettling, C., Gouin, A., Poulsen, K., Karihaloo, M., Rullamas, J., Evans, T., McMahon, S.B., Armanini, M.P., Berkemeier, L., Phillips, H.S. & Rosenthal, A. (1993) *Nature* **363**, 266–270.
Henderson, C.E., Phillips, H.S., Pollock, R.A., Davies, A.M., Lemeulle, C., Armanini, M., Simpson, L.C., Moffett, B., Vandlen, R.A., Koliatsos, V.E. & Rosenthal, A. (1994) *Science* **266**, 1062–1064.
Herero, M.T., Perez-Otano, I., Oset, C., Kastner, A., Hirsch, E.C., Agid, Y., Luqin, M.R., Obeso, J.A. & Del-Rio, J. (1993) *Neuroscience* **56**, 965–972.
Hofer, M., Pagliusi, S.R., Hohn, A., Leibrock, J. & Barde, Y.-A. (1990) *EMBO J.* **9**, 2459–2464.
Hou, J.G. & Mytilineou, C. (1994) *Soc. Neurosci. Abstr.* **20**, 1101.
Hudson, J., Granholm, A.-C., Gerhardt, G., Hoffman, A., Biddle, P., Leela, N.S., Collins, F. & Hoffer, B. (1993) *Soc. Neurosci. Abstr.* **19**, 652.
Hyman, C., Hofer, M., Barde, Y.-A., Juhasz, M., Yancopoulos, G.D., Squinto, S.P. & Lindsay, R.M. (1991) *Nature* **350**, 230–232.
Hyman, C., Juhasz, M., Jackson & Lindsay, R.M. (1994) *J. Neurosci.* **14**, 335–347.
Iacovetti, L. & Lyandvert, L. (1989) *Soc. Neurosci. Abstr.* **15** Pt 2, 277.
Ip, N.Y. & Yancopoulos, G.D. (1992) *Progr. Growth Factor Res.* **4**, 139–155.
Ip, N.Y., Li, Y., Vandestadt, I., Panayotatos, N., Alderson, R.F. & Lindsay, R.M. (1991) *J. Neurosci.* **11**, 3124–3134.
Ip, N.Y., McClain, J., Barrezueta, N.X., Aldrich, T.H., Pan, L., Li, Y., Wiegand, S.J., Friedman, B., Davis, S. & Yancopoulos, G.D. (1993a) *Neuron* **10**, 89–102.
Ip, N.Y., Li, Y., Yancopoulos, G.D. & Lindsay, R.M. (1993b) *J. Neurosci.* **13**, 3394–3405.
Johnson, J.E., Barde, Y.-A., Schwab, M. & Thoenen, H. (1986) *J. Neurosci.* **6**, 3031–3038.
Jolicoeur, F.B., Rivest, R., St.-Pierre, S. & Drumheller, A. (1991) *Brain Res.* **538**, 187–192.
Kearns, C. & Gash, D.M. (1994) *Soc. Neurosci. Abstr.* **20**, 1101.
Kikuchi, S., Muramatsu, H., Muramatsu, T. & Kim, S.U. (1993) *Neurosci. Lett.* **160**, 9–12.
Klein, R., Conway, D., Parada, L.F. & Barbacid, M. (1990) *Cell* **61**, 647–656.
Knüsel, B., Michel, P.P., Schwaber, S. & Hefti, F. (1990) *J. Neurosci.* **10**, 558–570.

Knüsel, B., Winslow, J.W., Rosenthal, A., Burton, L.E., Seid, D.P., Nikolics, K. & Hefti, F. (1991) *Proc. Natl Acad. Sci. USA* **88**, 961–965.
Kordower, J.H., Felten, D.L. & Gash, D.M. (1991) *The Scientific Basis for the Treatment of Parkinson's Disease*, pp. 175–224. Parthenon, Park Ridge, NJ.
Lapchak, P.A. & Hefti, F. (1992) *NeuroReport* **3**, 405–408.
Lärkfors, L., Lindsay, R.M. & Alderson, R. (1994) *Eur. J. Neurosci.* **6**, 1015–1025.
Lee, T., Seeman, P. & Rajput, A. (1978) *Nature* **273**, 59–61.
Leonard, S., Luthman, D., Logel, J., Luthman, J., Antle, C., Freedman, R. & Hoffer, B. (1993) *Mol. Brain Res.* **18**, 275–284.
Levi-Montalcini, R. & Angeletti, P.U. (1968) *Physiol. Rev.* **48**, 534–569.
Lieberman, A. & Goldstein, M. (1982) *Movement Disorders*, pp. 146–164. Butterworth, Boston.
Liebrock, J., Lottspeich, F., Hohn, A., Hofer, M., Hengerer, B., Masiakowski, P., Thoenen, H. & Barde, Y.-A. (1989) *Nature* **341**, 149–152.
Lin, L.-F.H., Mismer, D., Lile, J.D., Armes, L.G., Butler, E.T. Jr, Vannice, J.L. & Collins, F. (1989) *Science* **246**, 1023–1025.
Lin, L.-F.H., Doherty, D.H., Lile, J.D., Bektesh, S. & Collins, F. (1993) *Science* **260**, 1130–1132.
Lin, L.-F.H., Zhang, T.J., Collins, F. & Armes, L.G. (1994) *J. Neurochem.* **14**, 165–170.
Lindqvist, E., Tomac, A., Luthman, J., Hoffer, B., Bektesh, S. & Olson, L. (1994) *Abstr. Soc. Neurosci.* **20**, 1101.
Lindsay, R.M. (1993) In *Neurotrophic Factors* (eds Loughlin, S.E. & Fallon, J.H.), pp. 257–284. Academic Press, New York.
Lindsay, R.M., Thoenen, H. & Barde, Y.-A. (1985) *Dev. Biol.* **112**, 319–328.
Lindsay, R.M., Altar, C.A., Cedarbaum, J.M., Hyman, C. & Wiegand, S.J. (1993) *Exp. Neurol.* **124**, 103–118.
Lindsay, R.M., Wiegand, S.J., Altar, C.A. & DiStefano, P.S. (1994) *Trends Neurosci.* **17**, 182–190.
Lindvall, O. (1992) *Trends Neurosci.* **14**, 376–384.
Lindvall, O., Widner, H., Rhencrona, S. *et al.* (1992) *Ann. Neurol.* **31**, 155–165.
McGeer, P.L., Tooyama, I., Kawamata, T., Walker, D., Yamada, T., Hanai, K., Kimure, H., Igarashi, M. & McGeer, E.G. (1992) *Soc. Neurosci. Abstr.* **18**, 1248.
Mamounas, L.A., Blue, M.E., Siuciak, J.A. & Altar, C.A. (1994) *Soc. Neurosci. Abstr.* **20**, 441.
Maisonpierre, P.C., Belluscio, L., Squinto, S., Ip, N.Y., Furth, M.E., Lindsay, R.M. & Yancopoulos, G.D. (1990) *Science* **247**, 1446–1451.
Manthorpe, M. & Varon, S. (1985) *Growth and Maturation Factors*, pp. 77–117. Wiley, New York.
Manthorpe, M., Ray, J., Pettman, B. & Varon, S. (1989) *Nerve Growth Factors*. John Wiley, New York.
Manthorpe, M., Louis, J.C., Hagg, T. & Varon, S. (1992) *Neurotrophic Factors*. Academic Press, San Diego.
Martin-Iverson, M.T., Todd, K.G. & Altar, C.A. (1994) *J. Neurosci.* **14**, 1261–1270.
Mayer, E., Dunnett, S.B., Pellitteri, R. & Fawcett, J.W. (1993a) *Neuroscience* **56**, 379–388.
Mayer, E., Fawcett, J.W. & Dunnett, S.B. (1993b) *Neuroscience* **56**, 389–398.
Meakin, S.O. & Shooter, E.M. (1992) *Trends Neurosci.* **15**, 323–331.
Merlio, J.P., Ernfors, P., Jaber, M. & Persson, H. (1992) *Neuroscience* **51**, 513–532.
Mizuno, K., Carnahan, J. & Nawa, H. (1994) *Dev. Biol.* **165**, 243–256.
Morse, J.K., Wiegand, S.J., Anderson, K.A., You, Y., Cai, N., DiStefano, P.S., Altar, C.A., Lindsay, R.M. & Alderson, R.F. (1993) *J. Neurosci.* **13**, 4146–4156.
Mufson, E.J., Kroin, J.S., Sobreviela, T., Burke, M.A., Kordower, J.H., Penn, R.D. & Miller, J. (1995) *Exp. Neurol.* (in press).
Mytilineou, C., Park, T.H. & Shen, J. (1992) *Neurosci. Lett.* **135**, 62–66.
Niijima, K., Araki, M., Ogawa, M., Nagatsu, I., Sato, F., Kimura, H. & Yoshida, M. (1990) *Brain Res.* **528**, 151–154.
Nikkah, G., Odin, P., Smits, A., Tingström, A., Otrthberg, A., Brundin, P., Funa, K. & Lindvall, O. (1993) *Exp. Brain Res.* **92**, 516–523.
Nishino, H., Hashitani, T., Isobe, Y., Furuyama, F., Sato, H., Kumazaki, M., Horikomi, K. & Awaya, A. (1990) *Brain Res.* **534**, 73–82.

Numan, S. & Seroogy, K.B. (1994) *Abstr. Soc. Neurosci.* **20**, 1306.
O'Malley, E.K., Black, I.B. & Dreyfus, C.F. (1991) *Exp. Neurol.* **112**, 40–48.
Olson, L., Backlund, E., Ebendal, T., Freedman, R., Hamberger, B., Hansson, P., Hoffer, B., Lindholm, U., Meyerson, B., Strömberg, I., Sydow, O. & Seiger, A. (1991) *Arch. Neurol.* **48**, 373–381.
Oppenheim, R.W. (1991) *Annu. Rev. Neurosci.* **14**, 453–501.
Pardridge, W. (1991) *Peptide Drug Delivery to the Brain.* Raven Press, New York.
Park, T.H. & Mytilineou, C. (1993) *Brain Res.* **599**, 83–97.
Perschanski, M., Defer, G. & N'Guyen, J.P. (1994) *Brain* **117**, 487–499.
Perry, R., Curtis, M. & Dick, D. (1985) *J. Neurol Neurosurg. Psychiatry* **48**, 413.
Pezzoli, G., Fahn, S., Swork, A., Truong, D.D., de Yebenes, J.G., Jackson-Lewis, V., Herbert, J. & Cadet, J.L. (1988) *Brain Res.* **459**, 398–403.
Pezzoli, G., Zecchinelli, A., Ricciardi, S., Burke, R.E., Fahn, S., Scarlato, G. & Carenzi, A. (1991) *Mov. Dis.* **6**, 281–287.
Plunkett, R.J., Bankiewicz, K.S., Cummins, A.C., Miletich, R.S., Schwartz, J.P. & Oldfield, E.H. (1990) *J. Neurosurg.* **73**, 918–926.
Prochiantz, A., Daguet, M.C., Herbert, A. & Glowinski, J. (1981) *Nature* **293**, 570–572.
Przedborski, S., Levivier, M., Kostic, V., Jackson-Lewis, V., Dollison, A., Gash, D.M., Fahn, S. & Cadet, J.L. (1991) *Brain Res.* **550**, 231–238.
Purves, D. (1988) *Body and Brain: A Trophic Theory of Neural Connections*, pp. 123–141. Harvard University Press, Cambridge, MA.
Rotwein, P., Burgess, S.K., Milbrandt, J.D. & Krause, J.E. (1988) *Proc. Natl Acad. Sci. USA* **85**, 265–269.
Rudge, J.S., Li, Y., Pasnikowski, E.M., Mattson, K., Pan, L, Wiegand, S.J., Lindsay, R.M. & Ip, N.Y. (1994) *Eur. J. Neurosci.* **6**, 693–705.
Saadat, S., Sendtner, M. & Rohrer, H. (1989) *J. Cell Biol.* **108**, 1807–1816.
Sauer, H., Fischer, W., Nikkah, G., Wiegand, S.J., Brundin, P., Lindsay, R.M. & Björklund, A. (1993) *Brain Res.* **626**, 37–44.
Sauer, H., Campbell, K., Wiegand, S.J., Lindsay, R.M. & Bjorklund, A. (1994) *NeuroReport* **5**, 609–612.
Schmidt-Kastner, R., Tomac, A., Hoffer, B., Bektesh, S., Rosenzweig, B. & Olson, L. (1994) *Soc. Neurosci. Abstr.* **20**, 1101.
Schneider, J.A. & Yuwiler, A. (1989) *Exp. Neurol.* **105**, 171–181.
Schneider, J.A., Pope, A., Simpson, K., Taggart, J., Smith, M.G. & DiStefano, L. (1992) *Science* **256**, 843–846.
Schneider, J.S. (1992) *J. Neurosci. Res.* **31**, 112–118.
Schwab, M.E., Otten, U. & Thoenen, H. (1979) *Brain Res.* **168**, 473–483.
Schwaber, J.S., Due, B.R., Rogers, W.T., Junard, E.O., Sharma, A. & Hefti, F. (1991) *J. Comp. Neurol.* **309**, 27–39.
Segal, R.A., Takahashi, H. & McKay, R.D.G. (1992) *Neuron* **9**, 1041–1052.
Sendtner, M., Kreutzberg, G.W. & Thoenen, H. (1990) *Nature* **345**, 440–441.
Seroogy, K.B. & Gall, C.M. (1993) *Exp. Neurol.* **124**, 119–128.
Seroogy, K.B., Han, V.K.M. & Lee, D.C. (1991) *Neurosci. Lett.* **125**, 241–245.
Seroogy, K.B., Lee, D.C. & Gall, C.M. (1993a) *Abstr. Soc. Neurosci.* **20**, 1306.
Seroogy, K.B., Lundgren, K.H., Lee, D.C., Guthrie, K.M. & Gall, C.M. (1993b) *J. Neurochem.* **60**, 1777–1781.
Seroogy, K.B., Lundgren, K.H., Tran, T.M., Guthrie, K.M., Isackson, P.J. & Gall, C.M. (1994) *J. Comp. Neurol.* **342**, 321–334.
Seroogy, K.B., Numan, S., Gall, C.M., Lee, D.M. & Kornblum, H.I. (1995) *NeuroReport* **6**, 105–108.
Shen, R., Altar, C.A. & Chiodo, L.A. (1994) *Proc. Natl Acad. Sci USA* **91**, 8920–8924.
Shults, W., Matthews, R.T., Altar, C.A. & Langlais, P.J. (1994) *Exp. Neurol.* **125**, 183–194.
Siuciak, J.A., Altar, C.A., Wiegand, S.J. & Lindsay, R.M. (1994) *Brain Res.* **633**, 326–330.

Spencer, D.D., Robbins, R.J., Naftolin, F., Marek, K.L. *et al.* (1992) *N. Engl. J. Med.* **327**, 1548–1549.

Spenger, C., Hyman, C., Studer, L., Egli, M., Evtouchenko, L, Jackson, C., Dahl-Jorgensen, A., Lindsay, R.M. & Seiler, R.W. (1995) *Exp. Neurol.* (in press).

Spina, M.B., Squinto, S.P., Miller, J., Lindsay, R.M. & Hyman, C. (1992) *J. Neurochem.* **59**, 99–106.

Springer, J.E., Seeburger, J.L., Bergman, L.W. & Trojanowski, J.O. (1994) *Soc. Neurosci. Abstr.* **20**, 1311.

Squinto, S.P, Stitt, T.N., Aldrich, T.H., Davis, S., Bianco, S.M., Radziejewski, C., Glass, D.H., Masiakowski, P., Furth, M.E., Valenzuela, D.M., DiStefano, P.S. & Yancopoulos, G.D. (1991) *Cell* **65**, 885–893.

Stahl, N., Davis, S. & Wong, V. (1993) *J. Biol. Chem.* **268**, 7628–7631.

Stöckli, K.A., Lottspeich, F., Sendtner, M., Masiakowski, P., Carrol, P., Götz, R., Lindholm, D. & Thoenen, H. (1989) *Nature* **342**, 920–923.

Stöckli, K.A., Lillien, L.E., Näher-Noé, M., Breitfeld, G., Hughes, R.A., Raff, M.C., Thoenen, H. & Sendtner, M. (1991) *J. Cell Biol.* **115**, 447–459.

Strömberg, I., Herrera-Maraschitz, M., Ungerstedt, U., Ebendal, T. & Olson, L. (1985) *Exp. Brain Res.* **60**, 335–349.

Strömberg, I., Björklund, L., Johansson, M., Tomac, A., Collins, F., Olson, L., Hoffer, B. & Humpel, C. (1993) *Exp. Neurol.* **124**, 401–412.

Struder, L., Spenger, C., Seiler, R.W., Altar, C.A., Lindsay, R.M. & Hyman, C. (1995) *Eur. J. Neurosci.* **7**, 223–233.

Thoenen, H. (1991) *Trends Neurosci.* **14**, 165–170.

Tilson, H.A., Harry, G.J., Nanry, K., Hudson, P.M. & Hong, J.S. (1988) *J. Neurosci. Res.* **19**, 88–93.

Toffano, G., Savoini, G., Moroni, F., Lombardi, G., Calza, L. & Agnati, L.F. (1983) *Brain Res.* **261**, 163–166.

Toffano, G., Savoini, G., Moroni, F., Lombardi, G., Calza, L. & Agnati, L.F. (1984) *Brain Res.* **296**, 233–239.

Tomac, A., Lindqvist, E., Lin, L.-F.H., Ögren, S.O., Young, D., Hofer, B.J. & Olson, L. (1995) *Nature* **373**, 341–343.

Tomozawa, Y. & Appel, S. (1986) *Brain Res.* **399**, 111–124.

Tooyama, I., Walker, D., Yamada, T., Hanai, K., Kimurea, H., McGeer, E.G. & McGeer, P.L. (1992) *Brain Res.* **593**, 274–280.

van Horne, C.G., Strömberg, I., Young, D., Olson, L. & Hoffer, B. (1991) *Exp. Neurol.* **113**, 143–154.

Ventimiglia, R., Mather, P. & Lindsay, R.M. (1993) *Soc. Neurosci. Abstr.* **19**, 666.

Ventimiglia, R., Mather, P.E., Jones, B.E. & Lindsay, R.M. (1995) *Eur. J. Neurosci.* **7**, 213–222.

Wang, J., Bankiewicz, K.S., Plunkett, R.J. & Oldfield, E.H. (1994) *J. Neurosurg.* **80**, 484–490.

Weihmuller, F.B., Hadjiconstantinou, M., Bruno, J.P. & Neff, N.H. (1989) *Life Sci.* **45**, 2495–2502.

Wiegand, S.J., Anderson, K., Alexander, C., Criden, M., Altar, C.A., Lindsay, R.M. & DiStefano, P.S. (1992) *Mov. Dis.* **7** (Suppl. 1), 63.

Wiegand, S.J., Alexander, C., Jackson, C., Anderson, K.D., Lindsay, R.M., Altar, C.A. & Hyman, C. (1993) *Soc. Neurosci. Abstr.* **19**, 1681.

Williams, L.R. (1991) *Exp. Neurol.* **113**, 31–37.

Wilcox, J.N. & Derynck, R. (1988) *J. Neurosci.* **8**, 1901–1904.

Wong, V., Arriaga, R., Ip, N.Y. & Lindsay, R.M. (1993) *Eur. J. Neurosci.* **5**, 466–474.

Yurek, D.M., Collier, T.J. & Sladek, J.R. Jr (1990) *Exp. Neurol.* **109**, 191–199.

Yurek, D., Lu, W., Hipkins, S. & Wiegand, S. (1994) *Abstr. Am. Soc. Neural Transplant.* **1**, 18.

CHAPTER 11

IMMUNE-MEDIATED CELL DEATH AND NEUROPROTECTION IN NEURODEGENERATIVE DISEASES

Stanley H. Appel, R. Glenn Smith
and Wei Dong Le

*Department of Neurology,
Baylor College of Medicine,
Houston, TX 77030, USA*

Table of Contents

11.1 Introduction

Selective neuronal injury and cell death are the hallmarks of the neurodegenerative diseases amyotrophic lateral sclerosis (ALS), Parkinson's disease and Alzheimer's disease. In ALS, motor neurones are the major targets affected, while in Parkinson's disease, substantia nigra neurones are the predominant targets affected. In Alzheimer's disease, multiple networks are compromised involving the association cortex, connections to the hippocampus, and ascending projections of cholinergic,

Neurodegeneration and Neuroprotection
in Parkinson's Disease
ISBN 0-12-525445-8

noradrenergic, dopaminergic and serotinergic neurones. The specific factors initiating cell injury in these compromised neurones are known for only a small number of kindreds with inherited or familial forms of ALS and Alzheimer's disease. And while viral, toxin-mediated, trauma-related or immunological aetiologies have been proposed for the far larger proportion of sporadic cases, little evidence currently exists to support any single one of these theories as pathogenic for all neurodegenerative disease (Appel and Stefani, 1991).

Each of the neurodegenerative disorders has a varied clinical expression, and no single aetiology has been documented for all cases. As a result, these disorders are currently considered to be syndromes, having multiple diverse aetiologies. For example, in ALS, intoxications (e.g. lead), viral infections (e.g. enterovirus), endocrine dysfunction (e.g. hyperparathyroidism), genetic disturbances (e.g. hexosaminidase deficiency) (Johnson *et al.*, 1982) and other disturbances may simulate various facets of clinical disease. However, such cases comprise only a small percentage of the total number of patients presenting with symptoms of sporadic ALS, and do not appear to offer significant insight into the aetiology of most cases. Thus, it is still possible that a single aetiology may account for the majority of patients with ALS.

Even though our understanding of the aetiologies of these disorders is limited, our understanding of the biochemical mechanisms responsible for neuronal death has increased significantly. Several pathogenic mechanisms common to many types of neuronal injury and death have been implicated recently in neurodegenerative disease. Increased intracellular calcium (Choi, 1988; Orrenius *et al.*, 1989) and increased production of oxygen and nitrogen free radicals (Coyle and Puttfaroken, 1993; Lipton *et al.*, 1993) may directly contribute to neurodegenerative cell death, while altered production or response to trophic factors may hinder normal cellular homeostatic and protective mechanisms and facilitate neuronal death (Appel, 1981). These cytotoxic mechanisms may also interact, intensifyng cell injury. For example, increasing free radical production within the cell can raise intracellular calcium levels (Cantoni *et al.*, 1989), just as increasing intracellular calcium levels can induce the production of free radicals (Nicotera *et al.*, 1990). Thus, sustained cell injury and death may result from activation of a common cell death cascade that requires any or all of these cytotoxic processes, independent of aetiological determinants responsible for initiating cell injury in neurodegenerative disease. Further, diverse aetiological factors may initiate a similar cell death cascade, while the same aetiological factor may trigger diverse cytotoxic processes in different neurones, resulting in DNA fragmentation and apoptosis (Wyllie, 1980), or in cell necrosis.

This chapter focuses on selected aspects of two problems: (1) mechanisms of motor neurone cell death and the role of neurotrophic factors as neuroprotective agents, and (2) mechanisms of substantia nigra neurone cell death and the neuroprotective role of neurotrophic factors. First, studies in ALS are reviewed, to demonstrate that: (1) the pathogenesis of some cases of familial ALS may result from mutations in the gene coding for superoxide dismutase (SOD), producing increased free radical formation; (2) the pathogenesis of sporadic ALS may involve immune-mediated attack on voltage-gated calcium channels, producing increased levels of intracellular calcium;

and (3) neurotrophic factors can act as neuroprotective agents in motor neurone death with *in vitro* and *in vivo* models. Second, studies in Parkinson's disease suggest that free radical formation may play a significant role in substantia nigral neurone death. In a substantia nigra cell line, different forms of cell injury can lead to cell death either by increasing intracellular calcium or by increasing free radicals. Neurotrophic factors can act as neuroprotective agents in preventing cell injury due either to free radical- or calcium-dependent injury mechanisms.

11.2 Amyotrophic lateral sclerosis

The clinical presentation of ALS has been well characterized as one of painless motor weakness, usually accompanied by lower motor neurone findings of denervation atrophy and motor fasciculations, and by upper motor neurone findings of spasticity and hyperactive reflexes. Extremity weakness and bulbar dysfunction (initially involving speech, and subsequently affecting swallowing) are observed as initial symptoms of this disease, and worsening voluntary muscle weakness culminates in respiratory compromise and death. Sensory and autonomic systems are usually spared. The disorder has an incidence of 1–2 per 100 000, and a prevalence of 5–7 per 100 000. Ninety per cent of cases are sporadic, while 10% of cases are familial. Males are affected twice as commonly as females, and the mean age of onset is 57 years (Appel *et al.*, 1986). Disease progression varies with age at onset; patients with earlier onset appear to progress more slowly than do patients with later onset.

11.2.1 Familial ALS

The familial, or genetic, form of ALS differs from the classical sporadic form in several ways. The mean age of disease onset is 45.7 years, even though the range of ages at onset (20–72 years) is almost identical to that found for sporadic disease. Since the majority of patients have autosomal dominant inheritance, females are affected almost as often as males. Survival data for reported familial ALS cases suggest that, while the majority of patients have a median survival of 24 months, some families appear to have a median survival of 144 months. However, like sporadic disease, the rate of disease progression is faster in patients who are in their later decades at the time of symptom onset (Strong *et al.*, 1991).

11.2.1.1 Histopathological comparison of sporadic and familial ALS

Pathologically, hallmarks of sporadic and familial ALS include selective motor neurone involvement giving rise to specific features. Selective motor involvement is manifest in sporadic ALS by loss of large anterior horn motor neurones, loss of neurones in certain brainstem motor nuclei (i.e. hypoglossal nucleus, facial motor nucleus, and motor nucleus of cranial nerve V), and Betz cells of the motor cortex. In pathological studies of ALS cases early in the disease, motor neurones of the posterior horn, Clarke's column, intermediolateral column and Onuf's nucleus are spared in the spinal cord, while neurones of cranial nerves III, IV and VI are spared in the brainstem.

In familial ALS, while the pathology of motor neurones is similar or identical to that found for sporadic disease, other neurones are also variably affected. Some patients have had findings limited to the motor system, while others have had pathology more consistent with multisystem atrophies, with involvement of the middle root zone of the posterior columns, Clarke's nucleus and posterior spinocerebellar tracts, as well as significant neurone loss and gliosis of Onuf's nucleus, subthalamic, red and cerebellar dentate nuclei (Hirano, 1967; Terao *et al.*, 1991). One well-studied family had significant involvement of proprioceptive, general somatic afferent and spinocerebellar tracts, as well as loss of brainstem reticular formation neurones (Tanaka *et al.*, 1984; Kato and Hirano, 1992), findings almost never documented in sporadic disease.

However, with the use of assisted ventilation to prolong survival times of patients with ALS, more information has accumulated on the vulnerability of different motor neurone and non-motor neurone populations in sporadic ALS. While ventral spinal motor neurones and cortical motor neurones are affected early, and spinal motor neurone loss precedes clinical manifestations of disease, loss of oculomotor, trochlear and abducens motor neurones is a late finding in ALS (Hayashi and Kato, 1989; Mizutani *et al.*, 1992; Okamoto *et al.*, 1993). Likewise, neurones in Clarke's column are lost only after respiratory involvement requires ventilatory support (Takahashi *et al.*, 1992), and motor neurones of Onuf's nucleus are affected only after extensive loss of ventral horn motor neurones (Sasaki and Maruyama, 1993). Thus, patients with end-stage disease on respirators (and some patients with rapidly progressive disease) may develop ophthalmoplegia, autonomic motor dysregulation and bladder dysfunction, in apparent contradiction to early descriptions of classical disease. After complete loss of voluntary motor function in end-stage ALS, degeneration may also involve neurones in the globus pallidus, thalamus, subthalamic nucleus, red nucleus, substantia nigra, dentate nucleus, pontine tegmentum, and fibres of the median longitudinal fasciculus, dorsal spinocerebellar tract and central tegmental tract (Haysahi and Kato, 1989). While it is unclear that these changes reflect the natural history of all patients with sporadic ALS, significant similarities are observed for affected neuronal populations in these cases of endstage late sporadic ALS and in many patients with familial ALS. Different rates for neuronal loss suggest different initiating mechanisms in sporadic and familial ALS for cell injury, but common neuronal vulnerabilities suggest similar mechanisms for cell death.

11.2.1.2 Enzyme defects in familial ALS

A familial form of ALS has recently been documented to result from structural mutations in the gene for Cu^{2+}/Zn^{2+} SOD, with associated decreases in SOD enzyme activity (Deng *et al.*, 1993; Rosen *et al.*, 1993). These familial cases with SOD deficiency represent 15–20% of all familial cases, or 1.5–2% of the total ALS patient population. The presumption from this discovery is that decreased SOD activity leads to increased intracellular levels of superoxide anion radicals, which may subsequently be important for peroxynitrite and hydroxyl radical formation. However, because Cu^{2+}/Zn^{2+} SOD is present in many neurones, it is unclear why selective injury of upper and lower motor neurones results from this form of familial ALS. Given the extensive literature on mitochondrial abnormalities in Parkinson's disease (Shapira *et al.*, 1992; Mizuno, 1993) as well as the potential importance of free radical mechanisms in that disorder (Jenner *et al.*, 1992), it is surprising that the defect in Cu^{2+}/Zn^{2+} SOD would result primarily in motor neurone injury rather than substantia nigra injury.

Since the clinical syndrome of familial ALS is quite similar to the clinical syndrome of sporadic ALS, it is possible that the final common pathway of cell injury is identical in both familial and sporadic ALS. However, SOD enzyme activity is not decreased in patients with other forms of familial ALS or with any form of sporadic ALS, suggesting for these patients (1) that other mechanisms for free radical production may be important in motor neurone death, either directly altering activity of other enzymes responsible for the production of free radicals, or indirectly altering regulatory control over free radical production, or (2) other mechanisms such as increased intracellular calcium may be the more pertinent proximal event leading to subsequent motor neurone death.

11.2.2 Sporadic ALS: excitotoxicity

At present, the two major theories proposed to explain sporadic ALS involve excitotoxicity or autoimmunity. In the excitotoxic theory of ALS, motor neurone death is hypothesized to result from activation of ligand-gated ion channels. Evidence for the excitotoxic theory arises from the demonstration of decreased levels of glutamate in ALS spinal cord (Plaitakis *et al.*, 1988) and decreased high-affinity transport of glutamate in crude synaptosomal fractions derived from the spinal cord, motor cortex and somatosensory cortex of ALS patients at autopsy (Rothstein *et al.*, 1992). However the data from both experiments can be explained by a decrease in the number of glutamate terminals and glial activity, resulting from primary damage to the motor neurone and subsequent withdrawal of glutamatergic synaptic terminals and down-regulation of glutamate uptake by glia. Furthermore, neither the pattern of *N*-methyl-D-aspartate (NMDA) receptors nor the pattern of non-NMDA receptors on neurones in the spinal cord, brainstem and motor cortex corresponds to the pattern of selective neuronal vulnerability in ALS. Nevertheless, because activation by glutamate of ligand-gated calcium channels may increase intracellular calcium in model motor neurone systems and produce cell death, a theoretical argument exists for its role in the pathogenesis of motor neurone death in ALS.

11.2.3 Sporadic ALS: autoimmunity

The autoimmune theory of ALS suggests that antibodies to a motor neurone antigen may constitute the primary event leading to motor neurone destruction. Our own studies of sporadic neurodegenerative disease began with the development of two

Figure 1 Guinea-pig immunized with bovine ventral horn homogenate demonstrating bulbar compromise but no extremity weakness. This pattern is present in 25% of guinea-pigs with EAGMD. (Engelhardt *et al.*, 1990.)

animals models of immune-mediated motor neurone disease. Experimental autoimmune motor neurone disease (EAMND) is a model of lower motor neurone destruction induced by inoculation of purified bovine spinal cord motor neurones into guinea-pigs (Engelhardt *et al.*, 1989) (Figure 1). Experimental autoimmune grey matter disease (EAGMD) is a model of both upper and lower motor neurone degeneration induced by inoculation of bovine spinal cord ventral horn homogenates (Engelhardt *et al.*, 1990). In both models, high titres of anti-motor neurone antibodies are present in immune sera, and IgG is found at neuromuscular junctions and within motor neurones of immunized guinea pigs. Denervation is also present as evidenced by electromyographic and morphological criteria. Within the central nervous system of EAGMD animals, there is loss of both spinal cord motor neurones and large pyramidal cells in the motor cortex, with histological evidence of scattered perivascular inflammatory foci. The more extensive inflammatory reaction and involvement of upper motor neurones and lower motor neurones in EAGMD compared to EAMND may reflect the more extensive antigenic stimulus produced by inoculation of membranes from glia, vessel endothelium and motor neurones. Electrophysiological studies of EAGMD animals demonstrate increased resting release of acetylcholine from motor nerve terminals, without alteration of post-junctional membrane properties (Garcia *et al.*, 1990). Cyclophosphamide prevents or attenuates motor neurone destruction if administered prior to or immediately after inoculation of grey matter (Tajti *et al.*, 1991).

11.2.3.1 Comparison of the pathology of human ALS and guinea-pig EAGMD

Human ALS resembles guinea-pig EAGMD with respect to the loss of upper and lower motor neurones, the presence of inflammatory cells within the spinal cord (Engelhardt *et al.*, 1993), and the presence of immunoglobulin G (IgG) within motor neurones (Engelhardt and Appel, 1990). In 13 of 15 human ALS spinal cords, lower motor neurones stained positively for IgG; in 6 of 11 motor cortices, pyramidal cells stained positively for IgG (Engelhardt and Appel, 1990). Employing monoclonal antibodies directed against lymphocyte antigens, infiltrates of lymphocytes bearing T-cell markers could be demonstrated in the spinal cord of 18 of 27 consecutive ALS autopsies (Engelhardt *et al.*, 1993). T-helper cells were noted in the vicinity of degenerating corticospinal tracts while T-helper and T-suppressor/cytotoxic cells were present in ventral horns. Activated microglia were also prominent in ventral horns of affected patients. Using similar immunological markers, Kawamata *et al.* (1992) described the presence of significant numbers of CD8-reactive T cells and to a lesser extent CD4-reactive T cells in the spinal cord and brain parenchyma of 13 ALS patients. MHC class I and class II-reactive T cells were found, as were numerous reactive microglia.

11.2.3.2 Passive transfer of physiological changes with IgG

Most critical in establishing the importance of immune mechanisms in ALS is the ability of EAGMD IgG and human ALS IgG to passively transfer physiological changes to mouse neuromuscular junctions. Since increased acetylcholine release from motor nerve terminals was observed in guinea-pigs with EAGMD, this physiological change was used as a sensitive and early assay for immune-mediated injury. Mice injected with IgG purified from sera of guinea-pigs with EAGMD or from patients with ALS each demonstrated an increase in miniature end plate potential (MEPP) frequency at the neuromuscular junction (Appel *et al.*, 1991) (Figure 2).

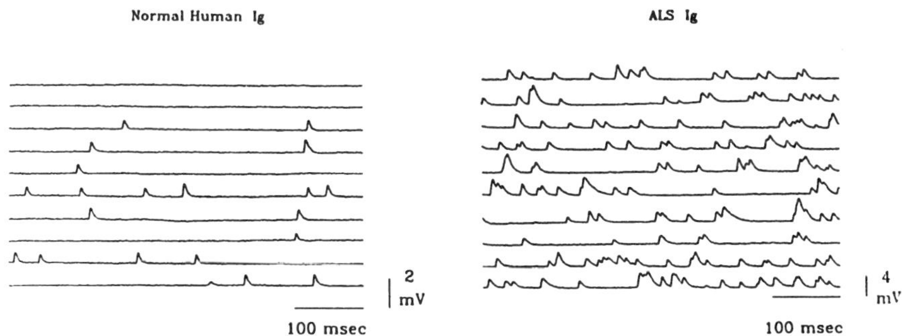

Figure 2 Increased frequency of MEPP in mouse neuromuscular junction, 24 hours after intraperitoneal injection of ALS IgG. (Appel *et al.*, 1991.)

Approximately 50% of the fibres had MEPP frequencies greater than 20 MEPP s^{-1}. No alteration in the normal histogram of MEPP frequency was noted when IgG from sera of preimmune guinea-pigs or of normal individuals or patients with other diseases were similarly injected. Chronic ALS IgG application produced axonal degeneration, in addition to altered spontaneous and evoked acetylcholine release (Uchitel *et al.*, 1992).

The demonstration that the passive transfer of ALS IgG increases acetylcholine release from the presynaptic motor terminal can be explained by an effect on ion channels (either opening a calcium channel or closing a potassium channel), resulting in increased intracellular calcium.

11.2.3.3 Effects of ALS IgG on calcium channels

Early studies tested the effects of ALS IgG on single mammalian skeletal muscle fibre ion channels and demonstrated that ALS IgG influenced the calcium current. However, observed effects were the opposite of what was anticipated or predicted. ALS IgG reduced the macroscopic L-type voltage-gated calcium current as well as the charge movement, but had no demonstrable effects on the sodium-dependent action potential (Delbono *et al.*, 1991). These actions of ALS IgG on L-type voltage-gated calcium channels (VGCCs) were similar to the action of nifedipine, an inhibitor of L-type calcium VGCCs. Fab fragments from ALS IgG also reduced the total calcium current, similar to effects of whole ALS IgG (Delbono *et al.*, 1993). ALS IgG or Fab fragments subjected to heating to over 90°C, or to adsorption with T tubules (which contain high concentrations of L-type VGCCs), lost VGCC inhibiting activity. Control IgG prepared from sera of normal individuals or of patients with familial ALS, myasthenia gravis, multiple sclerosis and chronic relapsing inflammatory polyneuropathy showed no alteration in calcium current kinetics, although IgG from patients with the Lambert–Eaton myasthenic syndrome (a disorder previously associated with anti-VGCC antibodies) also inhibited the calcium current.

In phospholipid bilayers containing skeletal muscle nifedipine-sensitive L-type VGCCs, ALS IgG reduced the mean calcium channel open time and diminished the amplitude of the calcium current (Magnelli *et al.*, 1993). Normal control IgG and myasthenia gravis IgG had no effects on calcium channel activity. After ALS IgG addition, a reduction in the time constant for the calcium channel open time was observed, indicating a reduced probability of VGCC opening. Furthermore, ALS IgG reduced calcium channel activity only when applied to the extracellular side of the calcium channel, as had already been documented in single mammalian skeletal muscle fibre studies. These results suggested that ALS IgG could interact directly with the dihydropyridine-sensitive VGCCs to alter calcium channel activity.

11.2.3.4 Immunological assays of calcium channel antibodies in ALS

To determine whether calcium channel antibodies could be identified immunochemically, we developed an enzyme-linked immunosorbent assay (ELISA) technique to detect the reaction of purified immunoglobulins with skeletal muscle L-type calcium channel complexes (Smith *et al.*, 1992). Using this assay, 75% of sera from ALS patients had significantly higher titres of VGCC-binding immunoglobulins than could be demonstrated with sera from control patients, including patients with familial ALS and spinal muscular atrophies. Of greatest interest was the fact that antibody titres appeared to correlate with the rate of ALS disease progression, rather than with the stage of disease. However, the presence of antibodies to L-type calcium channels was not absolutely specific for patients with ALS, since antibodies were also noted in the majority of patients with the Lambert–Eaton myasthenic syndrome, and in 15% of patients with the Guillain–Barré syndrome.

By immunoblot, ALS IgG appeared to interact selectively with the ionophore-forming α_1 subunit of the VGCC (Kimura *et al.*, 1994). A similar percentage of IgG from patients with ALS and the Lambert–Eaton myasthenic syndrome reacted against VGCC by immunoblot and ELISA, although Lambert–Eaton myasthenic syndrome IgGs could be distinguished from ALS IgG by the additional presence in Lambert–Eaton myasthenic syndrome IgG of antibodies to the VGCC β subunit. The binding of a mouse monoclonal antibody directed against the α_1 subunit, which itself could produce reductions in the L-type VGCC calcium current, was blocked by pre-incubation with ALS IgG.

11.2.3.5 Effects of ALS IgG on neuronal calcium channels

Because skeletal muscle is not a usual target in ALS, and because the effect of ALS IgG on VGCCs was anticipated from passive transfer studies to produce channel opening and increased intracellular calcium, rather than the observed decreased channel opening and decreased calcium current, studies were performed to test the effect of ALS IgG on neuronal calcium channels. A motor neurone line was developed by fusion of murine and N18TG2 neuroblastoma cells with dissociated embryonic rat ventral spinal cord, employing techniques which had previously yielded a substantia nigra cell line in our laboratory (Crawford *et al.*, 1992). Selection for inducible choline acetyl transferase (ChAT) provided a cell clone, VSC 4.1, which can differentiate in the presence of dibutyryl-cAMP and 0.1 μg ml^{-1} aphidocolin (a DNA polymerase inhibitor). After cAMP treatment, VSC 4.1 cells extended long branched processes and increases somatic size by 3–10-fold. This differentiated line appeared morphologically similar to a motor neurone cell line described by Cashman *et al.* (1992), but also possessed high levels of immunohistochemically identified neurone-specific enolase, 200 kDa neurofilament protein, synaptophysin and ChAT, cAMP-inducible ChAT activity, and radioligand binding for L-type, N-type and P-type VGCCs.

In physiological experiments, calcium currents in the VSC 4.1 motor neurone cell line were studied using whole-cell patch-clamp techniques (Mosier *et al.*, 1993). Our studies documented the presence of high threshold calcium currents. In the presence of ALS IgG, a prominent increase in peak calcium amplitude was noted within 10 minutes of addition. Boiling ALS IgG completely prevented effects in calcium cur-

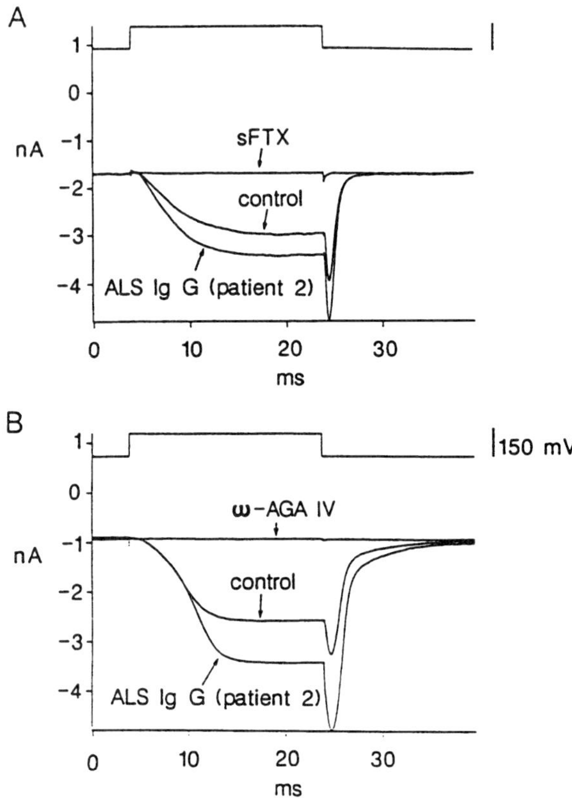

Figure 3 Barium current in Purkinje cells. (A) Control I_{Ba} is increased after exposure to ALS IgG. I_{Ba} is blocked by sFTX. (B) Similar recordings as in part (A) from a different neurone using the same IgG but blocked by another P-channel antagonist, ω–Aga IV A. (Llinas *et al.*, 1993.)

rents, and these effects were relatively specific for ALS IgG. The same ALS IgG significantly increased calcium current in P-type calcium channels in Purkinje cells and lipid bilayers (Llinas *et al.*, 1993) (Figure 3). Of greatest interest was the fact that the same ALS IgGs which were inhibitory for L-type VGCCs from skeletal muscle were found to be stimulatory in neuronal-type calcium channels, both in the VSC 4.1 motor neurone cell line and in Purkinje cells.

11.2.3.6 Cytotoxic effects of ALS IgG

When ALS IgG was added to differentiated VSC 4.1 cells, significant cytotoxicity was noted (Smith *et al.*, 1994). The cytotoxicity was time-dependent, beginning 6–12 hours after the addition of ALS IgG, and reaching maximal effect after 48–60 hours. The effect was not observed with normal or disease control IgG, and was not mediated by complement. Boiling or protease treatment of the IgG completely prevented cell loss. Employing direct daily cell counts, 40–60% of cells were lost within 3 days. A close correlation was observed between cell loss measured with whole-cell counts or cell viability assayed with vital dyes or by lactate dehydrogenase release. No significant effect of ALS IgG was noted on the same VSC 4.1 cells prior to cAMP treatment. Furthermore, no cytotoxic effects of ALS IgG were noted on our substantia nigra cell line, MES 23.5, or on the parental N18TG2 mouse neuroblastoma, either in the presence or absence of cAMP and aphidocolin.

After the addition of EGTA sufficient to buffer extracellular calcium concentrations from 10 μM to 10 nM, cells could be maintained without appreciable loss for several days (Smith *et al.*, 1994). Addition of ALS IgG in the presence of EGTA treatment no longer produced cytotoxicity, with complete protection observed at 1 μM extracellular calcium concentration. The cytotoxicity of ALS IgG could also be prevented by preincubating with purified L-type VGCC or with purified α_1 subunits of the L-type VGCC. However neither α_2 nor β subunits removed cytotoxicity from the ALS IgG. Nifedipine, an inhibitor of DHP-sensitive L-type VGCCs, had no effect on ALS IgG-induced cytotoxicity, while ω-conotoxin completely blocked ALS IgG-induced cell loss. Furthermore Aga IVA (an inhibitor of P-type channels) also completely blocked ALS IgG-induced cytotoxicity. No direct role for glutamate receptors could be demonstrated in the ALS IgG-mediated cytotoxicity in this motor neurone cell line, since inhibitors of NMDA receptors and α-amino-3-hydroxy-5-methyl-4-isoxazoleproprionate (AMPA)/kainic receptors did not block the cytotoxic effects of ALS IgG.

These data suggest that motor neurone cell death in ALS may be initiated either by multiple antibodies acting on different voltage-gated calcium channels, or by a single antibody acting on epitopes common to multiple calcium channels. The addition of antibodies to VGCCs to this motor neurone cell line leads to cytotoxicity *in vitro* which is totally dependent upon the entry of calcium into the cells. A similar mechanism for cell destruction (namely antibody activation of a calcium channel resulting in increased cytoplasmic calcium and cell death) has recently been documented in type I diabetes (Juntti-Berggrren *et al.*, 1993). In that study, serum IgM from patients with type I diabetes increases L-type calcium channel activity of insulin-producing cells and of GH_3 pituitary cells, leading to increased intracellular calcium and DNA fragmentation. In our motor neurone line, the ALS IgG interaction with neuronal calcium channels leads to increased calcium entry and similarly results in cell death by an apoptotic process.

11.2.3.7 Increased intracellular calcium mediates neuronal cell death in ALS models

Sicsjo (1981) originally suggested that elevations of cytosolic calcium consequent to energy failure may initiate a number of calcium-activated intracellular processes, causing cell death. He and others later expanded this view to include calcium entry through ligand-gated ion channels (Choi, 1988; Orrenius *et al.*, 1989; Siesjo and Bengtssar, 1989). Thus, alterations in intracellular calcium can be closely linked to the process of cell death following ischaemia or excitotoxic injury (Dubinsky, 1993). Our own studies on ALS now suggest that increases in intracellular calcium may involve an immune-mediated process whereby IgG from ALS patients interacts with neuronal VGCCs to increase calcium entry into the cell. Clearly, in different disease states, alterations in intracellular calcium and subsequent cell death may be caused by different triggering events, but the question must also be raised as to whether calcium entry *per se* is necessary, or whether parallel events may instead trigger cell injury. In our system, the strongest evidence for the importance of calcium entry in causation is provided by our ability to prevent ALS IgG-mediated cell death by removing extracellular calcium, or by blocking neuronal calcium channels with N- and P-type VGCC antagonists. Thus, calcium entry by this immune-dependent process may trigger cell injury, just as calcium entry by excitotoxic mechanisms triggers cell injury.

While we are only now examining intracellular processes after increased calcium entry that are important to IgG-mediated cytotoxicity, it is also likely that some of the same processes involved in calcium-mediated cell injury may be found in free radical-mediated cell injury. The similarity may be due to the fact that calcium entry can increase free radical production following activation of phospholipase A_2 (Glaser *et al.*, 1993), nitric oxide synthase (Schmidt *et al.*, 1992) or xanthine oxidase (Johnson *et al.*, 1987), as well as directly activate calcium-dependent kinases, proteases and endonuclease. Therefore, as indicated earlier, similar mechanisms of cell injury or death may yet be found for sporadic and familial ALS, even though they appear to be initiated by different aetiological processes.

11.3 Neurotrophic factors as neuroprotective agents

Neurotrophic factors are prominently proposed therapeutic agents for neurodegenerative diseases, including ALS (Appel, 1981). Specifically, insulin-like growth factor I (IGF-I) and ciliary neurotrophic factor (CNTF) have been documented to promote motor neurone cell survival during embryonic development (Oppenheim *et al.*, 1990; Lewis *et al.*, 1993), and to repair postnatal motor neurone injury (Sendtner *et al.*, 1990; Thoenen *et al.*, 1993). IGF-I has also been reported to increase intramuscular nerve sprouting 10-fold when administered subcutaneously to normal adult rats (Caroni and Grandes, 1990), while infusion of the IGF blocking protein (BP-4) inhibits muscle cell proliferation and neuronal sprouting in mice injected with botulinum toxin (Caroni, 1993). Thus, IGF-I or -II (both of which appear to act through IGF-I receptors) may be endogenous, physiological mammalian motor neurone sprouting factors.

The fact that these neurotrophic factors may serve a neuroprotective role as well as a repair role provides a cogent rationale for their use in ALS. However, there is no clear demonstration that any of the neurotrophic factors previously shown to enhance motor neurone repair are deficient in ALS. Indeed, while the rapidly progressive course of many ALS cases suggests that destruction triggered by autoimmunity may exceed the repair capacity of any neurotrophic factors, there is clear evidence of reinnervation and sprouting in more slowly progressive cases of ALS.

Such data instead suggest that destructive and repair processes in ALS may be simultaneous events: if the destructive process is excessive, then the repair capacity may not be sufficient, even with added neurotrophic factors. However, if the destructive process is slow, then repair may be of significant benefit and further trophic input may be of value. Although most animal models in which the neurotrophic molecules have been tested are not models of sporadic ALS, and the use of neurotrophic factors rests only on circumstantial evidence for neuroprotection, there are no other treatments known to influence the devastating course of ALS. Therefore, the current testing of these neuroprotective agents in clinical trials appears reasonable and appropriate, with general use awaiting direct demonstration that ALS-mediated motor neurone destruction can be reversed by these therapies.

11.4 Parkinson's disease

In Parkinson's disease, autopsy samples have documented that basal lipid peroxidation and levels of lipid hydroperoxide are increased in the substantia nigra (Dexter *et al.*, 1989a, 1994a). Although direct evidence for free radical overproduction is still lacking, decreased levels of reduced and total glutathione (Perry and Yong, 1986; Sofic *et al.*, 1992), increased iron, decreased ferritin, increased particulate SOD activity, and decreased activity of complex I of the mitochondrial respiratory chain all suggest that oxidative stress is high in the parkinsonian substantia nigra. However, whether these changes are primary or secondary to some other initiating pathological process is not clear. In brain tissue from unaffected individuals with incidental autopsy findings of Lewy bodies and pigmented substantia nigra neuronal loss, levels of reduced glutathione in the substantia nigra were only 65% of the levels found in normal subjects without such findings (Dexter *et al.*, 1994b). No statistically significant differences were observed in iron or mitochondrial complex I activity, suggesting that the oxidative stress may be a secondary rather than a primary process.

11.4.1 Autoimmunity in Parkinson's disease

Among aetiological theories of Parkinson's disease, the role of exogenous toxins has received the most support (Langston *et al.*, 1983; Tanner and Langston, 1990), but endogenous toxins (such as free radical production triggered by environmental factors or mitochondrial abnormalities) have also been implicated (Dexter *et al.*, 1989b; Shapira *et al.*, 1992; Mizuno, 1993). While immune processes are also involved in some aspects of Parkinson's disease (McGeer *et al.*, 1988; Yamada *et al.*, 1992), it has been unclear whether these serve an aetiological or secondary role. In studies that parallel our work on ALS, we have developed two animal models of immune-mediated destruction of substantia nigra neurones, to document that autoimmune mechanisms might also initiate substantia nigral injury.

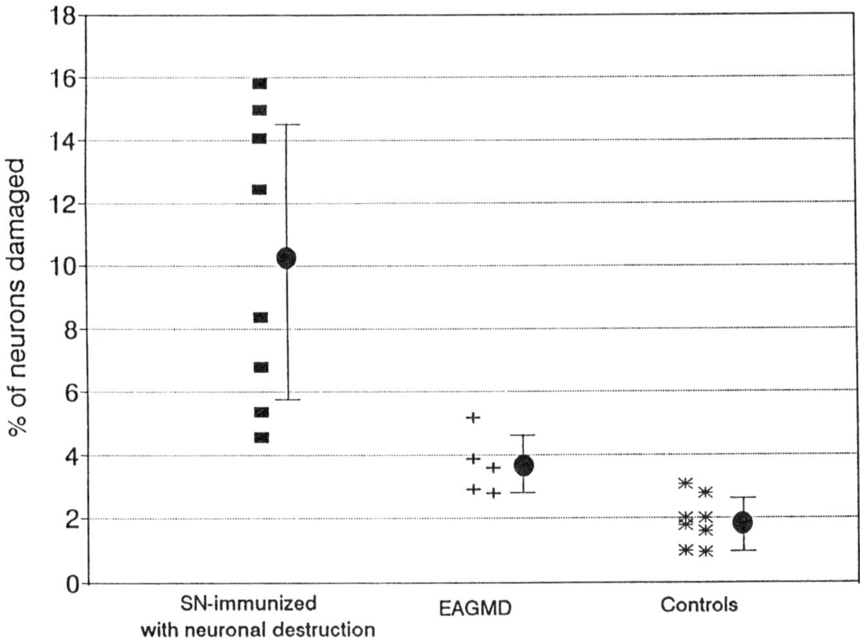

Figure 4 Substantia nigral (SN) neurone viability in guinea-pigs immunized with bovine mesencephalon, bovine ventral horn grey matter, or Freund's adjuvant alone. (Appel *et al.*, 1992.)

These studies began with the immunization of guinea-pigs with substantia nigra-containing bovine mesencephalon (Appel, 1992). Although no clinical signs of basal ganglia dysfunction appeared after bovine mesencephalon treatment, subsequent pathological examination revealed evidence of neuronal damage in the substantia nigra in 8 of 17 immunized guinea-pigs (Figures 4–6). No nigral pathology was noted in animals immunized with spinal cord grey matter or Freund's adjuvant alone. Accompanying the substantia nigra damage in mesencephalon-immunized guinea-pigs was a 25% decrease in nigral tyrosine hydroxylase activity, and a 27% decrease in striatal dopamine content. Immunohistochemically identified deposits of IgG were detected in sections of substantia nigra from mesencephalon-immunized guinea-pigs, and in sections of human substantia nigra after exposure to serum from mesencephalon-immunized guinea-pigs. However, since many non-neuronal and non-substantia nigra neuronal constituents also contained IgG deposits, and non-specific inflammatory responses were present within other parts of the central nervous system, this model did not establish that substantia nigra-specific disease could result by autoimmune processes.

Another immunological model has recently been produced by immunization of guinea-pigs with a pure neuronal cell line of hybrid dopaminergic cells (Crawford *et al.*, 1992). Immunized animals developed significant hypokinesia, and at autopsy showed a loss or damage to 50% of substantia nigra neurones, together with a 43%

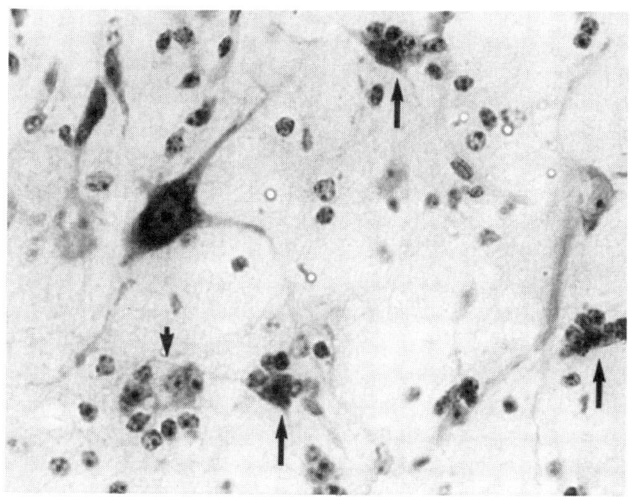

Figure 5 Degenerated shrunken neurones (arrows) surrounded by glial cells in the substantia nigra of a guinea-pig immunized with bovine mesencephalon. Arrowhead indicates a 'glial coffin' (cresyl violet stain, × 400). (Appel *et al.*, 1992.)

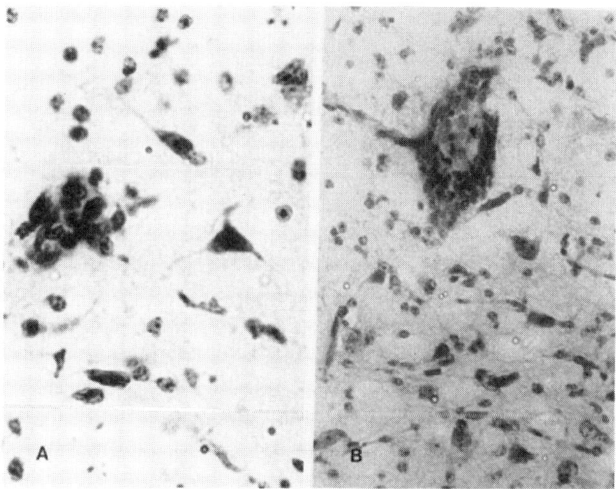

Figure 6 (A) Glial nodule in the substantia nigra of a guinea-pig immunized with bovine mesencephalon (cresyl violet stain, × 400). (B) Perivascular cuff consisting of mononuclear inflammatory cells at the border zone of the substantia nigra (cresyl violet stain, × 200). (Appel *et al.*, 1992.)

decrease in nigral tyrosine hydroxylase activity (Le, 1995). This model produced damage largely limited to the substantia nigra, suggesting that the structures involved in Parkinson's disease can potentially be targeted and damaged by an immune attack. Ongoing studies are attempting to determine whether an immunological process may contribute to the aetiology and/or pathogenesis of human Parkinson's disease.

11.4.2 Hypoglycaemia causes cell death by a calcium-mediated mechanism

As an additional approach to studying the mechanisms of substantia nigra neuronal injury and cell death, the substantia nigra MES 23.5 cell line has been subjected to hypoglycaemic injury (Le *et al.*, 1993). Hypoglycaemia was produced by replacing modified Sato's cell growth medium with glucose-free Locke's solution. In the first 6 hours following glucose deprivation, MES 23.5 cells undergo a remarkable change in morphology, with cell body shrinkage and neurite extension. The cell division rate transiently increased, possibly suggesting early division of cells in the G_1 phase. Subsequent cell damage and cell loss ensued, with almost 90% cell loss after 20 hours of glucose starvation. This cell loss was dependent on the presence of extracellular calcium, because removal of extracellular calcium significantly retarded the rate of cell death. Components of Locke's solution other than calcium did not appear to influence the cytotoxicity of MES 23.5 cells.

In this system, we were able to document the neuroprotective effect of IGF-I or cyclic AMP on hypoglycaemic injury (Figure 7). Saturating concentrations of 200 ng ml^{-1} of IGF-I increased cell survival during hypoglycaemia by 3.8-fold, while cAMP at a saturating concentration of 0.5 mM augmented cell survival by 2.6-fold. Basic fibroblast growth factor (bFGF), epidermal growth factor (EGF) and nerve growth factor (NGF) all failed to prevent MES 23.5 cell damage or loss induced by glucose starvation.

11.4.3 Increased free radicals mediate cell death

As an approach to understanding the effects of enhanced free radicals on neuronal cell death, the toxic effects of 6-hydroxydopamine (6-OHDA) and dopamine were tested with MES 23.5 cells (Le *et al.*, 1993). Although these agents did not produce significant cell damage to a number of cell lines, including the parental N18TG2 neuroblastoma, it did result in massive cell damage to MES 23.5 cells (which contain the high-affinity dopamine transporter). The cytotoxicity of either 6-OHDA or dopamine was concentration-dependent, with concentrations of 200 μM or greater for 20 hours resulting in 100% cell loss. Treatment with a 50 μM concentration of 6-OHDA or dopamine resulted in viable cell numbers which were 18 and 27%, respectively, of untreated controls. While the removal of extracellular calcium had no effect on cell survival, the removal of iron from the growth medium reduced cell death by more than two-fold.

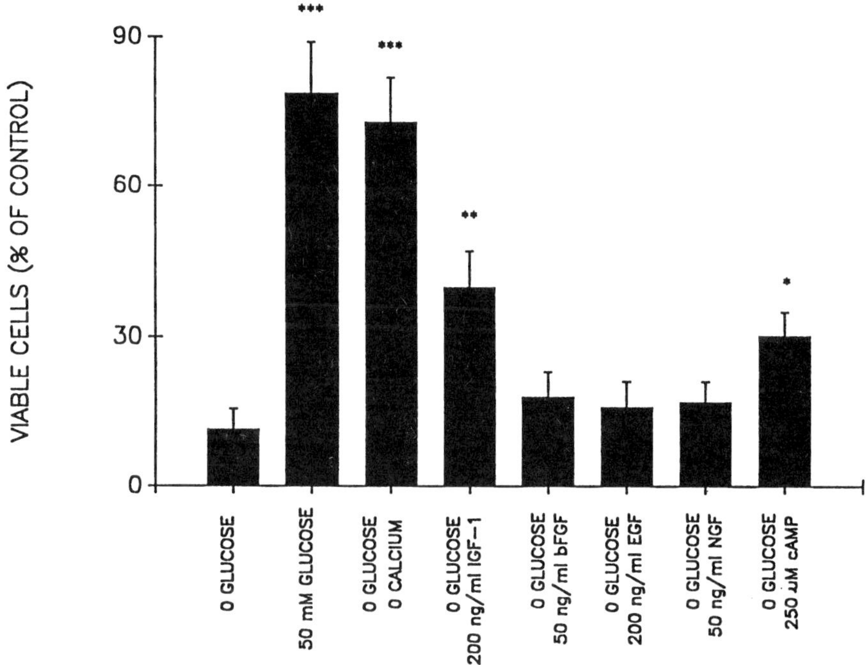

Figure 7 Protective effect of growth factors against calcium-mediated hypoglycaemic damage in MES 23.5 cells. The percentage of viable cells as compared to control cells grown in Sato's medium was determined in nine individual wells in three experiments. Each point represents the mean ± SD. ***$p<0.001$, **$p<0.01$, *$p<0.05$, compared to control with Student t test. (Le *et al.*, 1993.)

11.5 Neuroprotective action of neurotrophic factors

When cells were pretreated with cAMP (0.2 µM to 1 mM), bFGF (0.2–200 ng ml^{-1}) or IGF-I (1–1000 ng ml^{-1}) for 24 hours prior to treatment with 6-OHDA (50 µM) or dopamine (50 µM), the cytotoxicity was significantly reduced (Figure 8). A concentration of 25 ng ml^{-1} of bFGF increased the viable cell number by 2.8-fold after 6-OHDA treatment, and 2.3-fold after dopamine exposure. IGF also increased the viability by 2.6-fold; while EGF was less effective in enhancing the viability. NGF had no effect on MES 23.5 cell survival after dopamine or 6-OHDA additions. The neuroprotective effect of these factors appeared to involve a change in intracellular metabolism instead of an effect on uptake of 6-OHDA or dopamine, since each of these factors had previously been shown to increase rather than decrease the specific uptake of dopamine into dopaminergic neurones (Knusel *et al.*, 1990).

The major effect of these neurotrophic factors was presumably mediated through an antioxidant mechanism (Perez-Polo and Werrbach-Perez, 1987; Jackson *et al.*, 1990). Recently, brain-derived neurotrophic factor (BDNF) has been demonstrated

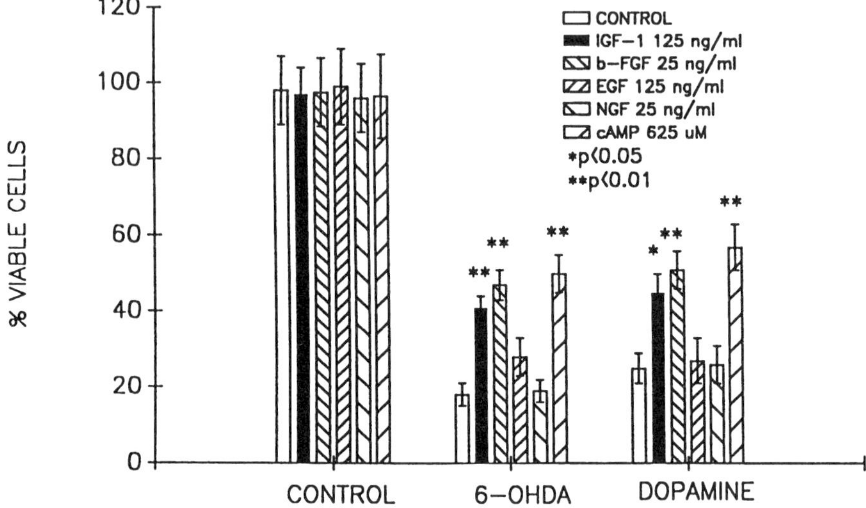

Figure 8 Protective effect of growth factors against 6-OHDA or dopamine-induced damage in MES 23.5 cells. The percentage of viable cells as compared to control cells grown in Sato's medium was determined in nine individual wells in three experiments. Each point represents the mean ± SD. $**p<0.01$, $*p<0.05$, compared to control with Student *t* test. (Le *et al.*, 1993.)

to protect dopaminergic neurones against 6-OHDA and 1-methyl-4-phenylpyridinium (MPP^+) toxicity, probably through similar antioxidant mechanisms (Spina *et al.*, 1992). Of considerable interest is the fact that the same neurotrophic factor, NGF, which could protect SY5Y cells against 6-OHDA-induced cytotoxicity (Perez-Polo and Werrbach-Perez, 1987), had no effect on 6-OHDA-induced cytotoxicity in our dopaminergic MES 23.5 cells. Thus, even when identical mechanisms of cell injury and death are involved, multiple factors clearly contribute to neuroprotective effects in different neurones. At present, it is unclear whether the extent of neuroprotection is dictated by changes at the level of the cell membrane neurotrophic factor receptor, intracellular messenger, or at a transcriptional or translational level.

Over a decade ago, neurotrophic factors were proposed as target-derived polypeptides whose relative deficiency would contribute to neuronal cell death in Parkinson's disease and the other disorders described in this chapter (Appel, 1981). Over the ensuing years, there has been no clear demonstration that any of these factors are deficient in neurodegenerative disorders. Nevertheless, a neuroprotective role for trophic factors has been documented under a variety of cytotoxic conditions. NGF and basic NGF can prevent damage to cholinergic neurones following brain lesions (Anderson *et al.*, 1988; Hefti *et al.*, 1989). bFGF has a similar protective effect on central dopaminergic neurones in animals treated with the parkinsonism-producing toxin 1-methyl-4-phenyl-1,2,5,6-tetrahydropyridine (MPTP) (Otto and Unsicker, 1990). Spina (1992) reported that BDNF also protects central dopaminergic neurones against 6-OHDA and MPP^+ toxicity *in vitro*, while Perez-Polo and Werrbach-Perez (1987) indicate that NGF protects SH-SY5Y tumour cells *in vitro* from 6-OHDA-induced cell death. The cytoprotective mechanism underlying these BDNF- and NGF-dependent effects appears to involve the glutathione antioxidant system. Cheng and Mattson (1992) reported that IGF-I, IGF-II, NGF and bFGF protect cultured central neurones against hypoglycaemic damage by an alternative mechanism, namely stabilizing calcium homeostasis. cAMP, in addition to its roles in trophic factor-dependent and -independent central neuronal differentiation (Rydel and Greene, 1988), also protects mesencephalic dopaminergic neurones from MPP^+-induced degeneration *in vitro* (Hartikka *et al.*, 1992). Our own studies with MES 23.5 cells provide additional evidence that increased intracellular calcium and free radicals are toxic for neurones, and that neurotrophic factors, especially IGF-I, can be neuroprotective. Thus, neuronal injury and cell death can be viewed as the net consequence of a relative imbalance of destructive and repair processes. If a specific type of cellular damage proceeds at a relatively slow rate, an active repair process assisted by a specific, relevant trophic factor might potentially reverse and maintain normal cellular structure and function.

What is eminently clear from these studies is that the specific mechanism of cellular injury differs from neuronal system to neuronal system, and from one toxic insult to another. As a result, varying efficiencies of neurone protection by different neurotrophic factors will also vary, and the therapeutic potential in neurodegenerative disease will depend critically upon a detailed understanding of the mechanisms underlying the aetiology and pathogenesis.

Disturbances of intracellular calcium and/or free radical-dependent mechanisms are clearly involved in neuronal cell death. In sporadic ALS, increased intracellular calcium mediated by ALS IgG, and possibly exacerbated by excitotoxic mechanisms, may be critical factors. In familial ALS, circumstantial evidence supports the role of free radicals in initiating processes leading to cell death. In substantia nigra neurones, dopamine or 6-OHDA mediate cytotoxicity by free radical mechanisms which can be ameliorated with IGF-I or bFGF, while hypoglycaemia-induced increases in intracellular calcium produce cell injury and death which can be prevented by IGF-I. Questions which must still be answered include whether the portal of calcium entry, the magnitude of the calcium rise and the duration of the rise are more pertinent variables in dictating cell injury, and whether increased intracellular calcium leads to increased free radical production (or increased free radical production leads to increased calcium) as the final common pathway of the cell death cascade. Nevertheless, it is clear that immunological mechanisms, at least in ALS, and possibly in other neurodegenerative diseases, may initiate the neurone death cascade by targeting ion channels and enhancing calcium entry. Furthermore, it is also clear that efforts directed towards neuroprotection of both free radical and intracellular calcium-mediated injury may well have significant therapeutic value for the neurodegenerative diseases, regardless of the initiating aetiologies.

Acknowledgements

Supported in part by grants from the T.L.L. Temple Foundation, the Muscular Dystrophy Association and Cephalon, Inc.

References

Anderson, K.J.D., Dam, D., Lee, S. & Cotman, C.W. (1988) *Nature* **332**, 360–362.

Appel, S.H. (1981) *Ann. Neurol.* **10**, 499–505.

Appel, S.H. & Stefani, E. (1991) In *Current Neurology* (ed. Appel, S.H.), Vol. 11, pp. 287–310. Mosby Year Book, St Louis.

Appel, S.H., Stockton-Appel, V., Stewart, S.S. & Kerman, R.H. (1986) *Arch. Neurol.* **43**, 234–238.

Appel, S.H., Engelhardt, J., Garcia, J. & Stefani, E. (1991) *Proc. Natl Acad. Sci. USA* **88**, 647–651.

Appel, S.H., Le, W.-D., Tajti, J., Haverkamp, L.J. & Engelhardt, J.I. (1992) *Ann. Neurol.* **32**, 494–501.

Cantoni, O., Sestili, P., Cattabeni, F., Bellomo, G., Pou, S., Cohen, M. & Cerutti, P. (1989) *Eur. J. Biochem.* **182**, 209–212.

Caroni, P. (1993) *Ann. NY Acad. Sci.* **692**, 209–222.

Caroni, P. & Grandes, P. (1990) *J. Cell Biol.* **110**, 1307–1317.

Cashman, N.R., Durham, H.D., Blusztajn, J.K., Oda, K., Tabira, T., Shaw, I.T., Dahrouge, S. & Antel, J.P. (1992) *Dev. Dynamics* **194**, 209–221.

Cheng, B. & Mattson, M.P. (1992) *J. Neurosci.* **12**(4), 1558–1566.

Choi, D.W. (1988) *Trends Neurosci.* **11**, 465–469.

Coyle, J.T. & Puttfarcken, P. (1993) *Science* **262**, 689–695.

Crawford, G.D., Le, W.-D., Smith, R.G., Xie, W.J., Stefani, E. & Appel, S.H. (1992) *J. Neurosci.* **12**, 3392–3398.

Delbono, O., Garcia, J., Appel, S.H. & Stefani, E. (1991) *J. Physiol.* **444**, 723–742.

Delbono, O., Magnelli, V., Sawada, T., Smith, R.G., Appel, S.H. & Stefani, E. (1993) *Am. Physiol.* **264**, C537–C543.

Deng, H.-X., Hentati, A., Tainer, J.A., Iqbal, Z., Cayabyab, A., Hung, W.-Y., Getzoff, E.D., Hu, P., Herzfeldt, B., Roos, R.P., Warner, C., Deng, G., Soriano, E., Smyth, C., Parge, H.E., Ahmed, A., Roses, A.D., Hallewell, R.A., Pericak-Vance, M.A. & Siddique, T. (1993) *Science* **261**, 1047–1051.

Dexter, D.T., Carter, C.J., Wells, F.R., Javoy-Agid, F., Agid, Y., Lees, A., Jenner, P. & Marsden, C.D. (1989a) *J. Neurochem.* **52**, 381–389.

Dexter, D.T., Wells, F.R., Lees, A.J., Agid, F., Agid, Y., Jenner, P. & Marsden, C.D. (1989b) *J. Neurochem.* **52**, 1830–1836.

Dexter, D.T., Holley, A.E., Flitter, W.D., Slater, T.F., Wells, F.R., Daniel, S.E., Lee, A.J., Jenner, P. & Marsden, C.D. (1994a) *Mov. Disord.* **9**, 92–97.

Dexter, D.T., Sian, J., Rose, S., Hindmarsh, J.G., Mann, V.M., Cooper, J.M., Wells, F.R., Daniel, S.E., Lees, A.J. & Schapira, A.H. (1994b) *Ann. Neurol.* **35**, 38–44.

Dubinsky, J.M. (1993) *Ann. NY Acad. Sci.* **679**, 34–42.

Engelhardt, J.I. & Appel, S.H. (1990) *Arch. Neurol.* **47**, 1210–1216.

Engelhardt, J.I., Appel, S.H. & Killian, J.M. (1989) *Ann. Neurol.* **26**, 368–376.

Engelhardt, J.I., Appel, S.H. & Killian, J.M. (1990) *J. Neuroimmunol.* **27**, 21–31.

Engelhardt, J.I., Tajti, J. & Appel, S.H. (1993) *Arch. Neurol.* **50**, 30–36.

Garcia, J., Engelhardt, J., Appel, S.H. & Stefani, E. (1990) *Ann. Neurol.* **47**, 1210–1216.

Glaser, K.B., Mobilio, D., Chang, J.Y. & Senko, N. (1993) *Trends Pharmacol. Sci.* **14**, 92–98.

Hartikka, J., Staufenbiel, M. & Lubbert, H. (1992) *J. Neurosci. Res.* **32**, 190–201.
Hayashi, H. & Kato, S. (1989) *J. Neurol. Sci.* **93**, 19–35.
Hefti, F., Hartikka, J. & Knusel, B. (1989) *Neurobiol. Aging* **10**, 515–533.
Hirano, A. (1967) *Arch. Neurol.* **16**, 232–243.
Jackson, G.R., Apffel, L., Werrbach-Perez, K. & Perez-Polo, J.K. (1990) *J. Neurosci. Res.* **25**, 360–368.
Jenner, P., Dexter, D.T., Sian, J., Shapira, A.H. & Marsden, C.D. (1992) *Ann. Neurol.* **32**, S82–S87.
Johnson, J.D., Conroy, W.G., Burrio, K.D. & Isom, G.E. (1987) *Toxicology* **46**, 21–28.
Johnson, W.G., Wigger, H.J. & Karp, H.R., Glaubiger, L.M. & Rowland, L.P. (1982) *Ann. Neurol.* **11**, 11–16.
Juntti-Berggren, L., Larsson, O., Rorsman, P., Ammala, C., Bokvist, K., Wåhlander, K., Nicotera, P., Dypbukt, J., Orrenuis, S., Hallberg, A. & Berggren, P.-O. (1993) *Science* **261**, 86–90.
Kato, S. & Hirano, A. (1992) *Clin. Neuropathol.* **11**, 41–44.
Kawamata, T., Akiyama, H., Yamada, T. & McGeer, P.L. (1992) *Am. J. Pathol.* **140**, 691–707.
Kimura, F., Smith, R.G., Nyormoi, O., Schneider, T., Nastainczyk, W., Hofmann, F., Stefani, E. & Appel, S.H. (1994) *Ann. Neurol.* **35**, 164–171.
Knusel, B., Michel, P.P., Schwaber, J.S. & Hefti, F. (1990) *J. Neurosci.* **10**(2), 558–570.
Langston, J.W., Ballard, P., Tetrud, J.W. & Irwin, I. (1983) *Science* **219**, 979–980.
Le, W.-D., Xie, W.J., Smith, R.G. & Appel, S.H. (1993) *Neurodegeneration J.* **2**, 227–236.
Lewis, M.E., Neff, N.T., Contreras, P.C., Strong, D.B., Oppenheim, R.W., Grebow, P.E. & Vaught, J.E. (1993) *Exp. Neurol.* **124**, 73–88.
Le, W.D., Engelhardt, J., Xie, W.J., Schneider, L., Smith, R.G. & Appel, S.H. (1995) Experimental autoimmune nigral damage in guinea pigs. *J. Neuroimmunol.* **57**, 45–53.
Lipton, S.A., Choi, Y.-B., Pan, Z.-H., Lei, S.Z., Chen, H.-S.V., Sucher, N.J., Loscalzo, J., Singel, D.J. & Stamler, J.S. (1993) *Nature* **364**, 626–632.
Llinas, R., Sugimori, M., Cherksey, B.D., Smith, R.G., Delbono, O., Stefani, E. & Appel, S.H. (1993) *Proc. Natl Acad. Sci. USA* **90**, 11743–11747.
McGeer, P.L., Itagake, S., Boyes, B.E. & McGeer, E.G. (1988) *Neurology* **38**, 1283–1291.
Magnelli, V., Sawada, T., Delbono, O., Smith, R.G., Appel, S.H. & Stefani, E. (1993) *J. Physiol.* **461**, 103–118.
Mizuno, Y., Ikebe, S., Hattori, N., Kondo, T., Tanaka, M. & Ozawa, T. (1993) *Adv. Neurol.* **60**, 282–287.
Mosier, D.R., Smith, R.G., Delbono, O., Appel, S.H. & Stefani, E. (1993) *Soc. Neurosci. Abstr.* **19**, 196.
Mizutani, T., Sakamaki, S., Tsuchiya, N., Kamei, S., Kohzu, H., Horuichi, R., Ida, M., Shiozawa, R. & Takasu, T. (1992) *Acta Neuropathol.* **84**, 372–377.
Nicotera, P., Bellamo, G. & Orrenius, S. (1990) *Chem. Res. Toxicol.* **3**, 484–494.
Okamoto, K., Hirai, S., Amari, M., Iizuka, T., Watanabe, M., Murakami, N. & Takatama, M. (1993) *Acta Neuropathol.* **85**, 458–462.
Oppenheim, R.W., Prevette, D., Qin-Wei, Y., Collins, F. & MacDonald, J. (1990) *Science* **251**, 1616–1618.
Orrenius, S., McConkey, D.J., Bellomo, G. & Nicotera, P. (1989) *Trends Pharmacol. Sci.* **10**, 281–285.
Otto, D. & Unsicker, K. (1990) *J. Neurosci.* **10**, 1912–1921.
Perez-Polo, J.R. & Werrback-Perez, K. (1987) In *Model Systems of Development and Aging of the Nervous System* (ed. Vernadakis, A.), pp. 433–442. Martinus Nijhoff, Boston.
Perry, T.L. & Yong, (1986) *Neurosci. Lett.* **67**, 269–274.
Plaitakis, A., Constantakakis, E. & Smith, J. (1988) *Ann. Neurol.* **24**, 446–449.
Rosen, D.R., Siddique, T., Patterson, D., Figlewicz, D.A., Sapp, P., Hentati, A., Donaldson, D., Goto, J., O'Regan, J.P., Deng, H.-X., Rahmani, Z., Krizus, A., McKenna-Yasek, D.,

Cayabyab, A., Gaston, S.M., Berger, R., Tanzi, R.E., Halperin, J.J., Herzfeldt, B., Van den Bergh, R., Hung, W.-Y., Bird, T., Deng, G., Mulder, D.W., Smyth, C., Laing, N.G., Soriano, E., Pericak-Vance, M.A., Haines, J., Rouleau, G.A., Gusella, J.S., Horvitz, H.R. & Brown, R.H. (1993) *Nature* **362**, 59–62.
Rothstein, J.D., Martin, L.J. & Kuncl, R.W. (1992) *N. Engl. J. Med.* **326**, 1464–1468.
Rydel, R.E. & Greene, L.A. (1988) *Proc. Natl Acad. Sci. USA* **85**, 1257–1261.
Sasaki, S. & Maruyama, S. (1993) *J. Neurol. Sci.* **119**, 28–37.
Schmidt, H.H.H.W., Pollock, J.S., Nakane, M., Forstermann, U. & Murad, F. (1992) *Cell Calcium* **13**, 427–434.
Sendtner, M., Kreutzberg, G.W. & Thoenen, H. (1990) *Nature* **345**, 440–441.
Schapira, A.H., Mann, V.M., Cooper, J.M., Krige, D., Jenner, P.J. & Marsden, C.D. (1992) *Ann. Neurol.* **32**, S116–S124.
Siesjo, B.K. (1981) *J. Cereb. Blood Flow Metab.* **1**, 155–185.
Siesjo, B.K. & Bengtssar, F. (1989) *J. Cereb. Blood Flow Metab.* **9**, 127–140.
Smith, R.G., Hamilton, S., Hofmann, F., Schneider, T., Nastainczyk, W., Birnbaumer, L., Stefani, E. & Appel, S.H. (1992) *N. Engl. J. Med.* **327**, 1721–1728.
Smith, R.G., Alexianu, M.E., Crawford, G., Nyormoi, O., Stefani, E. & Appel, S.H. (1994) *Proc. Natl Acad. Sci. USA* (in press).
Sofic, E., Lange, K.W., Jellinger, K. & Riederer, P. (1992) *Neurosci. Lett.* **142**, 128–130.
Spina, M.B., Squinto, S.P., Miller, J., Lindsay, R.M. & Hyman, C. (1992) *J. Neurochem.* **59**, 99–106.
Strong, M.J., Hudson, A.J. & Alvod, W.G. (1991) *Can. J. Neurol. Sci.* **18**, 45–58.
Tajti, J., Appel, S.H. & Stefani, E. (1991) *J. Neuroimmunol.* **34**, 143–151.
Takahashi, H., Oyanagi, K., Ohama, E. & Ikuta, F. (1992) *Acta Neuropathol.* **84**, 465–470.
Tanaka, J., Nakamura, H., Tabuchi, Y. & Takahashi, K. (1984) *Acta Neuropathol. (Berlin)* **64**, 22–29.
Tanner, C.M. & Langston, J.W. (1990) *Neurology* **40**(Suppl. 3), 17–30.
Terao, S., Sobue, G., Mukai, E., Murakami, N. & Hashizume, Y. (1991) *Clin. Neurol. (Tokyo)* **31**, 960–969.
Thoenen, H., Hughes, R.A. & Sendtner, M. (1993) *Exp. Neurol.* **124**, 47–55.
Uchitel, O.D., Scornik, F., Protti, D.A., Fumberg, C.G., Alvarez, V. & Appel, S.H. (1992) *Neurology* **42**, 2175–2180.
Wyllie, A.H. (1980) *Nature* **284**, 555–556.
Yamada, T., McGeer, P.L. & McGeer, E.G. (1992) *Acta Neuropathol.* **84**, 100–104.

CHAPTER 12

REDUCTION OF NEURONAL APOPTOSIS BY SMALL MOLECULES: PROMISE FOR NEW APPROACHES TO NEUROLOGICAL THERAPY

William G. Tatton[*,†,§], William Y.H. Ju[*,§], J. Wadia[*] and Nadine A. Tatton[‡,§]

Departments of [*]Physiology/Biophysics, [†]Psychology and [‡]Anatomy/Neurobiology and [§]the Institute for Neuroscience, Dalhousie University, Halifax, Nova Scotia, Canada

Table of Contents

12.1 Nerve cell death and nervous system disorders

The death of functionally specific nerve cells or the impairment of their interconnections with other nerve cells accounts for almost all neurological and psychiatric disease. Two fundamental cellular mechanisms appear to underlie neuronal death: necrosis and apoptosis. Necrosis usually is completed in minutes or hours after an insult and therefore might be termed immediate cell death. Three major features of necrosis have been emphasized (see Olanow, 1993): (1) failure of mitochondrial respiration with failure of ATP-dependent cellular functions, (2) increased cytosolic levels of oxidative radicals, due to a reduction in mitochondrial oxidative respiration or a failure of radical scavenging systems, causing lipid peroxidation and lysis of plasma membranes together with oxidation of DNA and proteins, and (3) increased intracellular Ca^{2+} levels with the activation of calcium-dependent proteolytic enzymes. Histologically, necrosis features plasma membrane fracture and cytoplasmic organelle dissolution with a relative maintenance of nuclear integrity. An inflammatory cell reaction due to the extrusion of cytoplasmic contents accompanies the neuronal changes.

Neurodegeneration and Neuroprotection
in Parkinson's Disease
ISBN 0-12-525445-8

A typical example of necrotic neuronal death was believed to be brain ischaemia/hypoxia causing oxygen deprivation and acidic tissue pH. Oxygen deprivation and intracellular acidosis caused mitochondrial failure and a loss of membrane electrical potential, which induced massive release of synaptic glutamate (Choi and Rotham, 1990). The glutamate activated receptors on nearby neurones, causing an immense influx of Ca^{2+} into their cytoplasm (Dugan and Choi, 1994). Necrotic neuronal death has been modelled by the application of excitotoxic glutaminergic agonists (Choi and Hartley, 1993), 1-methyl-4-phenyl-1,2,5,6-tetrahydropyridine (MPTP) toxicity (Tipton and Singer, 1993), 6-hydroxydopamine (6-OHDA) (Custodio *et al.*, 1994) and iron infusion (Sengstock *et al.*, 1992) or by direct mechanical trauma (Halland McCall, 1994).

Neuronal death by apoptosis may require a number of hours or days, and was initially only believed to involve developing neurones which did not compete adequately for trophic support from their target cells (other neurones or muscle) or from nearby non-neuronal cells (Hamburger and Oppenheim, 1982; Oppenheim, 1991). Programmed neuronal death or developmental apoptosis was responsible for the reduction of neuronal numbers found during embryogenesis or in early postnatal life.

Apoptosis was believed to have five characteristics which differentiated it from necrosis: (1) fragmentation of DNA, principally by Ca^{2+}-dependent endonucleases, initially into large pieces (5–300 kilobases) and then into oligonucleosomal sized pieces with the formation of 185 kilobase 'ladders' visible on agarose gels; (2) nuclear condensation and fractionation with the formation of nuclear bodies; (3) dependency on new protein synthesis; (4) a maintenance of external membrane integrity with the formation of membrane-wrapped cytoplasmic bodies or vacuoles; and (5) no inflammatory reaction (Wylie *et al.*, 1984; McConkey *et al.*, 1989).

It now has been shown that apoptosis can be initiated by a number of insults other than just trophic withdrawal or deprivation. Insults previously thought only to cause neuronal necrosis can induce apoptosis, particularly when delivered at low levels or at a slow rate (Cotman and Anderson, 1995). Figure 1 schematically shows a number of insults that can induce apoptosis, a number of which have also been shown to induce necrosis. For example, apoptosis has been shown to accompany nervous tissue damage caused by mitochondrial respiratory chain inhibitors (Hartley *et al.*, 1994), MPTP/1-methyl-4-phenylpyridinium (MPP^+) (Dipasquale *et al.*, 1991; Hartley *et al.*, 1994; Mochizuki *et al.*, 1994), hypoxia/ischaemia (Dragunow *et al.*, 1993; Linnik *et al.*, 1993; Rosenbaum *et al.*, 1994), exposure to excitotoxins (Samples and Dubinsky, 1993; Behl *et al.*, 1993; Joseph *et al.*, 1993; Kaku *et al.*, 1993; Montpied *et al.*, 1993; Zhong *et al.*, 1993; Csernasky *et al.*, 1994; Dessi *et al.*, 1994; Yan *et al.*, 1994; Copani *et al.*, 1994). and 6-OHDA or dopamine (Ziv *et al.*, 1994). Apoptosis of cultured neurones has been shown to result after exposure to the protein implicated in Alzheimer's disease, β-amyloid (Forloni, 1993; Forloni *et al.*, 1993; Loo *et al.*, 1993; Rabizadeh *et al.*, 1994) or gp120, the acquired immune deficiency syndrome (AIDS) protein (Muller *et al.*, 1992). Evidence is now accumulating which indicates that apoptotic neuronal death is an important component of a number of common neurological and psychiatric disorders (Altman, 1992; Margolis *et al.*, 1994; Thompson, 1995).

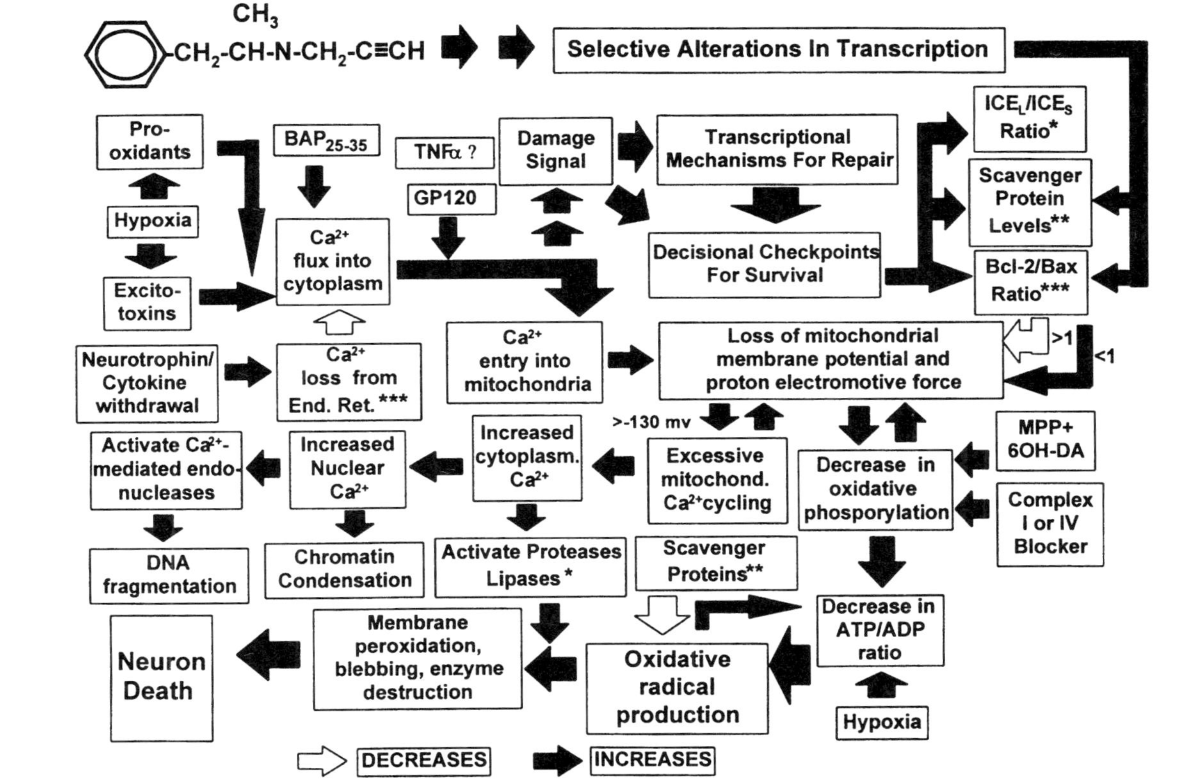

Figure 1 Schematic for the action of (–)-deprenyl on neuronal death. See text for details and abbreviations.

Stigmata of apoptosis have been found in the nervous systems of humans suffering with Alzheimer's disease (Su *et al.*, 1994; Cotman and Anderson, 1995), amyotrophic lateral sclerosis (Yoshiyama *et al.*, 1994), Huntington's disease (Yoshiyama *et al.*, 1994), several hereditary retinal degenerations (Shahinfar *et al.*, 1991; Chang *et al.*, 1993; Lolley *et al.*, 1994; Portera *et al.*, 1994; Steinberg, 1994; Tso *et al.*, 1994), glaucoma (Y.M. Buys, personal communication; Buchi, 1992; Berkelaar *et al.*, 1994; Garcia *et al.*, 1994; Rabacchi *et al.*, 1994; Silvera *et al.*, 1994) and spinal muscular atrophy (LeFebvre *et al.*, 1995; Roy *et al.*, 1995). Apoptosis of glial cells, particularly oligodendrocytes and Schwann cells, may also be important in the pathogenesis of demyelinating diseases and some peripheral neuropathies (Pender *et al.*, 1991; Raff *et al.*, 1993).

The schematic shown in Figure 1 is intended to illustrate the steps involved in neuronal death, including some of those that are common to apoptosis and necrosis. The schematic shows the features that are unique to apoptosis: the entry into transcriptional mechanisms for repair, the 'evaluation' of 'checkpoints' for survival (see Oltvai and Korsmeyer (1994) for details of the proposed decisional process), nuclear chromatin clumping (Deckwerth and Johnson, 1993), the activation of Ca^{2+}-dependent endonucleases, and DNA fragmentation (Altman, 1992). The schematic is meant to show that if the external membranes of the cell are rapidly fractured and the cytoplasmic contents lost, then there will be insufficient time for entry into transcriptional mechanisms for repair or the 'evaluation' of 'checkpoints' for survival. Work in our laboratory has shown that H_2O_2 treatment of nerve growth factor (NGF)-supported PC12 cells will induce transcription changes associated with apoptosis at low concentrations where the major onset of neuronal death is at about 8 hours. In contrast, high H_2O_2 concentrations induce death within 3 hours without any stigmata of apoptosis (W.Y.H. Ju, D.P. Holland & W.G. Tatton, unpublished observations).

12.2 Trophic and trophic-like agents and neuronal apoptosis

Several members of the neurotrophin, mitogen and cytokine/neurokine families have the capacity to reduce or slow neuronal apoptosis. Each of these neurotrophic factors acts on specific receptors which activate a number of cellular processes, some of which lead to a reduction in neuronal death. These agents seem to have promise in treating neurological disorders. Yet animal studies and several clinical trials indicate that systemic use of these powerful molecules can be complicated by severe side-effects. For example, animal studies (Henderson *et al.*, 1994; Zang *et al.*, 1995) and human clinical trials (Baringaga, 1994), using either systemic or intrathecal delivery of the neurotrophic cytokine ciliary neurotrophic factor (CNTF) have shown that doses sufficient to reduce neuronal apoptosis also cause disabling weight loss. Hence the practical use of neurotrophic factors may require new methods for the local delivery of agents that only activate neurotrophic receptors in the immediate region of dying or damaged neurones. Targeted delivery would then prevent side-effects caused by actions on cells that are not involved in the neurological disorder, particularly those lying outside of the nervous system.

Agents that can easily enter the nervous system and which are selectively targeted on the subcellular mechanisms mediating neuronal apoptosis or involved in process regrowth would seem ideal for the treatment of many nervous system disorders. A monoamine oxidase inhibitor, (–)-deprenyl, has been found to have 'trophic-like' effects in terms of neuronal survival and process growth, and may lead us to the development of an ideal agent (Tatton and Seniuk, 1994).

(–)-Deprenyl was synthesized as a 'psycho-energizer' that combined (–)-methamphetamine with a pargene chain and was subsequently found to selectively inhibit the B form of monoamine oxidase (MAO-B). Levodopa and (–)-deprenyl were used in combination to treat Parkinson's disease in the hope that levodopa would be converted to dopamine by nigrostriatal neurones, and (–)-deprenyl would cause an acute decrease in dopamine metabolism thereby increasing dopamine availability in the striatum. It was also postulated that (–)-deprenyl might decrease the rate of death of nigrostriatal neurons in Parkinson's disease by decreasing H_2O_2 production from dopamine metabolism and therefore oxidative radical damage to the cells. The celebrated finding that MAO-B inhibition protected nigrostriatal neurones from damage caused by the toxin MPTP, presumably by blocking the conversion of MPTP to MPP^+ in astroglia, reinforced the view that (–)-deprenyl could have clinical utility as an MAO-B inhibition-dependent neuroprotectant. Clinical trials were undertaken in both parkinsonism and Alzheimer's disease using oral doses of (–)-deprenyl which were sufficient to inhibit MAO-B. Definite but modest improvement was found in both diseases (see Tatton, 1993; Tatton *et al.*, 1993). Because of the absence of direct experimental evidence linking MAO-B inhibition to decreased nigrostriatal neuronal dysfunction or death, controversy has arisen as to whether the clinical improvements found with (–)-deprenyl monotherapy were due to increased neuronal dopamine levels or to decreased neuronal oxidative radical damage (Lieberman, *et al.*, 1992; Schulzer *et al.*, 1992; Olanow, 1994).

Recently, it has been shown that (–)-deprenyl has four different groups of actions which are independent of MAO-B inhibition. That is, they can be induced by (–)-deprenyl doses or concentrations that are insufficient to induce inhibition of either MAO-B or MAO-A. The MAO-independent actions include: (1) increased neuronal survival, either *in vivo* or *in vitro*, after damage or degeneration caused by hypoxia (Barber *et al.*, 1993), toxic exposure (Finnegan *et al.*, 1990; Tatton and Greenwood, 1991; Tatton *et al.*, 1993; Koutsiliere *et al.*, 1994), axonal transection (Salo and Tatton, 1992; Ansari *et al.*, 1993; Ju *et al.*, 1994; Oh *et al.*, 1994; Buys *et al.*, 1995) or developmental factors (Roy and Bedard, 1993); (2) modifications in gene expression or protein synthesis involving a variety of genes/proteins such as those for glial fibrillary acidic (Biagini *et al.*, 1993, 1994; Li *et al.*, 1993; Ju *et al.*, 1994), CNTF, basic fibroblast growth factor (Biagini *et al.*, 1994), neurotrophin receptor (TrkC) (Ekblom *et al.*, 1993), superoxide dismutases (SOD1, SOD2) (Carrillo *et al.*, 1990; Thiffault *et al.*, 1994) and aromatic amino acid decarboxylase (Li *et al.*, 1992); (3) increases in process growth of some glial cells and neurones (Iwasaki *et al.*, 1994; Seniuk *et al.*, 1994); and (4) decreases in hydroxyl radical levels after mitochondrial damage (Wu *et al.*, 1993).

The increases in neuronal survival induced by (–)-deprenyl *in vivo* are not trivial and approximate those achieved with some trophic factors. For example, a direct comparison showed that parenterally delivered (–)-deprenyl is more effective than

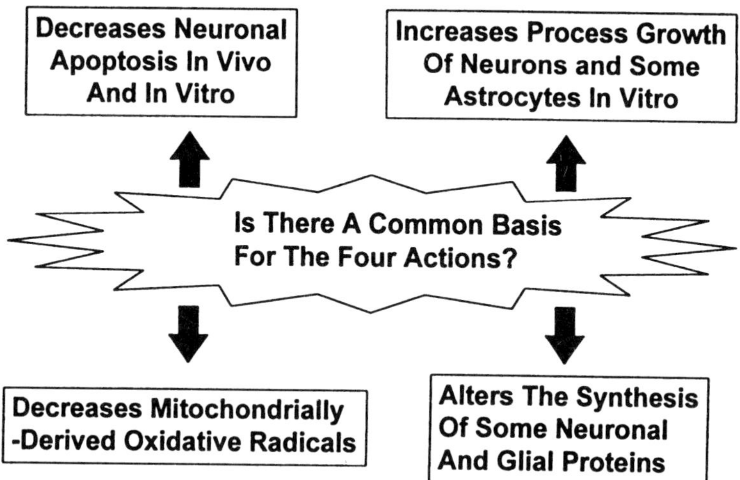

Figure 2 Four groups of MAO-independent actions of (–)-deprenyl. All of the actions have been shown to be mediated by concentrations or doses of (–)-deprenyl too small to inhibit MAO.

intracranially delivered CNTF in increasing the survival of immature facial motoneurones after axotomy (Zang *et al.*, 1995). Importantly, the capacity of (–)-deprenyl to promote neuronal survival seems more circumscribed than many neurotrophic factors since it does not induce the cachexia when given systemically or intrathecally (Baringaga, 1994; Henderson *et al.*, 1994; Zang *et al.*, 1995).

The neuronal death in most of the *in vivo* or *in vitro* models in which (–)-deprenyl increases survival is due to apoptosis (e.g. see Wilcox *et al.*, 1993) that has been linked to changes in protein synthesis and oxidative radical levels (see above). We therefore wondered whether the four different MAO-independent actions of (–)-deprenyl were linked by a common mechanism (Figure 2). In order to pursue a possible common mechanism, we examined the apoptotic death of PC12 cells caused by trophic withdrawal (Tatton *et al.*, 1994a) as a means of determining the capacity of (–)-deprenyl, (+)-deprenyl, (–)-deprenyl metabolites, and other MAO-B or -A inhibitors to mediate the four actions. We found that PC12 cells partially differentiated by serum and NGF treatment offered more reproducible results with (–)-deprenyl than undifferentiated PC12 cells treated with serum alone (Rukenstine *et al.*, 1991).

The PC12 cells died gradually over 5 days after trophic withdrawal with about 50% dying in the first 24 hours. *In situ* marking of cut DNA 3′ ends and DNA electrophoresis revealed that the internucleosomal DNA fragmentation characteristic of apoptosis began in most cells by 8–12 hours after trophic withdrawal. (–)-Deprenyl markedly reduced both the PC12 cell death and internucleosomal DNA fragmentation at concentrations of 10^{-5}–10^{-11} m, with 10^{-9} m inducing the maximum survival. Similar to our findings in axotomized facial motoneurones (Ansari *et al.*, 1993), (+)-deprenyl did not increase PC12 cell survival, indicating that (–)-deprenyl increased survival by interacting with a stereospecific site other than the FAD site of MAO-B. Most MAO-A and

MAO-B inhibitors (iproniazid, phenelzine, semicarbazide, tranylcypramine, nialamide, MDL 72974A, RO-16-6491, clorgyline and brofaromine) did not increase PC12 cell survival. Pargyline increased the survival with an IC_{50} 1000-fold greater than for (–)-deprenyl. Two of the major metabolites of (–)-deprenyl, (–)-methamphetamine and (–)-amphetamine, only increased the PC12 cell apoptosis at very high concentrations (10^{-3}–10^{-5} M) that are unlikely to be reached *in vivo*. More importantly, (–)-methamphetamine and (–)-amphetamine antagonized the capacity of (–)-deprenyl to increase PC12 cell survival in a dose-dependent fashion. Parallel studies in immature axotomized facial motoneurones showed a similar profile to other MAO inhibitors on motoneurone survival and also revealed the antagonism of (–)-methamphetamine and (–)-amphetamine (K.S. Ansari, F. Zhang & W.G. Tatton, unpublished observations).

Non-specific blockers of P_{450} enzymes inhibited the capacity of (–)-deprenyl to increase the survival of the trophically withdrawn PC12 cells and the axotomized immature facial motoneurones, but did not increase the death of either cell type when delivered without (–)-deprenyl (K.S. Ansari, D.P. Holland. & W.G. Tatton, unpublished findings). The P_{450} blockers did not alter the capacity of (–)-desmethyldeprenyl, the other major metabolite of (–)-deprenyl, to increase cell survival. The capacity of (–)-desmethyldeprenyl to increase PC12 cell survival was similar to that of (–)-deprenyl. Hence we believe that (–)-desmethyldeprenyl, not (–)-deprenyl itself, is the active molecule in inducing a decrease in neuronal apoptosis.

Translational or transcriptional blockade with cycloheximide, actinomycin or camptothecin showed that the increased PC12 cell survival depended on new protein synthesis induced by (–)-deprenyl (Tatton et al., 1994a). Kinetic experiments established that unblocked gene transcription for 4 hours or mRNA translation for 6 hours after (–)-deprenyl addition was necessary to increase cell survival (Tatton *et al.*, 1994a). Studies using differential display polymerase chain reaction (PCR) and two-dimensional protein gels showed that (–)-deprenyl alters the transcription/synthesis of 40 or more genes/proteins in the partially differentiated PC12 cells by 6 hours after trophic withdrawal (W.Y.H. Ju & W.G. Tatton, unpublished observations). To date, five of those genes/proteins have been identified using both reverse transcription–PCR and Western blots: Cu/Zn superoxide dismutase (SOD1), Mn superoxide dismutase (SOD2), Bcl-2, Bax and c-Fos (Tatton *et al.*, 1994b). Most genes/proteins that were examined, for example neurofilament light protein and tubulin, were not affected by (–)-deprenyl, showing that the transcriptional changes are selective.

Overexpression of the oncoprotein Bcl-2 makes a variety of neurones resistant to insults that induce apoptosis (Kane *et al.*, 1993), including PC12 cells (Mah *et al.*, 1993). In contrast, overexpression of the oncoprotein Bax increases vulnerability to apoptosis. The two proteins form a heterodimer, and their ratio is thought to be critical to the progression of apoptosis (Oltvai and Korsmeyer, 1994). Overexpression of SOD1 in nerve cells decreases apoptosis, presumably by reducing oxidative radical levels (Greenlund *et al.*, 1995). c-Fos is known to contribute to early events in changes in gene expression (Robertson *et al.*, 1995). Therefore, the changes in the synthesis of all five of the proteins may be linked as part of a gene expression programme to defend neurones from moderately damaging insults.

Bcl-2 is located in the outer membranes of mitochondria and is also found in nuclear membranes and the membranes of the endoplasmic reticulum (Lithgow *et al.*, 1994). Mitochondria lose their transmembrane potential (MMP) as one of earliest events in the progression of apoptosis and before DNA fragmentation is evident (Vayssiere *et al.*, 1994) or chromatin condensation can be detected (W.Y.H. Ju & W.G. Tatton, unpublished observations). The loss of MMP is known to correlate almost linearly with a loss of mitochondrial energy production (Richter and Kass, 1991; Bernardes *et al.*, 1994). Overexpression of Bcl-2 prevents or reduces the loss of MMP (Hennet *et al.*, 1993). The capacity of Bcl-2 overexpression to reduce apoptosis can be blocked by mitochondrial dysfunction caused by mitochondrial respiratory chain inhibitors and those inhibitors can induce apoptosis in cells that express normal levels of Bcl-2 (Smets *et al.*, 1994; Wolvetang *et al.*, 1994).

In order to determine if the changes in Bcl-2 gene expression induced by (–)-deprenyl were reflected in changes in MMP, we used confocal microscopy and the potentiometric dyes, chloromethyl tetramethylrosamine (CMTMR), a rhodamine derivative, and 5,5′,6,6′-tetrachloro-1,1′,3,3′-tetraethylbenzimidazolocarbocyanine (JC-1), a carbocyanine derivative, with fluoresence and confocal microscopy to examine MMP in PC12 cell after trophic withdrawal (J. Wadia & W.G. Tatton, unpublished findings). A significant reduction in MMP was apparent by 3 and 6 hours after trophic withdrawal and increased progressively after that time. Cells which did not show the nuclear stigmata of apoptosis did not show marked reductions in MMP. (–)-Deprenyl prevented any significant decrease in MMP in most mitochondria after trophic withdrawal. We have interpreted these results to indicate that the alteration in the Bcl-2/Bax ratio in the cells induced by (–)-deprenyl acted to maintain MMP and therefore ATP production.

The fluorochrome RHOD-AM was used to measure mitochondrial free Ca^{2+} levels. At 6 and 12 hours after trophic withdrawal, mitochondrial Ca^{2+} levels were significantly increased in the PC12 cell mitochondria, while addition of (–)-deprenyl significantly lowered free Ca^{2+} levels in the mitochondria. The maintenance of MMP and therefore mitochondrial energy production by (–)-deprenyl therefore seems to depend at least in part on a capacity to reduce intramitochondrial Ca^{2+} levels. Finally, we used dichlorofluorescein fluoresence to estimate oxidative radical levels in the PC12 cell cytosol after trophic withdrawal. The studies have shown that oxidative radical levels are markedly increased at 12 and 24 hours after trophic withdrawal and that (–)-deprenyl significantly reduces those increases. Therefore, the failure of mitochondrial potential caused by trophic withdrawal is accompanied by increased cytosolic levels of oxidative radicals. These results suggest that a maintenance of mitochondrial function coupled with increased radical scavenging by SOD1 and SOD2 accounts for the decrease in oxidative radicals with concentrations of (–)-deprenyl too low to inhibit MAO-B (10^{-9} M). These findings appear to explain the finding that concentrations of (–)-deprenyl as low as 10^{-12} M reduce striatal hydroxyl radical levels after MPP^+ infusion (Wu *et al.*, 1993).

Figure 3 schematically summarizes our findings in the partially differentiated PC12 cells to date. These results are incomplete and are estimated from a limited

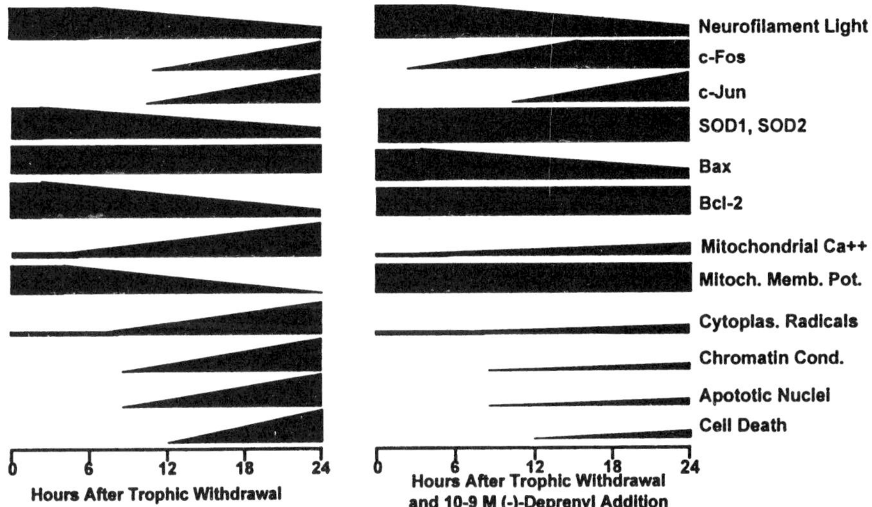

Figure 3 Estimation of temporal sequence of some events in PC12 cell apoptosis caused by trophic withdrawal and the alterations induced by (–)-deprenyl. The schematic on the left-hand side shows events in untreated apoptosis while that on the right-hand side shows the effect of (–)-deprenyl treatment. See text for details and abbreviations.

number of time points (3, 6, 9, 12, 18 and 24 hours after trophic withdrawal). Further time points, particularly in the first 6 hours after trophic withdrawal, will be required to determine an exact picture of the temporal relationships between different events. Despite its limitations, the schematic allows us to begin to draw together the hypothetical picture offered in Figure 1.

Figure 1 schematizes our preliminary findings and emphasizes the role of intramitochondrial Ca^{2+}, MMP and oxidative radical levels in the progression of apoptosis. It proposes that transcriptional modulation by (–)-desmethyldeprenyl acts to modulate the levels of some scavenger proteins as well as other genomically derived proteins that are critical to a maintenance of mitochondrial function.

Our results indicate that three of the four MAO-independent actions of (–)-deprenyl detailed above result from selective alterations in gene expression induced by (–)-desmethyldeprenyl. To date, we have not found any capacity of (–)-deprenyl to alter the growth of PC12 cell processes similar to that found for spinal motoneurones in organotypic culture (Iwasaki *et al.*, 1994). This is surprising since Bcl-2 has been reported to promote the differentiation of PC12 cells (Sato *et al.*, 1994). The induction of increased process growth of neurones by (–)-deprenyl may require the presence of glial cells in the environment similar to that in organotypic culture and may result then from a transcriptional action of (–)-deprenyl on reactive astrocytes (Seniuk *et al.*, 1994).

The relevance of the capacity of (–)-deprenyl to selectively alter gene expression through (–)-desmethyldeprenyl to the treatment of human nervous system disorders is uncertain. Yet a full understanding of mechanisms mediating those alterations may provide for the development of new therapeutic agents with the capacity to activate similar cellular mechanisms.

References

Altman, J. (1992) *TINS* **15**, 278–280.

Ansari, K.S., Yu, P.H., Kruck, T.X. & Tatton, W.G. (1993) *J. Neurosci.* **13**, 4042–4053.

Barber, A.J., Paterson, I.A., Gelowitz, D.L. & Voll, C.L. (1993) *Soc. Neuroscience Abstr.* **19**, 1646.

Baringaga, M. (1994) *Science* **264**, 772–774.

Behl, C., Hovey, L.E., Krajewski, S., Schubert, D. & Reed, J.C. (1993) *Biochem. Biophys. Res. Commun.* **197**, 949–956.

Berkelaar, M., Clarke, D.B., Wang, Y.C., Bray, G.M. & Aguayo, A.J. (1994) *J. Neurosci.* **14**, 4368–4374.

Bernardes, C.F., Meyer-Fernandes, J.R., Basseres, D.S., Castilho, R.F. & Vercesi, A.E. (1994) *Biochem. Biophys. Acta* **1188**, 93–100.

Biagini, G., Zoli, M., Fuxe, K. & Agnati, L.F. (1993) *NeuroReport* **4**, 955–958.

Biagini, G., Frasoldati, A., Fuxe, K. & Agnati, L.F. (1994) *Neurochemistry* **25**, 17–22.

Buchi, E.R. (1992) *Exp. Eye Res.* **55**, 605–613.

Buys, Y.M., Trope, G.E. & Tatton, W.G. (1995) *Current Eye Res.* **14**, 119–126.

Carrillo, M., Kanai, M., Nokubo, M. & Kitani, K. (1990) *Life Sci.* **48**, 517–521.

Chang, G.-Q., Hao, Y. & Wong, F. (1993) *Neuron* **11**, 595–605.

Choi, D.W. & Hartley, D.M. (1993) *Res. Publ. Assoc. Res. Nerv. Ment. Dis.* **71**, 23–34.

Choi, D.W. & Rotham, S.M. (1990) *Annu. Rev. Neurosci.* **13**, 171–182.

Copani, A., Bruno, V.M., Barresi, V., Battaglia, G., Condorelli, D.F. & Nicoletti, F. (1995) *J. Neurochem.* **64**, 101–108.

Cotman, C.W. & Anderson, A.J. (1995) *Mol. Neurobiol.* **10**, 19–45.

Csernansky, C.A., Canzoniero, L.M., Sensi, S.L., Yu, S.P. & Choi, D.W. (1994) *J. Neurosci. Res.* **38**, 101–108.

Custodio, J.B., Dinis, T.C., Almeida, L.M. & Madeira, V.M. (1994) *Biochem. Pharmacol.* **47**, 1989–1998.

Deckwerth, T.L. & Johnson, E.M. (1993) *J. Cell Biol.* **123**, 1207–1222.

Dessi, F., Ben-Ari, Y. & Charriaut-Marlangue, C. (1994) *Brain Res.* **654**, 27–33.

Dipasquale, B., Marini, A.M. & Youle, R.J. (1991) *Biochem. Biophys. Res. Commun.* **181**, 1442–1448.

Dragunow, M., Young, D., Hughes, P., MacGibbon, G., Lawlor, P. & Singleton, K. (1993) *Mol. Brain Res.* **18**, 347–352.

Dugan, L.L. & Choi, D.W. (1994) *Ann. Neurol.* **41**, 321–324.

Ekblom, J., Jossan, S.S., Ebendal, T., Soderstrom, S., Oreland, L. & Aquilonius, S.-M. (1993) *Neurol. Scand.* **84**(Suppl.), 79–86.

Finnegan, K.T., Skratt, J.J., Irwin, I. DeLanney, L.E. & Langston, J.W. (1990) *Eur. J. Pharmacol.* **184**.

Forloni, G. (1993) *Funct. Neurol.* **8**, 211–225.

Forloni, G., Chiesa, R., Smiroldo, S., Verga, L., Salmona, M. & Tagliavini, F. (1993) *NeuroReport* **4**, 523–526.

Garcia, V.E., Gorczyca, W., Darzynkiewecz, A. & Sharma, S.C. (1994) *J. Neurobiol.* **25**, 431–438.

Greenlund, L.J.S., Deckwerth, T.L. & Johnson, E.M. (1995) *Neuron* **14**, 303–315.

Hall, E.D. & McCall, J.M. (1994) *Adv. Pharmacol.* **28**, 221–268.

Hamburger, V. & Oppenheim, R.W. (1982) *Neurosci. Comment* **1**, 38–55.
Hartley, A., Stone, J.M., Heron, C., Cooper, J.M. & Schapira, A.H.V. (1994) *Neurochem.* **63**, 1987–1990.
Henderson, J.T., Seniuk, N.A., Richardson, P.M., Gaudie, J. & Roder, J.C. (1994) *J. Clin. Invest.* **93**, 2632–2638.
Hennet, T.G., Bertoni, G., Richter, C. & Peterhans, E. (1993) *Cancer Res.* **53**, 1456–1460.
Iwasaki, Y., Ikeda, K., Shoijima, T., Kobayashi, T., Tagaya, N. & Konoshita, M. (1994) *J. Neurol. Sci.* **125**, 11–13.
Joseph, R., Wei, L. & Han, E. (1993) *Mol. Brain Res.* **17**, 70–76.
Ju, W.J.H., Holland, D.P. & Tatton, W.G. (1994) *Exp. Neurol.* **126**, 233–246.
Kaku, D.A., Giffard, R.G. & Choi, D.W. (1993) *Science* **260**, 1516–1518.
Kane, D.J., Sarafian, T.A., Anton, R., Hahn, H., Gralia, E.B., Valentine, J.S., Ord, T. & Bresdesen, D.E. (1993) *Science* **262**, 1274–1276.
Koutsiliere, E., O'Callaghan, J.F.X., Chen, T.-S., Reiderer, P. & Rausch, W.-D. (1994) *Eur. J. Pharmacol.* **269**, R3–R4.
LeFebvre, S., Burglen, L., Reboullet, S., Clermont, O., Burlet, P., Viollet, L., Benichou, B., Cruad, C., Millasseau, P., Zeviani, M., LePaslier, D., Frezal, J., Cohen, D., Weissenback, J., Munnich, A. & Melki, J. (1995) *Cell* **80**, 155–165.
Li, X.M., Juorio, A.V., Paterson, I.A., Zhu, M.Y. & Boulton, A.A. (1992) *J. Neurochem.* **59**, 2324–2327.
Li, X.M., Qi, J., Juorio, A.V. & Boulton, A.A. (1993) *J. Neurochem.* **61**, 1573–1576.
Lieberman, A., Fahn, S., Olanow, C.W., Tetrud, J.W., Koller, W.C., Calne, D., Fazzini, E.A. & Muenter, M.D. (1992) *Neurology* **42**, 41–48.
Linnik, M.D., Zobrist, R.H. & Hatfiled, M.D. (1993) *Stroke* **24**, 2002–2008.
Lithgow, T., van Driel, R., Bertram, J.F. & Strasser, A. (1994) *Cell Growth Differ.* **5**, 411–417.
Lolley, R.N., Rong, H. & Craft, C.M. (1994) *Invest. Opthalmol. Vis. Sci.* **35**, 358–362.
Loo, D.T., Copain, A., Pike, C.J., Whittemore, E.R., Walencewicz, A.J. & Cotman, C.W. (1993) *PNAS* **267**, 7951–7955.
McConkey, D.J., Nicotera, P., Hartzell, P., Bellomo, G., Wyllie, A.H. & Orrenius, S. (1989) *Arch. Biochem. Biophys.* **269**, 365–370.
Mah, S.P., Zhong, L.T., Liu, Y., Roghani, A., Edwards, R.H. & Bredesen, D.E. (1993) *J. Neurochem.* **60**, 1183–1186.
Margolis, R.L., Chuang, D.M. & Post, R.M. (1994) *Biol. Psychiatry* **35**, 946–956.
Mochizuki, H., Nakamura, N., Nishi, K. & Mizuno, Y. (1994) *Neurosci. Lett.* **170**, 191–194.
Montpied, P., Weller, M. & Paul, S.M. (1993) *Biochem. Biophys. Res. Commun.* **195**, 623–629.
Muller, W.E., Schroder, H.C., Ushijima, H., Dapper, J. & Bormann, J. (1992) *Eur. J. Pharmacol.* **226**, 209–214.
Oh, C., Murray, B., Bhattacharya, N., Holland, D. & Tatton, W.G. (1994) *J. Neurosc. Res.* **38**, 64–74.
Olanow, W. (1993) *TINS* **16**, 439–444.
Olanow, W. (1994) *Mov. Disord.* **9**, 3.
Oltvai, Z.N. & Korsmeyer, S.J. (1994) *Cell* **79**, 189–192.
Oppenheim, R.W. (1991) *Annu. Rev. Neurosci.* **14**, 453–501.
Pender, M.P., Ngyen, K.B., McCombe, P.A. & Kerr, J.F.R. (1991) *J. Neurol. Sci.* **104**, 81–87.
Portera, C.C., Sung, C.H., Nathans, J. & Adler, R. (1994) *PNAS* **91**, 974–978.
Rabacchi, S.A., Bonfanti, L., Liu, X.H. & Maffei, L. (1994) *J. Neurosci.* **14**, 5292–5301.
Rabizadeh, S., Bitler, C.M., Butcher, L.L. & Bredesen, D.E. (1994) *PNAS* **91**, 10703–10736.
Raff, M.C., Barres, B.A., Burne, J.F., Coles, H.S., Ishizaki, Y. & Jacobson, M.D. (1993) *Science* **262**, 695–700.
Richter, C. & Kass, G.E.N. (1991) *Chem.-Biol. Interact.* **77**, 1–23.
Robertson, L.M., Kerppola, T.K., Smeyne, R.J., Bocchairo, C., Morgan, J.I. & Curran, T. (1995) *Neuron* **14**, 241–252.

Rosenbaum, D.M., Michaelson, M., Batter, D.K., Doshi, P. & Kessler, J.A. (1994) *Ann. Neurol.* **36**, 864–870.

Roy, E. & Bedard, P.J. (1993) *NeuroReport* **4**, 1183–1186.

Roy, N., Mahadevan, M.S., McLean, M. *et al.* (1995) *Cell* **80**, 167–178.

Rukenstein, A., Rydel, R.E. & Greene, L.A. (1991) *J. Neurosci.* **11**, 2552–2563.

Salo, P.T. & Tatton, W.G. (1992) *J. Neurosci. Res.* **31**, 394–400.

Samples, S.D. & Dubinsky, J.M. (1993) *J. Neurochem.* **61**, 382–385.

Sato, N., Hotta, K., Waguri, S., Nitatori, T., Tohyama, K., Tsujimoto, Y. & Uchiyama, Y. (1994) *Neurobiol.* **25**, 1277–1234.

Schulzer, M., Mak, E. & Calne, D.B. (1992) *Ann. Neurol.* **32**, 795–798.

Sengstock, G.J., Olanow, C.W., Dunn, A.J. & Arendash, G.W. (1992) *Brain Res. Bull.* **28**, 645–649.

Seniuk, N.A., Henderson, J.T., Tatton, W.G. & Roder, J.C. (1994) *J. Neurosc. Res.* **37**, 278–286.

Shahinfar, S., Edward, D.P. & Tso, M.O. (1991) *Curr. Eye Res.* **10**, 47–59.

Silveira, L.C., Russelakis, C.M. & Perry, H. (1994) *J. Neurocytol.* **23**, 75–86.

Smets, L.A., Van denBerg, J., Acton, D., Top, B., Van Rooij, H. & Verwijs-Janssen, M. (1994) *Blood* **84**, 1613–1619.

Steinberg, R.H. (1994) *Curr. Opin. Neurobiol.* **4**, 515–524.

Su, J.H., Anderson, A.J., Cummings, B.J. & Cotman, C.W. (1994) *Clin. Neurosci. Neuropath.* **5**, 2529–2533.

Tatton, W.G. (1993) *Mov. Disord.* **8**, 520–530.

Tatton, W.G. & Greenwood, C.E. (1991) *J. Neurosci. Res.* **30**, 666–672.

Tatton, W.G. & Seniuk, N.A. (1994) *Int. Acad. Biomed. Drug Res.* **7**, 238–248.

Tatton, W.G., Seniuk, N.A., Ju, W.Y.H. & Ansari, K.S. (1993) In *Monoamine Oxidase Inhibitors In Neurological Diseases* (eds Lieberman, A., Olanow, W., Youdim, M. & Tipton, K.), pp. 217–248. Raven Press, New York.

Tatton, W.G., Ju, W.Y.L., Holland, D.P., Tai, C.E. & Kwan, M.M. (1994a) *J. Neurochem.* **63**, 1572–1574.

Tatton, W., Ju, W., Wadia, J., Ansari, K., Zhang, F., Buys, Y. & Seniuk, N. (1994b) *Mov. Disord.* **9**, 4.

Thiffault, C., Aumont, N., Quirion, R. & Poirier, J. (1994) *Can. J. Physiol. Pharmacol.* **72**, 592.

Thompson, C.B. (1995) *Science* **267**, 1456–1462.

Tipton, K.F. & Singer, T.P. (1993) *J. Neurochem.* **61**, 1191–1207.

Tso, M.O., Zhang, C., Abler, A.S., Chang, C.J., Wong, F., Chang, G.Q. & Lam, T.T. (1994) *Invest. Ophthalmol. Vis. Sci.* **35**, 2693–2699.

Vayssiere, J.L., Petit, P.X., Risler, Y. & Mignotte, B. (1994) *Proc. Natl Acad. Sci. USA* **91**, 11752–11760.

Wilcox, B.J., Clatterbuck, R.E., Price, D.L. & Koliatsus, V.E. (1993) *Soc. Neurol. Abstr.* **19**, 441.

Wolvetang, E.F., Johnson, K.L., Krauer, K., Ralph, S.J. & Linnane, A.W. (1994) *FEBS Lett.* **339**, 40–44.

Wu, R.-M., Chiueh, C.C., Pert, A. & Murphy, D.L. (1993) *Eur. J. Pharmacol.* **243**, 241–248.

Wyllie, A.H., Morris, R.G., Smith, A.L. & Dunlop, D. (1984) *J. Pathol.* **142**, 67–77.

Yan, E., Ni, B., Weller, M., Wood, K.A. & Paul, S.M. (1994) *Brain Res.* **656**, 43–51.

Yoshiyama, Y., Yamada, T., Asanuma, K. & Asahi, T. (1994) *Acta Neuropathol. (Berlin)* **88**, 207–211.

Zhang, F., Richardson, P.M., Holland, D.P., Guo, G. & Tatton, W.G. (1995) *J. Neurosc. Res.* **40**, 564–570.

Zhong, L.T., Sarafian, T., Kanc, D.J., Charles, A.C., Mah, S.P., Edwards, R.H. & Bredesen, D.E. (1993) *Proc. Natl Acad. Sci. USA* **90**, 4533–4537.

Ziv, I., Melamed, E., Nardi, N., Luria, D., Achiron, A., Offen, D. & Barzilai, A. (1994) *Neurosci. Lett.* **170**, 136–140.

Index